Lecture Notes in Physics

Bisher erschienen/Already published

Vol. 1: J. C. Erdmann. Wärmeleitung in Kristallen, theoretische Grundlagen und fortgeschrittene experimentelle Methoden. II, 283 Seiten. 1969.

Vol. 2: K. Hepp, Théorie de la renormalisation. III, 215 pages. 1969.

Vol. 3: A. Martin, Scattering Theory: Unitarity, Analyticity and Crossing. IV, 125 pages. 1969.

Vol. 4: G. Ludwig, Deutung des Begriffs „physikalische Theorie" und axiomatische Grundlegung der Hilbertraumstruktur der Quantenmechanik durch Hauptsätze des Messens. 1970. Vergriffen.

Vol. 5: Schaaf, The Reduction of the Product of Two Irreducible Unitary Representations of the Proper Orthochronous Quantummechanical Poincare Group. IV, 120 pages. 1970.

Vol. 6: Group Representations in Mathematics and Physics. Edited by V. Bargmann. V, 340 pages. 1970.

Vol. 7: R. Balescu, J. L. Lebowitz, I. Prigogine, P. Résibois, Z. W. Salsburg, Lectures in Statistical Physics. V, 181 pages. 1971.

Vol. 8: Proceedings of the Second International Conference on Numerical Methods in Fluid Dynamics. Edited by M. Holt. 1971. Out of print.

Vol. 9: D. W. Robinson, The Thermodynamic Pressure in Quantum Statistical Mechanics. V, 115 pages. 1971.

Vol. 10: J. M. Stewart, Non-Equilibrium-Relativistic Kinetic Theory. III, 113 pages. 1971.

Vol. 11: O. Steinmann, Pertubation Expansions in Axiomatic Field Theory. III, 126 pages. 1976.

Vol. 12: Statistical Models and Turbulence. Edited by C. Van Atta and M. Rosenblatt. Reprint of the First Edition. VIII, 492 pages. 1975.

Vol. 13: M. Ryan, Hamiltonian Cosmology. VII, 169 pages. 1972.

Vol. 14: Methods of Local and Global Differential Geometry in General Relativity. Edited by D. Farnsworth, J. Fink, J. Porter, and A. Thompson. V, 188 pages.

Vol. 15: M. Fierz, Vorlesungen zur Entwicklungsgeschichte der Mechanik. V, 97 Seiten. 1972.

Vol. 16: H.-O. Georgii, Phasenübergang 1. Art bei Gittergasmodellen. IX, 167 Seiten. 1972.

Vol. 17: Strong Interaction Physics. Edited by W. Rühl and A. Vancura. V, 405 pages. 1973.

Vol. 18: Proceedings of the Third International Conference on Numerical Methods in Fluid Mechanics, Vol. I. Edited by H. Cabannes and R. Temam. VII, 186 pages. 1973.

Vol. 19: Proceedings of the Third International Conference on Numerical Methods in Fluid Mechanics, Vol. II. Edited by H. Cabannes and R. Temam. VII, 275 pages. 1973.

Vol. 20: Statistical Mechanics and Mathematical Problems. Edited by A. Lenard. VIII, 247 pages. 1973.

Vol. 21: Optimization and Stability Problems in Continuum Mechanics. Edited by P. K. C. Wang. V, 94 pages. 1973.

Vol. 22: Proceedings of the Europhysics Study Conference on Intermediate Processes in Nuclear Reactions. Edited by N. Cindro, P. Kulišic and Th. Mayer-Kuckuk. XIV, 329 pages. 1973.

Vol. 23: Nuclear Structure Physics. Proceedings 1973. Edited by U. Smilansky, I. Talmi, and H. A. Weidenmüller. XII, 296 pages. 1973.

Vol. 24: R. F. Snipes, Statistical Mechanical Theory of the Electrolytic Transport of Nonelectrolytes. V, 210 pages. 1973.

Vol. 25: Constructive Quantum Field Theory. The 1973 "Ettore Majorana" International School of Mathematical Physics. Edited by G. Velo and A. Wightman. III, 331 pages. 1973.

Vol. 26: A. Hubert, Theorie der Domänenwände in geordneten Medien. XII, 377 Seiten. 1974.

Vol. 27: R. K. Zeytounian, Notes sur les Ecoulements Rotationnels de Fluides Parfaits. XIII, 407 pages. 1974.

Vol. 28: Lectures in Statistical Physics. Edited by W. C. Schieve and J. S. Turner. V, 342 pages. 1974.

Vol. 29: Foundations of Quantum Mechanics and Ordered Linear Spaces. Advanced Study Institute, Marburg 1973. Edited by A. Hartkämper and H. Neumann. VI, 355 pages. 1974.

Vol. 30: Polarization Nuclear Physics. Proceedings 1973. Edited by D. Fick. IX, 292 pages. 1974.

Vol. 31: Transport Phenomena. Sitges International Schools of Statistical Mechanics, June 1974. Edited by G. Kirczenow and J. Marro. XIV, 517 pages. 1974.

Vol. 32: Particles, Quantum Fields and Statistical Mechanics. Proceedings 1973. Edited by M. Alexanian and A. Zepeda. V, 132 pages. 1975.

Vol. 33: Classical and Quantum Mechanical Aspects of Heavy Ion Collisions. Proceedings 1974. Edited by H. L. Harney, P. Braun-Munzinger, and C. K. Gelbke. VII, 311 pages. 1975.

Vol. 34: One-Dimensional Conductors GPS Summer School Proceedings, 1974. Edited by H. G. Schuster. VII, 371 pages. 1975.

Vol. 35: Proceedings of the Fourth International Conference on Numerical Methods in Fluid Dynamics, 1974. Edited by R. D. Richtmyer. V, 457 pages. 1975.

Vol. 36: R. Gatignol, Théorie Cinétique des Gaz à Répartition Discrète de Vitesses. II, 219 pages. 1975.

Vol. 37: Trends in Elementary Particle Theory. Proceedings 1974. Edited by H. Rollnik and K. Dietz. V, 472 pages. 1975.

Vol. 38: Dynamical Systems, Theory and Applications. Proceedings 1974. Edited by J. Moser. VI, 624 pages. 1975.

Vol. 39: International Symposium on Mathematical Problems in Theoretical Physics. Proceedings 1975. Edited by H. Araki. XII, 562 pages. 1975.

Vol. 40: Effective Interactions and Operators in Nuclei. Proceedings 1975. Edited by B. R. Barrett. XII, 339 pages. 1975.

Vol. 41: Progress in Numerical Fluid Dynamics. Proceedings 1974. Edited by H. J. Wirz. V, 471 pages. 1975.

Vol. 42: H II Regions and Related Topics. Proceedings 1975. Edited by D. Downes and T. L. Wilson. XII, 488 pages. 1975.

Vol. 43: Laser Spectroscopy. Proceedings 1975. Edited by S. Haroche, J. C. Pebay-Peyroula, T. W. Hänsch, and S. E. Harris. X, 466 pages. 1975.

Lecture Notes in Physics

77

Topics in Quantum Field Theory and Gauge Theories

Proceedings of the VIII International Seminar
on Theoretical Physics
Held by GIFT in Salamanca, June 13-19, 1977

Edited by J. A. de Azcárraga

Springer-Verlag
Berlin Heidelberg GmbH 1978

Editor
J. A. de Azcárraga
Facultad de Ciencias
Universidad de Salamanca
Salamanca
Spain

Library of Congress Cataloging in Publication Data

International Seminar on Theoretical Physics, 8th, Salamanca, Spain, 1977.
Topics in quantum field theory and gauge theories.

(Lecture notes in physics ; 77)
Bibliography: p.
Includes index.
1. Gauge fields (Physics)--Congresses. 2. Quantum field theory--Congresses. I. Azcárrage, J. A. de, 1941- II. Grupo Interuniversitario de Física Teórica. III. Title. IV. Series.
QC793.3.F5I59 1977 530.1'4 78-7978

ISBN 978-3-540-08841-7 ISBN 978-3-540-35820-6 (eBook)
DOI 10.1007/978-3-540-35820-6

Originally published by Springer-Verlag Berlin Heidelberg New York in 1978

2153/3140-543210

SALAMANCA - Façade of the Old University

INDEX

FOREWORD

This volume contains the Proceedings of the VIII G.I.F.T.* Seminar on Theoretical Physics, which took place in Salamanca during the third week of June, 1977.** During that eventful week for Spain, a distinguished group of lecturers*** (Dalitz, De Rújula, Goddard, Hey, Pati, Wess and Zinn-Justin) addressed an audience of about eighty participants and reported on several subjects under the general heading of the Seminar, 'Topics on Quantum Field Theory and Gauge Theories.' Their efforts to produce thorough and clear sets of lectures are here gratefully acknowledged.

Besides the J.E.N., which through the Instituto de Estudios Nucleares (Madrid) sponsors all G.I.F.T. activities, several persons and Institutions have contributed to make the 1977 Seminar possible. In particular, I wish to thank the I.C.E. and the Director of the Special Summer Courses of Salamanca University for the help granted to the Seminar. The collaboration of the British Council and the Banco de Santander as well as the extensive facilities given by the Rector of the University and the Dean of the Science Faculty are also acknowledged Finally, I wish to thank the members of the Department of Theoretical Physics and in particular V. Aldaya for helping me in the inevitable chores associated with the organization of the Seminar.

Faculty of Sciences, Salamanca University
1st February, 1978

José A. de Azcárraga

*Grupo Interuniversitario de Física Teórica.

**The other Seminars were held in Valencia (1970); Madrid (1971, 1972); Barcelona (1973); Zaragoza (1974); Jaca, Huesca (1975); L'Escala, Gerona (1976). This year's Seminar on 'Non Linear Problems in Theoretical Physics' will be held at Jaca (June, 1978).

***Prof. Hermann was unable to attend the Seminar, but the lectures he intended to give are included in this volume.

TOPICS IN QUANTUM FIELD THEORY AND GAUGE THEORIES

VIII INTERNATIONAL SEMINAR ON THEORETICAL PHYSICS

Held by GIFT at the Faculty of Sciences, Salamanca

June 13-19, 1977

List of Participants

Agapito, J.A.	(Salamanca)
Aldaya, V.	(Salamanca)
Alvarez, A.	(JEN,Madrid)
Asorey, M.	(Zaragoza)
Azcárraga, J.A. (Director of the Seminar)	(Salamanca)
Azcoiti, V.	(Zaragoza)
Baird, P.	(Oxford)
Benavent, F.	(Autónoma, Barcelona)
Boya, L.J.	(Salamanca)
Bramón, A.	(Valencia)
Bruce, D.	(Copenhagen)
Cariñena, J.F.	(Zaragoza)
Cerveró, J.M.	(Harvard-Salamanca)
Cid, L.	(Carabobo, Venezuela)
Cortés, J.L.	(Zaragoza)
Chaichian, M.	(Helsinki)
Dalitz, R.H. (Lecturer)	(Oxford)
Darby, D.	(Utrecht)
De Rújula, A. (Lecturer)	(Harvard)
Del Aguila, F.	(Aut. Barcelona)
Delgado, V.	(JEN,Madrid)
Elizalde, E.	(Central Barcelona)
Fernández-Rañada, A. (Director of G.I.F.T.)	(Madrid, Complutense)
Frère, J.M.	(Bruxelles)
Fustero, F.J.	(Aut. Barcelona)
García Alvarez, A.	(U.N.Méjico)
García Azcárate, A.C.	(Aut.Madrid)
García Estevez, J.V.	(Zaragoza)
García Gonzalo, L.	(Madrid, Compl.)
García Pérez, P.L.	(Salamanca)
Goddard, P. (Lecturer)	(Cambridge)
Gómez, J.M.	(Aut.Madrid)
Gomis, J.	(Central Barcelona)
González Arroyo, A.	(Aut. Madrid)
González Gascón, F.	(CSIC,Madrid)
Hernández, M.A.	(Aut. Madrid)
Hey, A.J.G. (Lecturer)	(Southampton)
Hüffel, H.	(Wien)
Ibáñez, L.	(Aut.Madrid)
León, J.	(CSIC,Madrid)
López Fraguas, A.	(JEN, Madrid)
López Martin, C.	(Aut. Madrid)
Lukierski, J.	(CEN Saclay)
Madurga, G.	(Sevilla)
Manton, N.S.	(Cambridge)
Martín, J.	(Aut. Madrid)
Martorell, J.	(Aut. Madrid)
Mateos, J.	(Salamanca)
McKinley, W.A.	(Rensselaer, N.Y.)
Mellado, I.	(JEN, Madrid)
Pajares, C.	(Aut. Barcelona)
Pati, J.C. (Lecturer)	(Imperial Coll. & Maryland Univ.)
Poves, A.	(Aut. Madrid)
Pujana, G.	(Bilbao)
Quirós, M.	(CSIC,Madrid)

Ramirez Cacho, F.	(Madrid Compl.)	Sanz, J.L.	(Aut. Madrid)
Ramirez Mittelbrunn, J.	(CSIC, Madrid)	Seguí, A.	(Zaragoza)
Ramón, M.	(Madrid Compl.)	Sesma, J.	(Valencia)
Ramos, J.	(Madrid Compl.)	Ugaz, E.	(Aut. Barcelona)
Rodriguez Espinosa, J.M.	(Madrid Compl.)	Usón, J.	(Madrid, Compl.)
Rodriguez González, M.A.	(Madrid Compl.)	Velasco, S.	(Salamanca)
Roig, F.	(Valencia)	Verdaguer, E.	(Aut. Barcelona)
Rothe, K.D.	(Montpellier)	Villalón, M.E.	(Madrid, Compl.)
Roy, L.J.	(Aut. Madrid)	Wess, J. (Lecturer)	(Karlsruhe)
Ruck, H.	(Frankfurt)	Wolff, U.	(München)
Sánchez Gómez, J.L.	(Aut. Madrid)	Yndurain, F.J.	(Aut. Madrid)
Sánchez Guillén, J.J.	(Zaragoza)	Zagury, N.	(Río de Janeiro)
		Zinn-Justin, J. (Lecturer)	(CEN, Saclay)

MAGNETIC MONOPOLES AND RELATED OBJECTS

Peter Goddard

Department of Applied Mathematics and Theoretical Physics,
University of Cambridge, U.K.

Table of contents

MAGNETIC MONOPOLES AND RELATED OBJECTS

Peter Goddard
Department of Applied Mathematics and Theoretical Physics,
University of Cambridge, U.K.

INTRODUCTION

The present intensive interest in extended objects in field theory started about three years ago. At that time two pieces of work appeared which provoked a lot of excitement. On the one hand, 't Hooft[1] and, independently, Polyakov[2] produced a solution to a classical non-Abelian gauge theory which could be interpreted as a magnetic monopole. On the other, Coleman[3] showed the equivalence of two two-dimensional quantum field theories: the sine-Gordon theory and the massive Thirring model. The sine-Gordon theory possesses an extended classical solution which becomes an elementary particle in the Thirring model description of the theory. Thus non-constant static solutions to classical theories can become elementary particles on quantization in the sense that they may be the quanta of a (local) quantum field in some other, equivalent description of the system. Objects which, in many respects, resemble particles at the classical level can indeed become particles in the usual sense after quantization. Coleman's results in one spatial dimension offer hope that extended solutions to classical field theories may give some information about the states of quantum theory even in three dimensions.

Although the 't Hooft-Polyakov solution may be a model for an elementary particle, none of the objects of this type which have been constructed so far seems to admit an interpretation which would give it direct physical relevance at the present, though the closely related idea of instantons may offer better prospects for a connection with the real world. But, in more general terms, these solutions teach us about a new way in which symmetries may be realised in a physical theory which is very different from the familiar Noether fashion. Further which symmetries are realised conventionally and which are realised in this new 'topological' fashion is a subjective question depending on the particular description of the system that has been chosen. This offers the possibility of a dual relationship between topological and conventional symmetries. Since such dual symmetries have inversely related couplings, forces of very different magnitudes may be related more intimately in this way than by some severely broken universal symmetry group. But perhaps we are ill-advised to speculate about the precise physical role these objects may eventually play in theoretical physics; if we cannot see exactly where these developments are leading we may console ourselves with the view expressed at the beginning of Dirac's famous first paper conjecturing the existence of magnetic monpoles[4]:

'There are at present fundamental problems in theoretical physics ... the solution of which will presumably require a more drastic revision of our fundamental concepts than any that have gone before. Quite likely these changes will be so great that it will be beyond the power of human intelligence to get the necessary new ideas by direct attempts to formulate the experimental data in mathematical terms. The theoretical worker in the future will therefore have to proceed in a more indirect way. The most powerful method of advance that can be suggested at present is to employ all the resorces of pure mathematics in attempts to perfect and generalise the mathematical formalism that forms the existing basis of theoretical physics, and after each success in this direction, to try to interpret the new mathematical features in terms of physical entities ...'

The central topic of these lectures is the 't Hooft-Polyakov solution and its relationship to the Dirac monopole. They are divided into three main parts. We begin by considering examples in one spatial and one temporal dimension where the discussion can be made more explicit. The second part deals with electromagnetism as the prototype gauge theory and Dirac's introduction of magnetic monopoles into it ending with a review of the formalism of non-Abelian gauge theories. The last and major section discusses extended objects in gauge theories, concentrating on the 't Hooft-Polyakov monopole but trying to understand its properties in a more general context.

At various points in this account I lean on Coleman's excellent review article[5]. Most of what I know about this subject has been acquired from lectures given in Cambridge in 1976 by Jeffrey Goldstone and from the patient explanations of David Olive and Edward Corrigan. I am also grateful to Nick Manton for interesting discussions. No attempt is made here at comprehensiveness; for a more complete treatment see Ref. 6.

1. EXTENDED OBJECTS IN TWO-DIMENSIONAL FIELD THEORIES.

1.1 Some examples.

We begin by discussing extended objects in field theories in a two-dimensional space-time. Consider the theory of a real scalar field φ given by the Lagrangian density

$$\mathcal{L} = \frac{1}{2}(\partial\varphi)^2 - U(\varphi) \tag{1.1}$$

giving equations of motion

$$\partial^2\varphi \equiv \frac{\partial^2\varphi}{\partial t^2} - \frac{\partial^2\varphi}{\partial x^2} = -U'(\varphi) \tag{1.2}$$

and the total energy or Hamiltonian is given by

$$H = \int_{-\infty}^{\infty} \left(\frac{1}{2}\left(\frac{\partial\varphi}{\partial t}\right)^2 + \frac{1}{2}\left(\frac{\partial\varphi}{\partial x}\right)^2 + U(\varphi) \right) dx. \tag{1.3}$$

It is clear that for the energy of the system to be bounded below the potential function $U(\varphi)$ must be. Without loss of generality we may take its mimimum value to be zero. So $U(\varphi) \geqslant 0$ and attains the value zero on a set of values of φ which we shall assume is discrete (in the sense of being countable and having no accumulation points).

We seek non-constant static solutions of finite energy. For definiteness first consider the φ^4 potential of Fig. 1,

$$U(\varphi) = \frac{\lambda}{2}\,(\varphi^2 - a^2)^2 \tag{1.4}$$

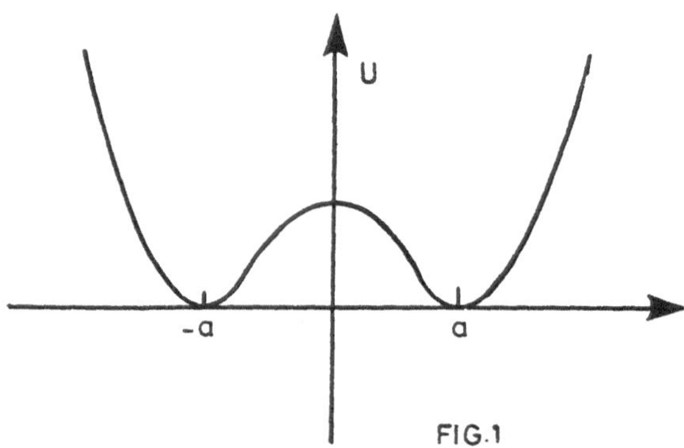

FIG.1

In this case we can find a time - independent static solution φ which tends to $\pm a$ as $x \to \pm\infty$. We integrate eq. (1.2) to give

$$\frac{1}{2}\left(\frac{\partial\varphi}{\partial x}\right)^2 = U(\varphi) \tag{1.5}$$

since $\frac{\partial\varphi}{\partial t} = 0$ by assumption. (No constant of integration can be introduced since we need both $\left(\frac{\partial\varphi}{\partial x}\right)^2$ and $U(\varphi)$ to tend to zero as $x \to \pm\infty$ to get a finite energy solution.) Thus

$$\int_0^{\varphi} \frac{d\varphi_1}{\sqrt{U(\varphi_1)}} = \pm(x-b), \qquad \text{b constant.} \tag{1.6}$$

This is illustrated in Fig. 2. For our explicit example of φ^4,

$$\frac{1}{\sqrt{\lambda}} \int_0^{\varphi} \frac{d\varphi_1}{a^2 - \varphi_1^2} = \pm(x-b),$$

giving

$$\frac{1}{2\sqrt{\lambda}a} \log\left(\frac{a+\varphi}{a-\varphi}\right) = \pm(x-b),$$

and hence

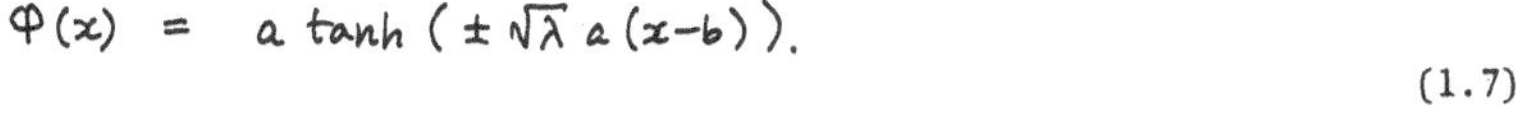

$$\varphi(x) = a \tanh\left(\pm\sqrt{\lambda}\, a(x-b)\right). \tag{1.7}$$

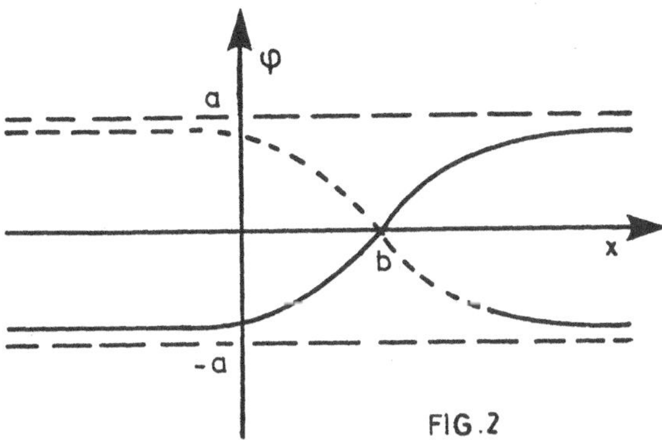

FIG.2

The solution with the plus sign in eq. (1.7) is shown as a solid line in Fig. 2, the minus sign corresponds to the dotted line. This solution is called a kink (for the plus sign), an antikink (for the minus sign). The sŏlutions move from one zero of U to an adjacent one as x goes from $-\infty$ to $+\infty$. This is a general feature not depending on the specific form of U; given any U with at least two zeros we can find such a solution. Let ξ_i, ξ_{i+1} be two successive zeros of U with $\xi_i < \xi_{i+1}$, as illustrated in Fig. 3. We can integrate eq. (1.5) to give

$$\int_{\varphi(0)}^{\varphi} \frac{d\varphi_1}{\sqrt{2U(\varphi_1)}} = \pm x \tag{1.8}$$

where $\varphi(0)$ is arbitrary subject to $\xi_i < \varphi(0) < \xi_{i+1}$. Eq. (1.8) implicitly defines two solutions sketched in Fig. 4, a lump (plus sign) and an antilump (minus sign, dotted), since as $\varphi \to \xi_{i+1}$, $x \to \pm\infty$ and as $\varphi \to \xi_i$, $x \to \mp\infty$ The solution φ is monotonically increasing or decreasing according to whether we take the plus or minus sign in eq. (1.8). The nomenclature is due to Coleman.[5]

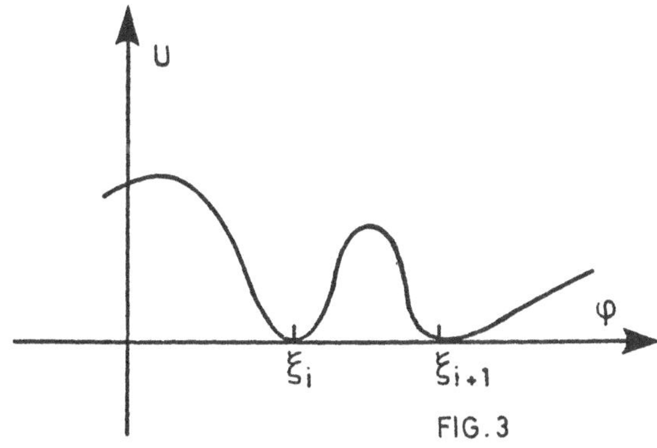

FIG.3

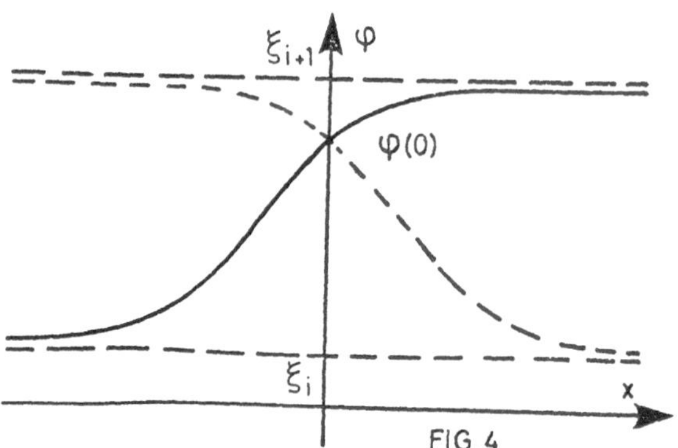

FIG.4

Other general properties of static solutions of eq. (1.2) for general U are that if $\varphi(x,t) = f(x)$ is a static solution so is f(x-b) and

$$\varphi(x,t) = f\left(\frac{x - vt}{\sqrt{1-v^2}}\right) \tag{1.9}$$

is also a solution of eq. (1.2). These are simple consequences of translation invariance and Lorentz invariance of that equation.

Thus in general lumps share some properties with particles: we can displace their centres of mass and we can boost them so that they move with any velocity less than that of light. To see what other properties they share in common we need to consider many 'lump' systems. Because the equations are non-linear we can not just add solutions to achieve this. To discuss this question further we look at another particular form for $U(\varphi)$, one that is particularly famous,

$$U(\varphi) = \frac{\alpha}{\beta^2}(1 - \cos\beta\varphi) \tag{1.10}$$

leading to the equation of motion

$$\partial^2\varphi + \frac{\alpha}{\beta}\sin\beta\varphi = 0, \tag{1.11}$$

which is usually refered to as the sine-Gordon equation. The corresponding finite energy static solutions are again easily evaluated from eq. (1.8), yielding

$$\varphi = \frac{4}{\beta}\tan^{-1}(\exp[\pm\sqrt{\alpha}\,x]) + \frac{2n\pi}{\beta} \tag{1.12}$$

where $n \in \mathbb{Z}$ (i.e. n is an integer) and the inverse tangent yields values between 0 and $\pi/2$. The solution (1.12) interpolates between the zeros of U at $\beta\varphi$ equals $2n\pi$ and $2(n+1)\pi$ The lump is called a <u>soliton</u> and the antilump an antisoliton. Usually, or at least often, the term soliton is used more loosely in particle physics to denote any lump. In applied mathematics the term soliton is frequently reserved for this sort of solution to a set of equations (of which the sine-Gordon equation is one) which possess remarkable properties. These equations may be completely solved through a series of <u>linear</u> problems (called the inverse scattering method) in terms of prescribed initial data. The system may thus be reduced to a normal coordinate Hamiltonian form. New solutions may be calculated from a given solution by a process which is somewhat analogous to a nonlinear version of Fourier analysis (called the Bäcklund transformation). For more details see Ref. 7.

To construct classical N soliton solutions in the sine-Gordon theory we must regard the solutions of eq. (1.12) for different n, but the same sign, as describing the same particle. This is because <u>if</u> we have a solution to an equation like eq. (1.2) φ say, such that for large t:

$$\varphi(x,t) \doteq f_1\left(\frac{x-a_1(t)}{\sqrt{1-V_1^2}}\right) \qquad \text{for } x \text{ near } a_1(t)$$

and

$$\varphi(x,t) \doteq f_2\left(\frac{x-a_2(t)}{\sqrt{1-V_2^2}}\right) \qquad \text{for } x \text{ near } a_2(t),$$

with $a_1(t) \to -\infty$, $a_2(t) \to \infty$ as $t \to \infty$ and the approximations get better in this limit, <u>then</u> we must have $f_1(\infty) = f_2(-\infty)$. Thus we have to follow the solution of (1.12), with the plus sign and n = r, with either the solution with the plus sign and n = r + 1 or the solution with the minus sign and n = r.

So we are forced to the above indentification to produce a two soliton state. This is physically permissable provided that we agree that the only functions of φ which are measurable are those which are periodic with period $2\pi/\beta$. Because of the remarkable properties of the sine-Gordon equation, N soliton solutions may be constructed explicitly in terms of elementary functions[7]. So soliton scattering may be studied. In this scattering the number of solitons and the number of antisolitons remains fixed, the velocities of the solitons are unchanged and all that happens to solitons is that they suffer a time delay. These are features which are special to the sine-Gordon equation depending on the details of its dynamics and connected with its exact solubility. The solution of eq. (1.11) which corresponds to a soliton-soliton collision is

$$\tan\frac{\beta\varphi}{4} = \frac{v\sinh\left(x(1-v^2)^{-\frac{1}{2}}\right)}{\cosh\left(vt(1-v^2)^{-\frac{1}{2}}\right)}$$

and that which corresponds to a soliton-anti-soliton collision is

$$\tan\frac{\beta\varphi}{4} = \frac{\sinh\left(vt(1-v^2)^{-\frac{1}{2}}\right)}{v\cosh\left(x(1-v^2)^{-\frac{1}{2}}\right)}$$

The sine-Gordon equation possesses one conservation law which can be deduced without reference to the detailed dynamics of the system. This law is topological in character and is a primitive version of other such laws which we will study in three spatial dimensions.

Suppose $\varphi(x,t)$ is any <u>finite energy</u> solution of the sine-Gordon equation. Then

$\phi(\infty,t) = \lim_{x\to\infty} \phi(x,t)$ must be a zero of U(ϕ). So $\phi(\infty,t) = 2n\pi/\beta$ for some integer n. Now an integer can not vary continuously with time unless it remains constant. Consequently $\phi(\infty,t)$ is independent of t. Likewise $\phi(-\infty,t)$ is independent of t, and the reason is essentially topological. We can not attach physical significance to the constants $\phi(\infty,t)$ and $\phi(-\infty,t)$ individually because we can only measure quantities invariant under $\phi \to \phi+2n\pi/\beta$, $m \in \mathbb{Z}$. But we can consider $\phi(\infty,t) - \phi(-\infty,t)$ which has this property. Now since ϕ increases by $2\pi/\beta$ as we pass a soliton and decreases by $2\pi/\beta$ as we pass an antisoliton,

$$\beta[\phi(\infty,t) - \phi(-\infty,t)]/2\pi = \text{soliton number}$$

$$= \text{(number of solitons) - (number of antisolitons)} \qquad (1.13)$$

is a constant of the motion. So without looking at the dynamics we can say that we have a conserved quantity, soliton number, which is like fermion number, but is conserved for a topological reason.

1.2 The Sine-Gordon equation and the Massive Thirring Model

All the discussion of section 1.1 has been classical. What happens when one of these scalar field theories is quantised? Do the lumps correspond in some sense to elementary particles? These questions have been answered in detail for the sine-Gordon theory by Coleman[3,5], whose results were foreshadowed to some extent by Skyrme[8] many years earlier. Coleman showed that the quantised Sine-Gordon equation was equivalent to another quantum field theory, the massive Thirring model, in which the quanta of the elementary field correspond to the solitons and that these are fermions. (To be more precise, he showed it equivalent to the zero charge sector of the Thirring model.) The massive Thirring model is described by the Langrangian density

$$\mathcal{L} = i\bar{\psi}\gamma_\mu\partial^\mu\psi - \tfrac{1}{2}g\, j_\mu j^\mu + m\bar{\psi}\psi\,; \quad j^\mu = \bar{\psi}\gamma^\mu\psi. \qquad (1.14)$$

Without the mass term the model is exactly soluble for $\hbar g > -\pi$. Coleman showed that perturbation theory in the mass term was equivalent to constructing the quantum theory of the sine-Gordon equation perturbatively in the potential U(ϕ) provided that

$$\frac{\beta^2\hbar}{4\pi} = \frac{1}{1+g\hbar/\pi} \qquad (1.15)$$

and $\beta^2\hbar < 8\pi$. Note that the classical limits of the models are not the same. This is not suprising because in the classical limit the Sine-Gordon theory has a particle-like structure corresponding to the fermions, i.e. the solitons, but the quanta of the ϕ field (the mesons) have 'dissolved'. It is only by going to quantum mechanics that we can expect to see this duality. The equivalence of the theories has subsequently made more rigorous by an existence proof given by Fröhlich[9].

So we have two equivalent descriptions of the same quantum field theory. In one there is a basic meson field and the fermion is an extended object, the soliton; in the other there is a basic fermi field and the meson is a fermion-anti-fermion bound state.

Finally consider fermion number conservation in the Thirring model. This is a consequence of the symmetry of the theory under $\psi \to e^{i\alpha}\psi$. Corresponding to this symmetry there is a conserved (Noether) current

$$j^\mu = \bar{\psi}\gamma^\mu\psi \tag{1.16}$$

The conservation of j^μ, $\partial_\mu j^\mu = 0$, follows from the equations of motion in the normal way. Now the relation with the sine-Gordon theory yields the correspondence

$$j^\mu = \frac{\beta}{2\pi}\epsilon^{\mu\nu}\partial_\nu\varphi$$

($\epsilon^{01} = -\epsilon^{10} = 1$; $\epsilon^{00} = \epsilon^{11} = 0$). In this description the conservation of j^μ does not depend on the equations of motion for φ. Also the charge, soliton number, equals

$$\int_{-\infty}^{\infty} j^0\,dx = \frac{\beta}{2\pi}\int_{-\infty}^{\infty}\frac{\partial\varphi}{\partial x}\,dx = \frac{\beta}{2\pi}(\varphi(\infty) - \varphi(-\infty)). \tag{1.18}$$

So we see that a topological conservation law in one description of the system is the result of a conventional symmetry in the other. One might speculate that this could occur in higher dimensions where one lacks such detailed results. Also notice that the divergence of the current associated with the topological conservation law vanishes purely 'kinematically'; this seems to be typical.

1.3 Higher Dimensions.

If we wish to find extended solutions to field theories in more than one spatial dimension we must look beyond theories containing only scalar fields. The reason for this is Derrick's Theorem[10] which states that if a field theory is defined by the Lagrangian density

$$\mathcal{L} = \tfrac{1}{2}(\partial_\mu\varphi\,\partial^\mu\varphi) - U(\varphi) \tag{1.19}$$

in D space and one time dimension, with the minimum value of U being zero, then there are no non-constant non-singular time-independent solutions with finite energy if $D \geqslant 2$. This is proved as follows. Suppose $\varphi_1(\underline{x})$ is a time independent non-singular solution with finite energy. For a time independent solution the action (per unit time) is

$$L[\varphi] = -\tfrac{1}{2}\int(\underline{\nabla}\varphi)^2 d^D\underline{x} - \int U(\varphi)\,d^D\underline{x} \equiv -V_1[\varphi] - V_2[\varphi] \tag{1.20}$$

If $\varphi_\lambda = \varphi_1(\lambda\underline{x})$,

$$L[\varphi_\lambda] = -\lambda^{(2-D)}V_1[\varphi_1] - \lambda^{-D}V_2[\varphi_1]. \tag{1.21}$$

Since $\frac{d}{d\lambda} L[\varphi_\lambda]\Big|_{\lambda=1} = 0, \quad (D-2)V_1[\varphi_1] + D V_2[\varphi_1] = 0.$

(1.22)

Since V_1 and V_2 are non-negative contributions to the energy (and so are finite), $V_2(\varphi_1) = 0$ and $V_1(\varphi_1) = 0$ if $D > 2$. If $D = 2$ we see immediately that $V_2[\varphi_1] = 0$. So since it minimises V_2 alone it must also make V_1 stationary by itself which yields φ_1 equals a constant (provided that the values of φ lie in some linear space).

To avoid this scaling argument we must prevent the division of the action into two separate positive terms which scale at different rates. This is acheived in gauge theories where $(\partial\varphi)^2$ is replaced by $(\partial\varphi + ieA\varphi)^2$ in the action. Thus motivated we will turn our attention to these.

2. GAUGE THEORIES AND ELECTROMAGNETISM.

2.1 Conventional Electromagnetism.

We begin to discuss gauge field theories by reviewing the formalism of the prototype gauge thoery: classical electromagnetism. Here the field tensor is related to the potential A^μ by the equation

$$F^{\mu\nu} = \partial^\mu A^\nu - \partial^\nu A^\mu \tag{2.1}$$

and this definition is invariant under the gauge transformation

$$A^\mu(x) \rightarrow A^\mu(x) + \frac{1}{e}\partial^\mu \Lambda(x) \tag{2.2}$$

From eq. (2.1) we obtain immediately the Bianchi identities:

$$\partial^\lambda F^{\mu\nu} + \partial^\mu F^{\nu\lambda} + \partial^\nu F^{\lambda\mu} = 0 \tag{2.3}$$

which are 'kinetic' in the sense of section 1 . They can be written more compactly as

$$\partial^\lambda F^*_{\mu\lambda} = 0 \quad \text{where} \quad F^*_{\lambda\mu} = -\tfrac{1}{2}\epsilon_{\lambda\mu\nu\rho}F^{\nu\rho} \tag{2.4}$$

is called the dual field tensor. ($\epsilon_{\lambda\mu\nu\rho}$ is totally antisymmetric with $\epsilon_{0123} = 1$.)

The equations of motion of the fields, with a given charge and current distribution $(j^\mu) = (\rho, \underline{j})$, are obtained from the action

$$-\frac{1}{4}\int F_{\mu\nu}F^{\mu\nu}d^4x - \int j_\mu A^\mu d^4x \tag{2.5}$$

The Euler-Lagrange equation $\partial^\mu\left(\frac{\partial\mathcal{L}}{\partial(\partial^\mu A^\nu)}\right) - \frac{\partial\mathcal{L}}{\partial A^\nu} = 0$ reads

$$\partial_\nu F^{\mu\nu} = -j^\mu \tag{2.6}$$

The electric and magnetic fields $\underline{E}$ and $\underline{B}$ are contained in F as shown in eq. (2.7),

$$(F^{\mu\nu}) = \begin{pmatrix} 0 & -E^1 & -E^2 & -E^3 \\ E^1 & 0 & -B^3 & B^2 \\ E^2 & B^3 & 0 & -B^1 \\ E^3 & -B^2 & B^1 & 0 \end{pmatrix} \tag{2.7}$$

$\underline{E} = (E^i)$, etc., and hence $E^i = F^{i0}, \quad F^{ij} = -\epsilon_{ijk} B^k$. So we have the following correspondence between eqs. (2.4) and (2.6) and the familiar Maxwell eqs., using $(x^\mu) = (t, \underline{x})$

$$\partial^\nu F^*_{\mu\nu} = 0 \qquad \begin{cases} \underline{\nabla} \wedge \underline{E} = -\dot{\underline{B}} \\ \underline{\nabla} . \underline{B} = 0 \end{cases} \tag{2.8}$$

$$\partial_\nu F^{\mu\nu} = -j^\mu \qquad \begin{cases} \underline{\nabla} . \underline{E} = \rho \\ \underline{\nabla} \wedge \underline{B} = \underline{j} + \dot{\underline{E}} \end{cases} \tag{2.9}$$

Since

$$(A^\mu) = (\varphi, \underline{A}), \qquad \underline{E} = -\underline{\nabla}\varphi - \dot{\underline{A}}, \quad \underline{B} = \underline{\nabla} \wedge \underline{A}. \tag{2.10}$$

Now consider the interaction between the electromagnetic field F and a real or complex scalar field φ. The form of the gauge transformation on φ is

$$\varphi \rightarrow e^{-iQ\Lambda} \varphi, \qquad \Lambda \equiv \Lambda(x), \tag{2.11}$$

where Q is a hermitian matrix ($Q^+ = Q$) and eQ is the charge operator. (φ is in general a vector in the internal space.) Under the gauge transformation,

$$\partial^\mu \varphi \rightarrow e^{-iQ\Lambda} (\partial^\mu \varphi - iQ\partial^\mu \Lambda \varphi) \tag{2.12}$$

not $e^{-iQ\Lambda} \partial^\mu \varphi$. But, because of eq. (2.2), if we define

$$\mathcal{D}^\mu \varphi = (\partial^\mu + ieA^\mu Q)\varphi \tag{2.13}$$

we have

$$\mathcal{D}^\mu \varphi \rightarrow e^{-iQ\Lambda} \mathcal{D}^\mu \varphi \tag{2.14}$$

and we can construct a gauge invariant Lagrangian density by setting

$$\mathcal{L} = -\frac{1}{4} F_{\mu\nu} F^{\mu\nu} + (\mathcal{D}_\mu \varphi)^+ \mathcal{D}^\mu \varphi - U(\varphi) \tag{2.15}$$

provided that U is invariant under global gauge transformations, $U(e^{i\alpha Q}\varphi) = U(\varphi)$ (since it involves no derivatives). This action yields the equations of motion,

$$\partial_\nu F^{\mu\nu} = -j^\mu \tag{2.16}$$

where now,

$$j^\mu = ie\,\varphi^+ Q \mathcal{D}^\mu \varphi - ie(\mathcal{D}^\mu \varphi)^+ Q \varphi, \tag{2.17}$$

and

$$\mathcal{D}_\mu \mathcal{D}^\mu \varphi = - U'(\varphi) \tag{2.18}$$

and these are still supplemented by the Bianchi identities, $\partial^\nu F^*_{\mu\nu} = 0$, which are purely kinematic. Note that if we write $g = e^{i\Lambda}$, $\in U(1)$, the group of unimodular complex numbers, the gauge transformations take the form,

$$\begin{aligned} A^\mu &\to A^\mu + \tfrac{i}{e}(\partial^\mu g)g^{-1} \qquad & F^{\mu\nu} &\to F^{\mu\nu} \\ \varphi &\to D(g)\varphi & \mathcal{D}^\mu\varphi &\to D(g)\mathcal{D}^\mu\varphi, \end{aligned} \tag{2.19}$$

where $D(g) = e^{iQ\Lambda}$ is a representation of U(1).

Note that the constant e which we have introduced has the dimensions of inverse charge rather than charge. When this theory is quantised we would expect the possible electric charges in the theory to be the eigenvalues of $e\hbar Q$. Since we have taken the velocity of light, c, to be unity, the dimensions of action, $[\hbar]$, $ML^2T^{-1} \equiv ML$. The dimensions of charge, $[q_0] = M^{\frac{1}{2}}L^{\frac{1}{2}}$ so that $[e] = M^{-\frac{1}{2}}L^{-\frac{1}{2}}$. There is no basic unit of electric charge in this theory until we make the passage to quantum mechanics.

2.2 Dirac Monopoles.

We wish to discuss Dirac's incorporation of magnetic monopoles in the theory of electromagnetism because the first example of an extended object in three dimensional field theory to be constructed[1,2] has an interpretation as a magnetic monopole. Dirac's original motivation in his 1931 paper was to understand the quantization of el ectric charge: he showed that this would result in a quantum theory of electric and magnetic charges. His original approach was to show that a quantum mechanics in which the wave functions had path-dependent phase factors led naturally to such a theory. Subsequently[11] in 1948, he gave a more comprehensive treatment starting from an action principle for a fixed number of electrically and magnetically charged particles interacting with the electromagnetic field. Here it is clear that the quantization condition which relates the electric and magnetic charges is only necessary in the quantum theory. (For a review of Dirac monopoles with particular reference to the experimental situation see Ref. 12. The theory is reviewed in Refs. 13 and 6.)

We can write down a classical theory involving magnetic charges if we relax the Bianchi identities which, through $\underline{\nabla}\cdot\underline{B} = 0$, dictate the absence of magnetic charge. We replace them by field equations, introducing a magnetic current k^μ as a source for $F^*_{\mu\nu}$ so that we have symmetry under the Dirac duality transformation $F^{\mu\nu} \to F^{*\mu\nu}$, $F^*_{\mu\nu} \to -F_{\mu\nu}$, $j^\mu \to k^\mu$, $k^\mu \to -j^\mu$.

$$\partial^\nu F_{\mu\nu} = -j_\mu, \qquad \partial^\nu F^*_{\mu\nu} = -k_\mu, \tag{2.20}$$

$$F^*_{\lambda\mu} = -\tfrac{1}{2}\epsilon_{\lambda\mu\nu\rho}F^{\nu\rho}$$

If k^μ is non-vanishing we cannot introduce a potential A^μ as in eq. (2.1). If we have particles labelled by an index i with electric charges q_i and magnetic charges g_i travelling along paths X_i^μ the currents are given by

$$j^\mu(x) = \sum_i q_i \int \delta_4(x - X_i)\, dX_i^\mu \tag{2.21}$$

$$k^\mu(x) = \sum_i g_i \int \delta_4(x - X_i)\, dX_i^\mu \tag{2.22}$$

All we lack is the potential A^μ. This is not essential for classical mechanics but it is necessary if we are to obtain the Schrödinger equation for a particle, possessing electric charge q, moving in the electromagnetic field, by simply replacing p^μ by ($p^\mu - qA^\mu$) that is $i\hbar\partial^\mu$ by $i\hbar\partial^\mu - qA^\mu \equiv i\hbar(\partial^\mu + ieA^\mu)$ where $q = \hbar e$. Dirac's approach was to construct a potential which is not uniquely specified by $F^{\mu\nu}$ even up to (non-singular) gauge transformations, but whose curl, $\partial^\mu A^\nu - \partial^\nu A^\mu$ yields $F^{\mu\nu}$ everywhere except on certain curves, one emanating from each magnetic pole. (A^μ is determined by $F^{\mu\nu}$ and these curves up to gauge transformations.) He then required that the quantum mechanics for different A^μ be equivalent and obtained a quantisation condition.

To describe his approach consider the field due to a magnetic pole of strength m at the origin:

$$\underline{B} = \frac{g}{4\pi r^3}\underline{r}, \qquad \underline{\nabla}.\underline{B} = g\,\delta_3(\underline{r}) \tag{2.23}$$

Dirac associates with B the potential which gives B modified by a magnetic flux fed in along an arbitrary curve $\mathcal{S}$ from the origin to infinity (see Fig. 5).

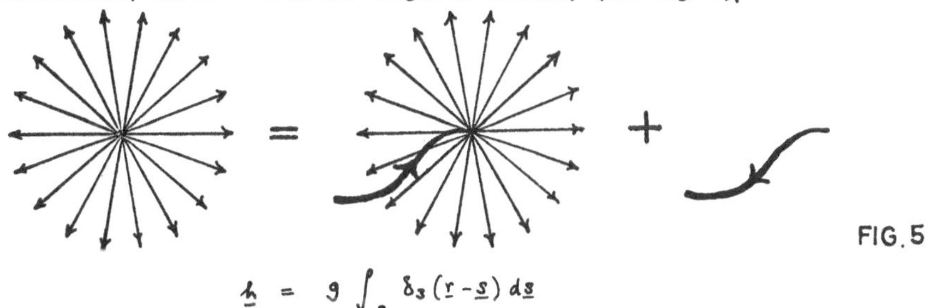

FIG. 5

If

$$\underline{h} = g\int_{\mathcal{S}} \delta_3(\underline{r} - \underline{s})\, d\underline{s} \tag{2.24}$$

$$\oiint_{S} \underline{h}.d\underline{S} = g \tag{2.25}$$

where S is any closed surface enclosing the origin. This implies

$$\underline{\nabla}.\underline{h} = g\,\delta_3(\underline{r}) \tag{2.26}$$

We split $\underline{B}$ up as indicated in Fig. 5,

$$\underline{B} = (\underline{B} - \underline{h}) + \underline{h}. \tag{2.27}$$

Then

$$\underline{\nabla}\,(\underline{B} - \underline{h}) = 0 \tag{2.28}$$

so that we may introduce $\underline{A}$ so that

$$\underline{B} - \underline{h} = \underline{\nabla} \wedge \underline{A} \tag{2.29}$$

So that

$$\underline{B} = \underline{\nabla} \wedge \underline{A} + \underline{h} \tag{2.30}$$

The curve $\mathcal{S}$ is called the Dirac string associated with the potential A^{μ}. We can give an explicit formula for $\underline{A}$ in terms of $\mathcal{S}$ and $\underline{B}$[13],

$$\underline{A} = -\int d\underline{s} \wedge \underline{B}(\underline{r} - \underline{s}). \tag{2.31}$$

From this we may verify eq. (2.29)

$$\underline{\nabla} \wedge \underline{A} = -\int_{\mathcal{S}} d\underline{s}\,(\underline{\nabla}_r . \underline{B}(\underline{r} - \underline{s})) + \int_{\mathcal{S}} (d\underline{s} . \underline{\nabla}_r)\, \underline{B}(\underline{r} - \underline{s}) \tag{2.32}$$

$$= -g \int_{\mathcal{S}} d\underline{s}\ \delta_3(\underline{r} - \underline{s}) - \int_{\mathcal{S}} (d\underline{s} . \underline{\nabla}_s)\, \underline{B}(\underline{r} - \underline{s})$$

$$= -\underline{h}(\underline{r}) + \underline{B}(\underline{r}). \tag{2.33}$$

So for the monopole field of eq. (2.23) with $\mathcal{S}$ the negative z-axis

$$\underline{A} = \frac{g}{4\pi r} \cdot \frac{\sin\theta}{1 + \cos\theta}\, \hat{\underline{\chi}} \tag{2.34}$$

where $\hat{\underline{\chi}}$ is a unit vector in the azimuthal direction about the (positive) z-axis. For $\mathcal{S}$ a straight line in the direction of the unit vector $\hat{\underline{n}}$

$$\underline{A} = \frac{g}{4\pi r}\ \frac{\hat{\underline{n}} \wedge \underline{r}}{r + \hat{\underline{n}} . \underline{r}} \tag{2.35}$$

The particular choice of $\underline{A}$ must not lead to any physical effects dependent on the Dirac string $\mathcal{S}$. Near $\mathcal{S}$ the only terms in the wave equation which are rapidly varying, are $\underline{A}$ and consequently the wave function ψ. The effects of this rapid variation may be absorbed into a phase factor u which satisfies $(\underline{\nabla} + ie\underline{A})\,u = 0$. To calculate the change in phase of ψ between two points both near to one another and near to the string we integrate along a path joining them

$$\psi(x_2) = \exp\left(-ie \int_{x_1}^{x_2} \underline{A} . d\underline{x}\right) \psi(x_1). \tag{2.36}$$

In particular if we are not to see interference effects (Bohm-Aharonov effect) we must have

$$\exp\left(-ie \oint \underline{A} . d\underline{x}\right) = 1, \tag{2.37}$$

where the integral is taken round a small closed loop encircling $\mathcal{S}$. Thus

$$e\oint \underline{A}'\, d\underline{x} = 2n\pi, \quad n \in \mathbb{Z}. \tag{2.38}$$

But for this loop $\oint \underline{A}.d\underline{x} = g$, the flux along $\mathcal{S}$. Consequently we have the quantisation condition

$$\frac{qg}{4\pi\hbar} = \frac{eg}{4\pi} = \frac{n}{2}, \qquad n \in \mathbb{Z}, \tag{2.39}$$

for a quantum mechanical particle of charge the moving in the (fixed) field of a magnetic monopole.

To see that this condition does ensure the equivalence of the quantum theories of the motion of an electrically charged particle in the field of a magnetic monopole for different choices of $\mathcal{S}$ we consider two choices of $\mathcal{S}$, $\mathcal{S}_1$ and $\mathcal{S}_2$. Corresponding to these we will have two decompositions of $\underline{B}$

$$\underline{B} = \underline{\nabla}\wedge\underline{A}_1 + \underline{h}_1 = \underline{\nabla}\wedge\underline{A}_2 + \underline{h}_2, \qquad \text{respectively,} \tag{2.40}$$

leading to

$$\underline{\nabla}\wedge(\underline{A}_2 - A_1) = \underline{h}_1 - \underline{h}_2 \tag{2.41}$$

$$= g\int_{\mathcal{C}} \delta_3(\underline{r}-\underline{s})\,d\underline{s} \tag{2.42}$$

where $\mathcal{C}$ is the curve $-\mathcal{S}_2$ followed by $\mathcal{S}_1$. $\mathcal{C}$ is effectively a loop closed at infinity. We can compare the problem of solving eq. (2.42) with the problem, from elementary electromagnetic theory, of finding the magnetic field $\underline{H}$ due to a current I flowing in a closed loop $\mathcal{C}$; that is of solving

$$\underline{\nabla}\wedge\underline{H} = \underline{j} \tag{2.43}$$

where

$$\underline{j} = I\int_{\mathcal{C}} \delta_3(\underline{r}-\underline{s})\,d\underline{s}\,.$$

The solution to this problem can be given in terms of the gradient of the solid angle subtended by the loop $\mathcal{C}$ at the point at which the field is being measured. This is a many-valued function of position changing by 4π as we encircle the loop $\mathcal{C}$. To pick a specific value we consider a particular spanning surface of $\mathcal{C}$, $\mathcal{S}$ say, and take the angle subtended by this surface, which, as given by the formula

$$\Omega(\underline{r}) = \int_{\mathcal{S}} \frac{d\underline{S}.(\underline{r}-\underline{s})}{|\underline{r}-\underline{s}|^3}, \tag{2.44}$$

is single-valued but jumps by 4π as we cross $\mathcal{S}$. Then $\underline{H}$ is given by

$$\underline{H} = -\frac{I}{4\pi}\underline{\nabla}\Omega + \left\{ I\int_{\mathcal{S}} \delta_3(\underline{r}-\underline{s})\,d\underline{S} \right\} \tag{2.45}$$

where the term in brackets contributes only on $\mathcal{S}$ and cancels the singularity of Ω

there. ($\underline{s}$ is the integration variable in eqs. (2.44) and (2.45)). In consequence the solution of eq. (2.42) is

$$\underline{A}_2 - \underline{A}_1 = -\frac{g}{4\pi}\underline{\nabla}\Omega_{12} + \left\{ g \int_S \delta_3(\underline{r}-\underline{s})\, d\underline{S} \right\} \tag{2.46}$$

where S is <u>any</u> surface spanning $\mathcal{C}$. Now suppose two electron wave functions are related by

$$\psi_2(\underline{r}) = \exp\left(-\frac{ieg}{4\pi}\Omega(\underline{r})\right)\psi_1(\underline{r}) \tag{2.47}$$

Ω depends on the choice of S but this only introduces an ambiguity of a multiple of 4π. Consequently the relation eq. (2.47) will be <u>independent</u> of the choice of S provided that

$$\frac{eg}{4\pi} \in \mathbb{Z} \tag{2.48}$$

(i.e. the Dirac quantization condition is satisfied). Given (2.48) we can always choose S so that the point under consideration $\underline{r} \notin S$. Then we can use eq. (2.46) ignoring the term in brackets and

$$\begin{aligned}(\partial^r + ieA_2^r)\psi_2 &= \exp\left(-i\frac{eg}{4\pi}\Omega\right)\left(\partial^r + ieA_2^r - \frac{ieg}{4\pi}\partial^r\Omega\right)\psi_1 \\ &= \exp\left(-\frac{ieg}{4\pi}\Omega\right)(\partial^r + ieA_1^r)\psi_1\end{aligned} \tag{2.49}$$

Thus given the Dirac quantization condition the wave functions corresponding to the two potentials will be related by a unitary transformation; if ψ_2 satisfies the Schrödinger equation with $p^r \to i\hbar(\partial^r + ieA_2^r)$, ψ_1 given by eq. (2.47) will satisfy it with $p^r \to i\hbar(\partial^r + ieA_1^r)$. This argument will easily extend to the situation of an electron moving in the prescribed field of a number of magnetic poles.

The following <u>heuristic</u> argument is given by Zumino.[12] Consider the motion of a particle of mass m in the field of a magnetic pole of strength g. The equation of motion is given by the Lorentz force law:

$$m\ddot{\underline{r}} = q\,\dot{\underline{r}} \wedge \frac{g}{4\pi}\,\underline{r}/r^3 \tag{2.50}$$

assuming the particle to have electric charge $q = e\hbar$ and no magnetic charge. The rate of change of orbital angular momentum,

$$\begin{aligned}\frac{d}{dt}(m\,\underline{r}\wedge\dot{\underline{r}}) &= \frac{qg}{4\pi}\,\underline{r}\wedge(\dot{\underline{r}}\wedge\underline{r})\frac{1}{r^3} \\ &= \frac{qg}{4\pi}\left(\dot{\underline{r}}/r - \frac{(\dot{\underline{r}}\cdot\underline{r})\underline{r}}{r^3}\right) \\ &= \frac{qg}{4\pi}\frac{d}{dt}(\underline{r}/r).\end{aligned} \tag{2.51}$$

This leads us to add $-qg\hat{\underline{r}}/4\pi$ to the orbital angular momentum to give a total angular momentum $\underline{J}$ which is conserved.

$$\frac{d}{dt}(\underline{J}) = \underline{0} \quad \text{if} \quad \underline{J} = m\underline{r}\wedge\dot{\underline{r}} - \frac{qg}{4\pi}\hat{\underline{r}} \tag{2.52}$$

Now the component of $\underline{J}$ in the radial direction:

$$\hat{\underline{r}}.\underline{J} = -\frac{qg}{4\pi} \tag{2.53}$$

and if we require this to be $-\frac{n\hbar}{2}$ for some $n \in \mathbb{Z}$, as is the component of angular momentum in any fixed direction in a quantum theory, we obtain Dirac's quantization condition

$$\frac{eg}{4\pi} = \frac{qg}{4\pi\hbar} = \frac{n}{2} \quad \text{for some} \quad n \in \mathbb{Z}. \tag{2.54}$$

A similar argument to this may be used to obtain quantization conditions on monopole solutions in non-Abelian gauge theories. However, in this case the theories are classical and the argument rigorous.[15] The additional term $-qg\hat{\underline{r}}/4\pi$ in $\underline{J}$ can be understood as the angular momentum residing in the electromagnetic field. (See e.g. Ref. 6.)

Note that Dirac's quantization condition is a quantum effect. It simultaneously implies the quantization of electric and magnetic charges. Electric charges come in units of $2\pi\hbar/g_0$ where g_0 is the smallest magnetic charge and magnetic charges come in units of $2\pi\hbar/q_0$, where q_0 is the smallest electric charge. It is this explanation of charge quantization provided naturally by the existence of a magnetic monopole that Dirac found interesting. Classically we can get no quantization between poles as the dimensions of $[qg] = ML = [\hbar]$ $(c=1)$. In non-Abelian gauge theories there is quantization at the classical level (see section 3.4) but it is a quantization of the magnetic charge in terms of the inverse coupling constant of the gauge theory. Note that, since eq. (2.54) cannot be maintained while performing a perturbation expansion in positive powers of e and g, difficulties may be expected in a perturbation theoretic development of quantum electrodynamics. This is indeed the case[16,13] but a consistent quantum field theory can be constructed if a quantization condition is maintained.[17] The quantization condition suggests that magnetic poles may be a non-perturbative phenomenon in terms of the electromagnetic field as usually formulated and, with hindsight, the analogy with the solitons in one spatial dimension, which are non-perturbative in terms of the φ field, is obvious.

2.3 Non-Abelian Gauge Theories.

We consider the generalisation of the gauge theory of section 2.1 from U(1) to a general compact Lie gauge group G. We can regard G as a group of matrices by taking any faithful (i.e. one to one) representation of G. Suppose $\{T^a\}$ are set of hermitian generators for G, that is a basis for the Lie algebra L(G) of G. Suppose $\varphi(x)$ is a n-dimensional, Lorentz-scalar, (real or complex) field on which there acts a unitary representation D of G,

$$\varphi \rightarrow D(g)\varphi \tag{2.55}$$

By taking $g \equiv g(x)$ we can define local gauge transformations. Under these $\partial^\mu\varphi$ transforms as follows (cf. eq. (2.12)),

$$\partial^\mu\varphi \rightarrow D(g)\,\partial^\mu\varphi + \{\partial^\mu D(g)\}\varphi . \tag{2.56}$$

To remove the unwanted second term in the right hand side of the transformation (2.56) we introduce gauge fields W^μ_a and associate with them a matrix in the Lie algebra

$$\underset{\sim}{W}^\mu = W^\mu_a T^a \in L(G) \tag{2.57}$$

If we specify that under a gauge transformation

$$\underset{\sim}{W}^\mu \rightarrow g\underset{\sim}{W}^\mu g^{-1} + \frac{i}{e}(\partial^\mu g)\,g^{-1} \tag{2.58}$$

(Compare eqs. (2.2) and (2.19)) then

$$\begin{aligned} \mathcal{D}^\mu\varphi \equiv \partial^\mu\varphi + ie\,D(\underset{\sim}{W}^\mu)\varphi \rightarrow D(g)\,(\partial^\mu + ieD(\underset{\sim}{W}^\mu))\,\varphi + \{[\partial^\mu D(g)]\,D(g^{-1}) \\ - D((\partial^\mu g)\,g^{-1})\}\,D(g)\varphi \end{aligned} \tag{2.59}$$

$$= D(g)(\partial^\mu + ie\,D(\underset{\sim}{W}^\mu))\varphi = D(g)\,\mathcal{D}^\mu\varphi \tag{2.60}$$

[Since $\quad D(g+\delta g)\,D(g^{-1}) = D(1+(\delta g)g^{-1}) = 1 + D((\delta g)g^{-1}) + O(\delta g)^2$

$\qquad (D(g+\delta g) - D(g))\,D(g^{-1}) = D((\delta g)g^{-1}) + O(\delta g)^2$]

To define the field tensor for the non-Abelian gauge fields in a way which makes its transformation properties clear consider

$$\begin{aligned} [\mathcal{D}^\mu, \mathcal{D}^\nu]\varphi &\equiv \mathcal{D}^\mu(\mathcal{D}^\nu\varphi) - \mathcal{D}^\nu(\mathcal{D}^\mu\varphi) \\ &= [\partial^\mu + ie\,D(\underset{\sim}{W}^\mu),\ \partial^\nu + ie\,D(\underset{\sim}{W}^\nu)]\,\varphi \\ &= ie\,\{D(\partial^\mu\underset{\sim}{W}^\nu) - D(\partial^\nu\underset{\sim}{W}^\mu) + ie\,D[\underset{\sim}{W}^\mu, \underset{\sim}{W}^\nu]\}\,\varphi \end{aligned} \tag{2.61}$$

So if we introduce gauge fields $G^{\mu\nu}_a$ such that

$$\underset{\sim}{G}^{\mu\nu} = G^{\mu\nu}_a T^a = \partial^\mu\underset{\sim}{W}^\nu - \partial^\nu\underset{\sim}{W}^\mu + ie[\underset{\sim}{W}^\mu, \underset{\sim}{W}^\nu] \tag{2.62}$$

we have

$$[\mathcal{D}^\mu, \mathcal{D}^\nu]\,\varphi = ie\,D(\underset{\sim}{G}^{\mu\nu})\,\varphi \tag{2.63}$$

From this equation the simple gauge transformation property of $G^{\mu\nu}_a$ may be deduced.

From eq. (2.60) it follows that $\mathcal{D}^\mu\mathcal{D}^\nu\varphi \rightarrow D(g)\,\mathcal{D}^\mu\mathcal{D}^\nu\varphi$, which equals $\{D(g)\mathscr{D}^\mu\mathscr{D}^\nu D(g^{-1})\}D(g)\varphi$. So from eq. (2.63) we see that

$$D(\underset{\sim}{G}^{\mu\nu})\varphi \rightarrow D(g\underset{\sim}{G}^{\mu\nu}g^{-1})D(g)\varphi \tag{2.64}$$

implying

$$\underset{\sim}{G}^{\mu\nu} \rightarrow g\underset{\sim}{G}^{\mu\nu}g^{-1} \tag{2.65}$$

since eq. (2.64) is true for arbitrary φ, in any representation. Notice that $\underset{\sim}{G}^{\mu\nu}$ transforms covariantly according to the adjoint representation of the group; it is only an invariant for an Abelian group. [The adjoint representation of a Lie group is the one with the same dimension as the group defined by $(\xi_a) \rightarrow \xi_a' = D_{ab}(g)\xi_b$ where $\underset{\sim}{\xi}' = \xi_a' T^a = g\underset{\sim}{\xi}g^{-1}$]

For an infinitessimal gauge transformation $g(x) = 1 - iT^a\epsilon_a(x)$ (2.66)

$$\delta\varphi_\alpha = i\epsilon_a D_{\alpha\beta}(T^a)\varphi_\beta, \qquad \delta W_a^\mu = -iC^{bc}{}_a\epsilon_b W_c^\mu + \frac{1}{e}\partial^\mu\epsilon_a \tag{2.67}$$

$$\delta G_a^{\mu\nu} = -i\epsilon_b C^{bc}{}_a G_c^{\mu\nu} \tag{2.68}$$

where $C^{ab}{}_c$ are the structure constants of the group (corresponding to the basis $\{T^a\}$ for $L(G)$),

$$[T^a, T^b] = iC^{ab}{}_c T^c \tag{2.69}$$

We can always arrange that $\mathrm{Tr}(T^aT^b) = \kappa\,\delta^{ab}$ and then $C^{ab}{}_c$ is totally antisymmetric in a, b and c.

There is a generalisation of the Bianchi identities of eqs. (2.3), (2.4) which follows from the Jacobi identity for the differential operators $\mathscr{D}^\lambda$.

$$[\mathscr{D}^\lambda,[\mathscr{D}^\mu,\mathscr{D}^\nu]] + [\mathscr{D}^\mu,[\mathscr{D}^\nu,\mathscr{D}^\lambda]] + [\mathscr{D}^\nu,[\mathscr{D}^\lambda,\mathscr{D}^\mu]] = 0. \tag{2.70}$$

Applying this to any φ , since

$$\begin{aligned}[\mathscr{D}^\lambda,[\mathscr{D}^\mu,\mathscr{D}^\nu]]\varphi &= ie\,[\mathscr{D}^\lambda, D(\underset{\sim}{G}^{\mu\nu})]\varphi \\ &= ie\,D(\mathscr{D}^\lambda\underset{\sim}{G}^{\mu\nu})\varphi\end{aligned} \tag{2.71}$$

where $\mathscr{D}^\lambda\underset{\sim}{G}^{\mu\nu} = \partial^\lambda\underset{\sim}{G}^{\mu\nu} + ie[\underset{\sim}{W}^\lambda,\underset{\sim}{G}^{\mu\nu}]$ is the covariant derivative (2.72) of $\underset{\sim}{G}^{\mu\nu}$; we obtain

$$\mathscr{D}^\lambda\underset{\sim}{G}^{\mu\nu} + \mathscr{D}^\mu\underset{\sim}{G}^{\nu\lambda} + \mathscr{D}^\nu\underset{\sim}{G}^{\lambda\mu} = 0. \tag{2.73}$$

We can rewrite these Bianchi identities by introducing the dual field tensor

$$\underset{\sim}{G}^*_{\lambda\mu} = -\frac{1}{2}\epsilon_{\lambda\mu\nu\rho}\underset{\sim}{G}^{\nu\rho}. \tag{2.74}$$

Eq. (2.73) then takes the form

$$\mathscr{D}^{\lambda}\underset{\sim}{G}^{*}_{\mu\lambda} = 0 \tag{2.75}$$

in exact parallel with eq. (2.4). The covariant derivative given in eq. (2.72) is the appropriate form of eq. (2.59) for the adjoint representation (eq. (2.65)). Under a gauge transformation $\mathscr{D}^{\lambda}\underset{\sim}{G}^{\mu\nu} \rightarrow g\,\mathscr{D}^{\lambda}\underset{\sim}{G}^{\mu\nu}g^{-1}$.

We can construct a manifestly gauge invariant Lagrangian density

$$\mathscr{L} = -\tfrac{1}{4}G_a^{\mu\nu}G_{a\mu\nu} + (\mathscr{D}^{\mu}\varphi)^{\dagger}\mathscr{D}_{\mu}\varphi - U(\varphi) \tag{2.76}$$

where the potential U is symmetric under G, $U(\mathcal{D}(g)\varphi) = U(\varphi)$ and $Tr\,(T^a T^b) = \kappa\,\delta_{ab}$
The invariance of the field tensor term follows from the cyclic property of traces:

$$\begin{aligned} G_a^{\mu\nu}G_{a\mu\nu} = \tfrac{1}{\kappa}Tr\,(\underset{\sim}{G}^{\mu\nu}\underset{\sim}{G}_{\mu\nu}) &\rightarrow \tfrac{1}{\kappa}Tr(g\underset{\sim}{G}^{\mu\nu}g^{-1}g\,\underset{\sim}{G}_{\mu\nu}g^{-1}) \\ &= \tfrac{1}{\kappa}Tr\,(\underset{\sim}{G}^{\mu\nu}\underset{\sim}{G}_{\mu\nu}) \end{aligned} \tag{2.77}$$

The Lagrangian density (2.76) leads to equations of motion

$$\mathscr{D}^{\nu}\underset{\sim}{G}_{\mu\nu} = -\underset{\sim}{j}_{\mu} \tag{2.78}$$

$$\mathscr{D}_{\mu}\mathscr{D}^{\mu}\varphi = -U'(\varphi) \tag{2.79}$$

where $\quad j_a^{\nu} = ie\,\varphi^{\dagger}T^a\mathscr{D}^{\nu}\varphi - ie\,(\mathscr{D}^{\nu}\varphi)^{\dagger}T^a\varphi$

which are supplemented by the Bianchi identities eq. (2.75); these are purely kinematic. (Eqs. (2.76) to (2.79) follow the conventions for a complex scalar field. For a real scalar field, $\mathscr{L} = -\tfrac{1}{4}G_a^{\mu\nu}G_{a\mu\nu} + \tfrac{1}{2}(\mathscr{D}_{\mu}\varphi)^{T}\mathscr{D}^{\mu}\varphi - U(\varphi)$ and $j_a^{\mu} = ie\varphi^{T}T_a\mathscr{D}^{\mu}\varphi$).

3. EXTENDED OBJECTS IN GAUGE THEORIES.

3.1 Generalities.

We will now discuss non-singular finite energy solutions to non-Abelian gauge field theories in three-dimensional space. Before considering specific solutions, we will make some general observations in an attempt to gain some preliminary understanding of the stranger features of those solutions, such as the correlations between directions in physical space and internal space.

It is always possible to choose a gauge in which the time components of the gauge potentials, W_a° , vanish (that is given any solution we can find a gauge transformation which takes it into a solution with $W_a^{\circ} = 0$; for a proof see Coleman's lectures[5] or Ref. 6.). In such a gauge the energy takes a simple form

$$E = \int d^3\underline{x} \left\{ \frac{1}{2} (\partial^0 W_a^i)^2 + |\partial^0 \varphi|^2 + \frac{1}{4} (G_{ij}^a)^2 + |\mathcal{D}^i \varphi|^2 + U(\varphi) \right\} \tag{3.1}$$

Assuming again that U has minimum value zero, for finite energy we clearly need

$$U(\varphi) \to 0 \quad \text{and} \quad |\mathcal{D}^i \varphi| = o\left(r^{-3/2}\right) \quad \text{as} \quad r = |\underline{x}| \to \infty . \tag{3.2}$$

We would like to infer that φ must tend to a limit, which would then have to be a zero of U, as $\underline{x}$ tends outwards radially to infinity (that is $\underline{x} = r\hat{\underline{x}}$, $r \to \infty$ with $\hat{\underline{x}}$ fixed). But such a statement is not gauge invariant, because, given a gauge in which it is true, we may perform a gauge transformation which is an oscillatory function of r and which will make φ oscillate at infinity (unless φ tends to a zero of U which remains fixed under the action of G). However at any given instant it is possible to find a gauge in which the radial components of W_a^i, namely $\hat{x}^i W_a^i$, vanish for all r greater than some non-zero radius[5]. In this gauge eq. (3.2) implies

$$\left|\frac{\partial \varphi}{\partial r}\right| = o\left(r^{-3/2}\right) \tag{3.3}$$

so that

$$\varphi_\infty(\hat{\underline{x}}) = \lim_{r \to \infty} \varphi(r\hat{\underline{x}})$$

exists for each unit vector $\hat{\underline{x}}$. We think of φ_∞ as defining a map from the set of unit vectors in three dimensions (called a 2-sphere by mathematicians and denoted by S^2) and the set of zeros of U which we will denote by $\mathcal{M}_0$. We assume that, as a result of the equation of motion, φ_∞ varies continuously with time and so its topological characteristics will be preserved. Much of the remainder of these lectures will be concerned with elaborating and explaining this statement. Notice in particular that φ_∞ relates directions in physical space with those in the internal space in which φ takes its values.

If $\varphi_\infty(\hat{\underline{x}})$ has a non-trivial dependence on the angular variables $\hat{\underline{x}}$ the transverse components of $\underline{\nabla}\varphi$ will behave like 1/r. Because of (3.3) we see that

$$|\hat{\underline{x}} \wedge \underline{\nabla}\varphi| = O(r^{-1}), \quad |\hat{\underline{x}} \wedge \underline{W}_a| = O(r^{-1}) \ \text{but} \ |\hat{\underline{x}} \wedge \underline{\mathcal{D}}\varphi| = o(r^{-3/2}) \tag{3.4}$$

Consequently a delicate cancellation has to occur.

Rather than discuss further general properties of such solutions at this stage we will see how the above considerations apply in a particular example.

3.2 The 't Hooft-Polyakov Monopole.

't Hooft[1] and Polyakov[2] considered the case of an SU(2) gauge theory with $\underline{\varphi}$ transforming according to the adjoint representation (i.e. as a vector) and with

$$U(\underline{\varphi}) = \frac{\lambda}{4} (\varphi^2 - a^2)^2 \tag{3.5}$$

Clearly we do not need the full gauge theory formalism, set up in section 2.3, to deal with this simple case. Here $\underline{\Phi}$, $\underline{G}^{\mu\nu}$ and $\underline{W}^{\mu}$ are all vectors in the internal space and the Lagrangian density is

$$\mathcal{L} = -\frac{1}{4}\underline{G}_{\mu\nu}\cdot\underline{G}^{\mu\nu} + \frac{1}{2}\mathcal{D}_{\mu}\underline{\Phi}\cdot\mathcal{D}^{\mu}\underline{\Phi} - \frac{\lambda}{4}(\Phi^2 - a^2)^2 \tag{3.6}$$

with $\mathcal{D}^{\mu}\underline{\Phi} = \partial^{\mu}\Phi - e\underline{W}^{\mu}\wedge\underline{\Phi}$ $\qquad$ $\underline{G}^{\mu\nu} = \partial^{\mu}\underline{W}^{\nu} - \partial^{\nu}\underline{W}^{\mu} - e\underline{W}^{\mu}\wedge\underline{W}^{\nu}$

(We may explicitly connect with section 2.3 by taking $T^a = \frac{1}{2}\sigma^a$, where σ^a are the Pauli matrices, and $D(T^a)_{\alpha\beta} = -i\epsilon_{a\alpha\beta}$.) The equations of motion are

$$\mathcal{D}_{\nu}\underline{G}^{\mu\nu} = e\underline{\Phi}\wedge\mathcal{D}^{\mu}\underline{\Phi} \equiv -\underline{j}^{\mu} \tag{3.7}$$

$$\mathcal{D}_{\mu}\mathcal{D}^{\mu}\underline{\Phi} = -\lambda\underline{\Phi}(\Phi^2 - a^2) \tag{3.8}$$

In the last section we argued that topological characteristics of $\underline{\Phi}_{\infty}$ would provide conservation laws. We are thus interested in cases in which $\underline{\Phi}$ has a non-trivial angular behaviour at infinity. In this model the set of zeros of U is itself a two-dimensional sphere.(i.e. the surface of a ball in three-dimensional space).

$$\mathcal{M}_0 = \{\underline{\Phi} : \Phi^2 = a^2\} \tag{3.9}$$

The simplest non-trivial angular behaviour is

$$\underline{\Phi}_{\infty}(\hat{\underline{x}}) = \lim_{r\to\infty}\underline{\Phi}(r\hat{\underline{x}}) = a\hat{\underline{x}} \tag{3.10}$$

An ansatz which leads to this is

$$\underline{\Phi} = \underline{x}\,\frac{H(aer)}{er^2} \tag{3.11}$$

$$W_a^i = \epsilon_{aij}x^j\,\frac{1 - K(aer)}{er^2} \tag{3.12}$$

$$W_a^0 = 0. \tag{3.13}$$

For this ansatz the energy takes the form

$$E = \frac{4\pi a}{e}\int_0^{\infty}\frac{d\xi}{\xi^2}\left\{\xi^2\left(\frac{dK}{d\xi}\right)^2 + \frac{1}{2}\left(\xi\frac{dH}{d\xi} - H\right)^2 + \frac{1}{2}(K^2-1)^2 + K^2H^2 + \frac{\lambda}{4e^2}(H^2-\xi^2)^2\right\} \tag{3.14}$$

using ξ for the argument, aer, of H and K.

The equations of motion are satisfied by the ansatz provided that

$$\xi^2\frac{d^2K}{d\xi^2} = KH^2 + K(K^2-1) \tag{3.15}$$

$$\xi^2\frac{d^2H}{d\xi^2} = 2K^2H + \frac{\lambda}{e^2}H(H^2-\xi^2) \tag{3.16}$$

(These equations can be obtained more easily by applying the variational principle directly to the functional of eq. (3.14). It is not obvious _a priori_ that we will obtain the correct equations in this way but it can be shown to follow by a symmetry argument.[5]).

The appropriate boundary conditions for a finite energy solutions are

$$K-1 \leq O(\xi), \quad H \leq O(\xi) \quad \text{as} \quad \xi \to 0, \tag{3.17}$$

$$K \to 0, \quad H \sim \xi \qquad \text{sufficiently fast as } \xi \to \infty.$$

The equations for H and K present no problems from the point of view of the existence of solutions or numerical analysis. The functions H and K tend to their asymptotic forms exponentially quickly at infinity. The characteristic length scale is 1/ae. The forms of H and K are sketched in Fig. 6.

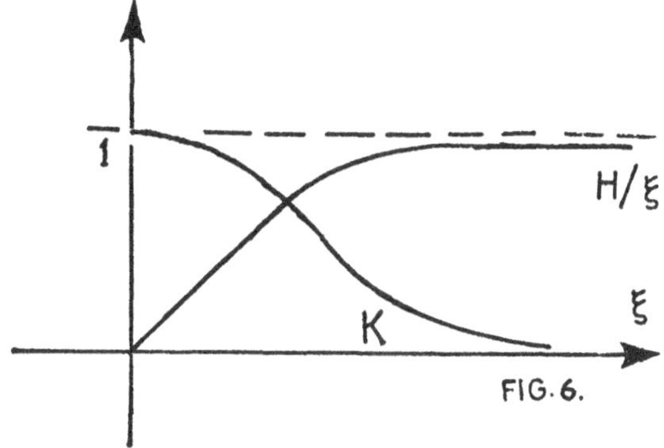

FIG. 6.

It is easy to estimate the total energy of the solution numerically using variational techniques. This energy would be interpreted as the classical mass of a corresponding particle. The integral of eq. (3.14) is a function of λ/e^2 and we may write

$$\text{Mass} = E = \frac{4\pi a}{e} f(\lambda/e^2) \tag{3.19}$$

The following values of f have been calculated in connection with various investigations:

$f(0) = 1$ (Prasad and Sommerfield[18])

$f(0{\cdot}1) - 1{\cdot}1$ ('t Hooft[1])

$f(0{\cdot}5) = 1{\cdot}42$ (Julia and Zee[19])

$f(10) = 1{\cdot}44$ ('t Hooft[1])

In the course of numerical investigations Prasad and Sommerfield[18] chanced upon an exact analytical form for the solution for $\lambda/e^2 = 0$. The limit $\lambda/e^2 \to 0$ corresponds to switching the potential U off, though a shadow remains in that $\varphi^2 = a^2$ at infinity still. Nothing is left of the potential except the boundary condition it enforces. Thus it is a limit in which we may expect to make precise statements uninfluenced by the detailed form of any potential. The solution of Prasad and

Sommerfield enables us to write.

$$K(\xi) = \xi/\sinh\xi + O(\lambda/e^2) \tag{3.20}$$

$$H(\xi) = \xi\coth\xi - 1 + O(\lambda/e^2) \tag{3.21}$$

We can rewrite eq. (3.63) in the following way:

$$\text{Mass} = \frac{M_W}{4\alpha} f(\lambda/e^2) \tag{3.22}$$

where $M_W = ae\hbar$ and $\alpha = q_0^2/4\pi\hbar$ is the fine structure constant if we interpret $q_0 = e\hbar/2$ as the basic unit of charge in the theory. Introducing M_W and α is really a bit bogus since they only have relevance after the theory has been quantised. The quantum theory of these extended solutions is still at a rudimentary stage and we will make no attempt to discuss it at any length. However we can make some guesses about the magnitude of the quantum of electric charge and the masses of the elementary particles associated with the fields which may serve to give an indication of orders of magnitude. The significance of M_W is then that it is the naive value for the mass of the heavy vector boson in the quantum field theory of the Lagrangian (3.6). In this theory the Higgs mechanism (see Ref. 20 for a review) operates. Thus rather than having three massless vector particles and three scalars, two of the scalars are absorbed to make two of the vector particles massive leaving a single vector particle massless and one other scalar. To lowest order the mass of the vector particles is $M_W = ae\hbar$. If we assume that one of the SU(2) directions corresponds to electromagnetism as is suggested by the presence of the massless vector particle, which will be the photon, the electric charges (of the particles associated with any other fields coupled into the theory) will be the eigenvalues of $e\hbar D(\sigma_3/2)$ in the various representations, D, of SU(2). Since the eigenvalues of $D(\sigma_3/2)$ are multiples of a half, electric charge will be quantised in units of $q_0 = e\hbar/2$. If this is the charge on the electron, α will be the fine structure constant.

Thus the mass of the solution is expected to be greater than $137M_W/4$ where M_W is the mass of a vector boson. Consequently in a physical relevant theory it would be very heavy indeed. This sadly limits the immediacy of its physical relevance.

't Hooft[1] and Polykaov[2] in proposing this solution pointed out that it would carry magnetic charge; it is a magnetic monopole. To understand this interpretation we next discuss how electromagnetion can be embedded in a non-Abelian gauge theory.

3.3 Embedding Electromagnetism.

In the SU(2) - gauge theory of the last section if ϕ tends sufficiently fast to constant vector at a infinity, $\underline{\phi} \to a\hat{\underline{z}}$, say (where $\hat{\underline{z}}$ is the unit vector in the third direction) we could interpret the corresponding component of $\underline{G}^{\mu\nu}$ (i.e. $G_3^{\mu\nu}$) as the electromagnetic tensor, at least asymptotically. This is consistent because if $\partial_\mu \varphi_3 \to 0$ faster than $r^{-3/2}$ we must have $\underline{W}^\mu \wedge \underline{\varphi} \to 0$ as fast to ensure that $(\mathcal{D}_\mu \varphi)_3 \to 0$ sufficiently fast to give finite energy. In consequence

$$(\mathcal{D}^\lambda \underline{G}^{\mu\nu})_3 \doteq \partial^\lambda G_3^{\mu\nu} \tag{3.23}$$

and the third components of the equations

$$\mathcal{D}^\nu \underline{G}_{\mu\nu} = -\underline{j}_\mu \qquad \mathcal{D}^\nu \underline{G}^*_{\mu\nu} = 0 \tag{3.24}$$

may be interpreted as the Maxwell equations for the elctromagnetic field tensor, $F^{\mu\nu} = G_3^{\mu\nu}$ at infinity.

This interpretation is also suggested by the effect of the spontaneous symmetry breaking through the Higgs mechanism on the theory as discussed in the last section. If φ tends to a constant, φ_0, at infinity, we may interpret this as the ground state. It is the gauge fields associated with elements of the Lie algebra which generate transformations which do not preserve φ_0 that correspond to massive vector particles (with mass $ae\hbar$ in the model of section 3.3). The vector fields associated with generators which leave φ_0 unchanged (that is those that generate the little group of φ_0) acquire no mass. In the specific case of SU(2) and if $\underline{\varphi} \to a\hat{\underline{z}}$, it is only rotations about the third axis that have a massless vector associated with them and it is W_3^μ which must be interpreted as the photon field A^μ.

For the 't Hooft-Polyakov solution the asymptotic direction of φ varies and one is led to identify this variable direction with electromagnetism. Thus we introduce an electromagnetic field tensor

$$F^{\mu\nu} = \frac{1}{a}\underline{\varphi}\cdot\underline{G}^{\mu\nu}. \tag{3.25}$$

To investigate Maxwell's equations consider

$$\begin{aligned}\partial^\lambda F^{\mu\nu} &= \frac{1}{a}(\partial^\lambda \underline{\varphi})\cdot\underline{G}^{\mu\nu} + \frac{1}{a}\underline{\varphi}\cdot\partial^\lambda \underline{G}^{\mu\nu} \\ &= \frac{1}{a}(\mathcal{D}^\lambda \underline{\varphi})\cdot\underline{G}^{\mu\nu} + \frac{1}{a}\underline{\varphi}\cdot(\mathcal{D}^\lambda \underline{G}^{\mu\nu})\end{aligned} \tag{3.26}$$

using the symmetry of the triple scalar product

$$(\underline{W}^\lambda \wedge \underline{\varphi})\cdot\underline{G}^{\mu\nu} = -\underline{\varphi}\cdot(\underline{W}^\lambda \wedge \underline{G}^{\mu\nu}).$$

Thus,

$$\partial_\nu F^{\mu\nu} = -j^\mu, \tag{3.27}$$

$$\partial_\nu F^{*\mu\nu} = -k^\mu \tag{3.28}$$

where, using the equations of motion and Bianchi identities,

$$j^\mu = -\frac{1}{a}(\mathfrak{D}_\nu\underline{\Phi}).\underline{G}^{\mu\nu} + \frac{1}{a}\underline{\Phi}.\underline{j}^\mu \tag{3.29}$$

$$k_\mu = \frac{1}{2}\epsilon_{\mu\nu\lambda\rho}\partial^\nu F^{\lambda\rho} \tag{3.30}$$

$$= \frac{1}{2}\epsilon_{\mu\nu\lambda\rho}(\mathfrak{D}^\nu\underline{\Phi}).\underline{G}^{\lambda\rho} \tag{3.31}$$

At infinity, where $\mathfrak{D}_\mu\underline{\Phi}$ vanishes exponentially fast, Maxwell's equations are satisfied in the conventional form with $k_\mu=0$ and j^μ being the appropriate component of the SU(2) current $\underline{j}^\mu$. (In fact, $j^\mu=0$ at infinity because $\underline{j}^\mu = -e\underline{\Phi}\wedge\mathfrak{D}^\mu\underline{\Phi}$ but this is a rather special feature of the 't Hooft-Polyakov model. If there were any other charged fields in the theory they would make a contribution to j^μ.) The total magnetic charge

$$\int k^0 d^3x = \oint\!\!\oint \underline{B}.d\underline{S}$$

taken over a large sphere, where

$$B^i = -\frac{1}{ea}\epsilon_{ijk}\underline{\Phi}.\underline{G}^{jk} \tag{3.32}$$

We emphasise that we have only been able to identify electromagnetism where $\mathfrak{D}^\mu\Phi$ vanishes, that is asymptotically. It is only there that the argument about the Higgs mechanism makes an unambiguous suggestion. When Φ is at a zero of U it determines a ground state and rotations which preserve this ground state we expect to be associated with massless vector mesons, candidates for the photon. But it is not so clear that this argument can be applied before Φ reaches a zero of U. Coleman[5] amplifies this point as follows. The magnetic field given by eq. (3.32) is certainly the quantity that a magnetometer would measure in region where Φ is constant at a zero of U or, more generally, in a region in which $U(\Phi)$ and $\mathfrak{D}^\mu\Phi$ are approximately zero in comparison with the characteristic length and energy scales in the theory. But if these conditions are not satisfied the detailed mechanism of the magnetometer (how it interacts with the various fields) must be known before it can be decided what it would measure. The mathematical reflection of this is that one can add any term to the definition of $F^{\mu\nu}$ which vanishes where $\mathfrak{D}^\mu\Phi$ is effectively zero. This corresponds to some definite choice of magnetometer mechanism.

All such magnetometers will measure the same total magnetic charge, as the asymptotic magnetic field is uniquely specified, but they will differ as to its distribution throughout the non-asymptotic region. The magnetic charge is confined to the region where $\mathfrak{D}^\mu\Phi$ differs appreciably from zero. We will see in the next section that the 't Hooft-Polyakov solution has non-zero magnetic charge.

The discussion of this section may be generalised to a situation where we have a compact gauge group G with a Higgs field Φ transforming under the adjoint representation

[20,21] Using the formalism developed in section 2.3, we may introduce an electromagnetic tensor $F^{\mu\nu}$ in a region in which Φ is covariantly constant,

$$\mathcal{D}^{\mu}\Phi = 0 \tag{3.33}$$

by,

$$F^{\mu\nu} = \frac{1}{K}\mathrm{Tr}(\underset{\sim}{\Phi}\underset{\sim}{G}^{\mu\nu}) \tag{3.34}$$

where $K^2 = \mathrm{Tr}(\underset{\sim}{\Phi}^2)$ is constant as a result of eq. (3.33). We may then establish the Maxwell equations (3.27) and (3.28) with $j^{\mu} = K^{-1}\mathrm{Tr}(\underset{\sim}{\Phi}\underset{\sim}{j}^{\mu})$ and $k^{\mu} = 0$ provided that eq. (3.33) holds.

3.4 Magnetic Charge and its Quantization.

Having identified the electromagnetic field at infinity in the last section we can calculate the total magnetic flux and hence deduce the magnetic charge of the 't Hooft-Polyakov solution. We will present the calculation of the asymptotic field in a sufficiently general form to enable us to discuss the possible values that the charge can take for an arbitrary solution. The condition that the covariant derivative of Φ vanishes tells us a lot about the structure of the field tensor $G_{\mu\nu}$. The condition

$$\mathcal{D}^{\mu}\underline{\Phi} = \partial^{\mu}\underline{\Phi} - e\underline{W}^{\mu}\wedge\underline{\Phi} \tag{3.35}$$

is satisfied by

$$\underline{W}^{\mu} = \frac{1}{a^2 e}\underline{\Phi}\wedge\partial^{\mu}\underline{\Phi} + \Lambda^{\mu}\underline{\Phi} \tag{3.36}$$

in a region in which $|\underline{\Phi}| = a$. It is in fact the general solution for $\underline{W}^{\mu}$ given Φ. The arbitrary function Λ^{μ} is zero for the 't Hooft-Polyakov solution. From eq. (3.36) we can calculate the asymptotic from of $\underline{G}_{\mu\nu}$,

$$\underline{G}^{\mu\nu} \doteq \frac{1}{a^2 e}\partial^{\mu}\underline{\Phi}\wedge\partial^{\nu}\underline{\Phi} + (\partial^{\mu}\Lambda^{\nu} - \partial^{\nu}\Lambda^{\mu})\underline{\Phi} \tag{3.37}$$

$$= \underline{\Phi}\,F^{\mu\nu}/a \tag{3.38}$$

where the electromagnetic tensor

$$F^{\mu\nu} = \frac{1}{a^3 e}\underline{\Phi}.(\partial^{\mu}\underline{\Phi}\wedge\partial^{\nu}\underline{\Phi}) + a(\partial^{\mu}\Lambda^{\nu} - \partial^{\nu}\Lambda^{\mu}) \tag{3.39}$$

For the specific case of the 't Hooft-Polyakov solution,

$$G_a^{ij} \doteq \frac{\epsilon_{ijk}\hat{x}^k}{er^2}\hat{x}^a, \qquad F^{ij} \doteq \frac{\epsilon_{ijk}\hat{x}^k}{er^2} \tag{3.40}$$

Since the magnetic field $B^i = -\frac{1}{2}\epsilon_{ijk}F^{jk}$ we see that the solution has a magnetic charge $-4\pi/e$.

The total magnetic charge for a general solution is given by

$$g = \oint\!\!\oint \underline{B}.d\underline{S} = -\tfrac{1}{2}\oint\!\!\oint \epsilon_{ijk} F^{jk} dS^i \tag{3.41}$$

$$= -\frac{4\pi}{e} N \tag{3.42}$$

where

$$N = \frac{1}{4\pi a^3}\oint\!\!\oint dS^i \tfrac{1}{2}\epsilon_{ijk}\,\underline{\Phi}.(\partial^j\underline{\Phi}\wedge\partial^k\underline{\Phi}). \tag{3.43}$$

Here all the integrals are taken over large spheres; the second term in eq. (3.37) does not contribute by the divergence theorem. The quantity N defined by eq. (3.42) is always an integer, the number of times the unit sphere in real space is mapped onto the sphere of minima of the potential U, $|\Phi| = a$, by Φ_∞. Thus the magnetic charge g is quantized in units of $4\pi/e = g_0$. This satisfies the Dirac condition

$$\frac{q_0 g_0}{4\pi\hbar} = \frac{1}{2} \tag{3.44}$$

where $q_0 = \frac{1}{2}\hbar e$, the quantum of charge we expect in the quantum theory of this model (see section 3.2).

It is important to notice that the quantization of magnetic charge obtained here is an entirely classical phenomenon. The appearance of $\hbar$ in eq. (3.44) is deceptive; it only enters when we speculate about the unit of electric charge in the corresponding quantum theory. In the classical theory studied here electric charge is <u>not</u> quantised. The constant e occurring in the Lagrangian is an electric coupling constant. The dimensions of e are those of inverse charge, while the dimensions of $\hbar$ are charged squared. Consequently g_0 and q_0 do have the correct dimensions for the quanta of charge. On the other hand, Dirac's quantization of charge is a quantum phenomenon as we explained in section 2.2. It is clear from dimensional considerations that Dirac's quantization must take place after the passage to quantum mechanics if it is to take the form qg is an integral multiple of a given constant. Dirac was dealing with point charges put in by hand as sources. The relation of the 't Hooft-Polakov monopole to the Dirac monopole is discussed further in section 3.6.

If $\underline{n}$ is a unit vector the asymptotic forms

$$\underline{\Phi} = a\underline{n}, \qquad \underline{W}^\mu = \frac{1}{e}\,\underline{n}\wedge\partial^\mu\underline{n}, \qquad \underline{G}^{\mu\nu} = \frac{1}{e}\,\partial^\mu\underline{n}\wedge\partial^\nu\underline{n} \tag{3.45}$$

exactly satisfy the equations of motion where $\underline{n}$ is non-singular as a function of x. However if $\underline{n}$ is nonsingular everywhere there can be no net magnetic charge. To see this consider the map $\hat{\underline{x}} \to \underline{n}(r\hat{\underline{x}})$, for fixed radius r. This defines a map of the sphere S^2 into itself. For large r the number of times this map covers S^2 as $\hat{\underline{x}}$ ranges over S^2 once determines the magnetic charge. But if $\underline{n}$ is non-singular this map changes continuously till we obtain a constant map for r = 0. During this continuous def-formation the number of times the sphere is covered does not change. Consequently if $\underline{n}$ is nonsingular this number is zero, the number we obtain for r = 0.

We may construct solutions of the form of eq. (3.45) with magnetic charge $-4\pi N/e$ for any integer N, which are singular only at the origin. For such a solution $\underline{n}(r\hat{\underline{x}})$ must cover the limit sphere N times. An example of such a map is given in polar coordinates (r, θ, χ) by

$$\underline{n}(r,\theta,\chi) = (\cos N\chi \sin\theta, \sin N\chi \sin\theta, \cos\theta). \tag{3.46}$$

The 't Hooft-Polakov solutions have this asymptotic behaviour for $N = \pm 1$, but are smoothed out in the non-asymptotic region so that they satisfy the equations of motion everywhere, even at the origin, and have finite energy. The trivial solution, φ constant, provides an example of N = 0. For $N \geqslant 2$ no explicit smooth solutions are known. It seems quite possible that there are no static non-singular finite energy solutions with these charges.

3.5 Topological Currents and the Bogomolny Bound.

We have seen in the last section that solutions to the 't Hooft-Polyakov model possess a quantized magnetic charge. Eq. (3.42) makes it plain that this charge is a topological property of the Higgs field $\underline{\varphi}$, and more particularly, of the asymptotic map $\underline{\varphi}_\infty: S^2 \to \mathcal{M}_0$. Since, presumably, $\underline{\varphi}_\infty$ changes continuously with time, this charge is conserved. In section 1.2 we show that the soliton number, which is conserved for topological reasons, could be obtained as a charge from a current whose divergence vanished independently of the dynamics. We can construct such a conserved current for the magnetic charge in the 't Hooft-Polyakov model, or, more generally, for the magnetic fields embedded in gauge theories, as discussed at the end of section 3.3 [23,24] Indeed defining k^μ as in eq. (3.30) we see immediately that

$$\partial_\mu k^\mu = 0$$

simply from the antisymmetry of $\epsilon_{\lambda\mu\nu\rho}$.

A major application of this current is to provide a lower bound for the mass of the monopole. We start with the trivial inequality,[23,25]

$$\left(\tfrac{1}{2}\epsilon_{ijk}\underline{G}^{jk} \pm \mathcal{D}^i\underline{\varphi}\right)^2 \geqslant 0 \tag{3.47}$$

$$\tfrac{1}{4}\left(\underline{G}^{jk}\right)^2 + \tfrac{1}{2}\left(\mathcal{D}^i\underline{\varphi}\right)^2 \geqslant \pm\tfrac{1}{2}\epsilon_{ijk}\left(\mathcal{D}^i\underline{\varphi}\right).\underline{G}^{jk} \tag{3.48}$$

Consequently the energy

$$E \geqslant |g|a \tag{3.49}$$

by using eqs. (3.1) and (3.31).

In calculating this lower bound we have considered only the kinetic contributions to the energy entirely neglecting the positive contribution of the potential energy, U.

For a solution with magnetic charge $4\pi N/e$, the Bogomolny bound (3.49) says that the mass is greater than or equal to

$$\frac{4\pi|N|a}{e} = \frac{|N|M_W}{4\alpha} \tag{3.50}$$

In particular the Prasad-Sommerfield solution, discussed in section 3.2, saturates this bound for the relevant value of N = 1. Since this exact solution only applies in the limit of $\lambda/e^2=0$ in which the potential is turned off, leaving only the boundary condition at infinity as a shadow, the fact that the potential has been neglected is no obstacle to saturation. Indeed it is only in this limit that we stand any chance of saturating the bound. Because of charge conservation we may infer that the 't Hooft-Polyakov solution is stable in the limit $\lambda/e^2 \to 0$.

The fact that the Bogomolny bound is saturated for the Prasad-Sommerfield solution means that it must satisfy the equation

$$\mathcal{D}^i\phi = -\tfrac{1}{2}\epsilon_{ijk}G^{jk} \tag{3.51}$$

Indeed this equation together with the Bianchi identities imply the equations for motion for a static solution, thus reducing the equations of motion to first order equations. This gives a systematic way of arriving at the Prasad-Sommerfield solution.

3.6 The Relationship between the Dirac and 't Hooft-Polyakov monopoles.

There are differences and similarities between the magnetic monopoles of 't Hooft[1] and Polyakov[2] on the one hand and Dirac[4] on the other. Obviously, the principal distinction is that the 't Hooft-Polyakov monopole exactly satisfies the SU(2) gauge theory equations without point sources whilst the Dirac monopole needs sources 'put in by hand'. In a sense, which it is hoped this section clarifies, this is accomplished by smoothing out the Dirac string into the other SU(2) directions so that the Bianchi identities do not prevent a net flux reaching infinity, and by leaving the electromagnetic tensor ambiguous near the origin, avoiding the necessity for a pole.

Coleman[5] stresses the differences between the monopoles saying that they could not be more different. He draws attention to the fact that additional degrees of freedom are needed to describe the point sources in Dirac's theory, but the 't Hooft-Polyakov type of monopole requires no extra degrees of freedom. But if we examine the 't Hooft-Polyakov monopole on a scale large compared with its dimensions, which are of order $1/ae$, it looks like for Dirac monopole. A detailed analysis of scattering on SU(2) magnetic monopoles is given by Boulware et al.[26] who show that only in deep scattering do deviations from the Dirac theory occur. Here we show the eqinvalence of the asymptotic fields; a similar calculation is performed in Ref. 26.

Consider rotating the SU(2) directions of φ and $G^{\mu\nu}$, which are parallel at each point to $\underline{x}$, so that everywhere they point in the same direction, that of the z-axis say. Actually this cannot be done continuously throughout all space; indeed it cannot be done continuously on any sphere containing the origin. It is impossible to find a rotation defined continuously over the sphere S^2 which rotates $\hat{\underline{x}}$ to the fixed direction $\hat{\underline{z}}$. But it can be done throughout the whole of space outside a cone, with arbitrarily small semi-vertical angle, surrounding the negative z-axis. Outside the cone we can make all the fields point in the third direction in SU(2) space. In the limit as the solid angle contained by the cone tends to zero we regain the Dirac monopole potential of eq. (2.34) and the expression for the radial magnetic field is just that of eq. (2.30) complete with the string. We see in detail how this comes about starting with the asymptotic fields,

$$\varphi = \frac{a}{2}\hat{\underline{x}}\cdot\underline{\sigma}, \quad W^\mu = \frac{1}{2e}\underline{\sigma}\cdot(\hat{\underline{x}}\wedge\partial^\mu\hat{\underline{x}}), \quad G^{ij} = \frac{1}{2e}\hat{\underline{x}}\cdot\underline{\sigma}\frac{\epsilon_{ijk}\hat{x}^k}{r^2} \tag{3.52}$$

Under a gauge transformation

$$u = \cos\frac{\psi}{2} + i\sin\frac{\psi}{2}\,\underline{k}\cdot\underline{\sigma} \in SU(2), \tag{3.53}$$

where $k^2 = 1$,

$$W^\mu \rightarrow uW^\mu u^{-1} + ie^{-1}(\partial^\mu u)u^{-1}, \tag{3.54}$$

$$G^{\mu\nu} \rightarrow uG^{\mu\nu}u^{-1}. \tag{3.55}$$

Now

$$u\underline{\sigma}u^{-1} = \underline{\sigma}\cos\psi + \underline{k}\wedge\underline{\sigma}\sin\psi + (1-\cos\psi)\,\underline{k}(\underline{k}\cdot\underline{\sigma}). \tag{3.56}$$

So that $\underline{k} = \hat{\underline{x}} = (\hat{\underline{z}}\wedge\hat{\underline{x}})/\sin\theta$,

$$u\underline{x}\cdot\underline{\sigma}u^{-1} = \underline{x}\cdot\underline{\sigma}\frac{\sin(\theta-\psi)}{\sin\theta} + r\hat{\underline{z}}\cdot\underline{\sigma}\frac{\sin\psi}{\sin\theta} \tag{3.57}$$

where $\hat{\underline{z}}\cdot\hat{\underline{x}} = \cos\theta$. Now choose $\psi(\theta)$ to be a suitable differentable function of θ with $\psi(\pi)=0$ and $\psi(\theta)=\theta$ for $0 \leq \theta \leq \pi-\varepsilon$. We will consider a sequence of such functions with $\varepsilon \rightarrow 0$, so that $\psi(\theta)\uparrow\theta$. (See fig. 7).

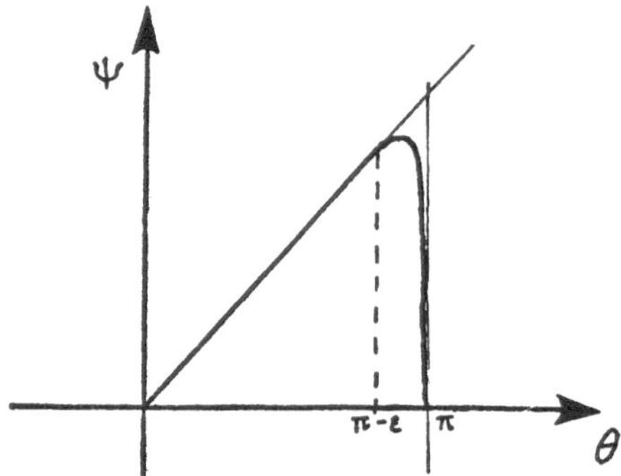

FIG.7.

We have $G^{ij}_a \to \frac{1}{e}\,\delta_{a3}\,\epsilon_{ijk} r^k / r^3$ in the limit. Under the gauge transformation

$$\begin{aligned} \underset{\sim}{W}_r &\to 0 \\ \underset{\sim}{W}_\theta &\to -\frac{1}{2er}(1-\psi')\,\hat{\underline{\chi}}.\underline{\sigma} \\ \underset{\sim}{W}_\chi &\to \frac{1}{2er}\,\frac{\sin(\theta-\psi)}{\sin\theta}\,\hat{\underline{\theta}}.\underline{\sigma} - \frac{1}{2er}\,\frac{(1-\cos\psi)}{\sin\theta}\,\hat{\underline{z}}.\underline{\sigma} \end{aligned} \tag{3.58}$$

Let us introduce the potential

$$A^\mu = \underline{\Phi}.\underline{W}^\mu/a \tag{3.59}$$

In the limit of $\psi \uparrow \theta$, A^μ becomes the Dirac potential

$$\underline{A} = -\frac{1}{2er}\,\frac{(1-\cos\theta)}{\sin\theta}\,\hat{\underline{\chi}} \tag{3.60}$$

The field tensor

$$\begin{aligned} F^{\mu\nu} &= \underline{\Phi}.\underline{G}^{\mu\nu}/a \\ &= \partial^\mu A^\nu - \partial^\nu A^\mu - 2ie\,\mathrm{Tr}(\underset{\sim}{\Phi}\,[\underset{\sim}{W}^\mu, \underset{\sim}{W}^\nu])/a \end{aligned} \tag{3.61}$$

In the limit the last term in eq. (3.61) generates the Dirac string

$$\begin{aligned} 2ie\,\mathrm{Tr}(\underset{\sim}{\Phi}\,[\underset{\sim}{W}^i, \underset{\sim}{W}^j]) &\sim -\frac{1}{er^2}(1-\psi')\,\frac{\sin(\theta-\psi)}{\sin\theta}\,\epsilon_{ijk}\hat{x}^k \\ &\to \frac{4\pi}{e}\,\epsilon_{ij3}\,\delta(x)\,\delta(y)\,\theta(-z) \end{aligned} \tag{3.62}$$

Thus we indeed regain the Dirac representation of the field, complete with string.

3.7 Introduction to some Topological Ideas

In section 3.1 we argued that each solution of the field equations defines at any time a map $\varphi_\infty : S^2 \to \mathfrak{M}_0$. In section 3.4 we showed that for the 't Hooft monopole, where $\mathfrak{M}_0$ is essentially the sphere S^2 again, the magnetic charge is a topological characteristic of this map, the number of times it covers the sphere (multiplied by

$4\pi/e$). Now we will try to elaborate the connection between monopoles and topological concepts. (For more detailes see Refs. 5 and 6).

First consider the structure of $\mathcal{M}_0$. For 't Hooft-Polyakov case it is a sphere and any one point of it may be obtained from any other by a gauge transformation. In general if this is the case we say G acts transitively on $\mathcal{M}_0$ and we may write

$$\mathcal{M}_0 = \{D(g)\varphi_0 : g \in G\} \tag{3.63}$$

for any fixed $\varphi_0 \in G$. Transitivity may be interpreted as saying that all of the ground state degeneracy is a reflection of symmetry under G; there are no further 'accidentally degenerate' minima.

Now we get the same φ by acting with g_1 and g_2 on φ_0 if and only if

$$D(g_1)\varphi_0 = D(g_2)\varphi_0$$

$$D(g_2^{-1}g_1)\varphi_0 = \varphi_0 \tag{3.64}$$

which we may write $g_2^{-1}g_1 \in H_{\varphi_0}$ where H_{φ_0} is a subgroup of G, called the little group of φ_0 .

$$H_{\varphi_0} = \{h \in G : D(h)\varphi_0 = \varphi_0\}. \tag{3.65}$$

Equation (3.64) says that g_1 and g_2 give the same point in $\mathcal{M}_0$ when acting on φ_0 if and only if they are in the same coset of H_{φ_0} in G. So we can identify $\mathcal{M}_0$ with G/H, the set of cosets of $H \equiv H_{\varphi_0}$ in G. For different φ_0 the subgroups H_{φ_0} are isomorphic. Thus φ defines a map $S^2 \to G/H$. For the 't Hooft-Polyakov case we may take $\varphi_0 = a\hat{\underline{z}}$ and H_{φ_0} is the group of rotations about the z-axis: $\{\exp(i\sigma_3\theta/2)\}$. So $S^2 \cong SU(2)/U(1)$.

As time progresses the map $\varphi_\infty : S^2 \to G/H$ varies continuously. In mathematical terminology the maps defined at different times are <u>homotopic</u>. [Two maps $f_1, f_2 : X \to Y$ are homotopic if and only if there exists a continuous map, $F : X \times [0,1] \to Y$ such that $F(x,0) = f_1(x)$, $F(x,1) = f_2(x)$]. We can devide up all the maps $S^2 \to G/H$ into equivalence classes according to whether they are homotopic. Which class a solution is in is gauge invariant [For if $\varphi_2(\underline{x}) = D(g(\underline{x}))\varphi_1(\underline{x})$, define $F(\hat{\underline{x}},t) = D(g(\lambda(t)\hat{\underline{x}}))\varphi_{1\infty}(\hat{\underline{x}})$ where $\lambda(t) = t/(1-t)$. Then $F(\hat{\underline{x}},1) = \varphi_{2\infty}(\hat{\underline{x}})$, $F(\hat{\underline{x}},0) = D(g(0))\varphi_{1\infty}(\hat{\underline{x}})$ so that $\varphi_{2\infty}$ is homotopic to $D(g_0)\varphi_{1\infty}$ for some constant g_0. But the gauge group G will be assumed connected. So we may continuously change g_0 to the identity giving $\varphi_{2\infty}$ homotopic to $\varphi_{1\infty}$]. So the homotopy class is a gauge invariant constant of the motion and we may get a conservation law if we find and label the homotopy classes of maps $S^2 \to G/H \cong \mathcal{M}_0$.

Usually mathematicians are interested in a slightly different concept: the homotopy classes of maps $S^2 \to \mathcal{M}_0$ where in the homotopies used to define the equivalence classes one point of S^2, $\hat{\underline{z}}$ say, is kept fixed, $\varphi_\infty(\hat{\underline{z}}) = \varphi_0$. It map be shown that provided H is connected these classes are in one-one correspondence with those in which no effort is made to fix a point. The set of homotopy classes $S^2 \to \mathcal{M}_0$ with $\varphi_\infty(\hat{\underline{z}}) = \varphi_0$ is densted by $\Pi_2(\mathcal{M}_0)$. An Abelian group structure can be defined on it and it is called the <u>second homotopy group</u> of $\mathcal{M}_0$.

In the case in which there are techniques for simplifying the calculations of $\Pi_2(\mathcal{M}_0)$. To describe this we try to 'lift' the map $\varphi_\infty : S^2 \to G/H$ to a map $S^2 \to G$, that is to find a $g(\hat{\underline{x}})$ such that $\varphi_\infty(\hat{\underline{x}}) = D(g(\hat{\underline{x}}))\varphi_0$. In general this will not be possible. Indeed it will only be possible if the map φ_∞ is homotopic to a constant map $S^2 \to \mathcal{M}_0$ (i.e. is trivial homotopically). This is because any map $S^2 \to G$ is trivial, by a theorem of E. Cartan; this holds for any Lie group G. Consequently if we can lift the map φ_∞ to G it must be trivial. However we could lift if it were defined over a square. To arrange this let $\hat{\underline{x}}(s,t)$ be a map from the unit square [0,1]x[0,1] which maps the whole boundary of the square to a single point, the fixed point $\hat{\underline{z}}$, but is otherwise one to one. So

$$\hat{\underline{x}}(0,t) = \hat{\underline{x}}(1,t) = \hat{\underline{x}}(s,0) = \hat{\underline{x}}(s,1) = \hat{\underline{z}} \tag{3.66}$$

Now we can find a g(s,t) such that

$$\varphi_\infty(\hat{\underline{x}}(s,t)) = D(g(s,t))\varphi_0 \tag{3.67}$$

by starting with g(s,0) = 1 and then determining g(s,t) for succesive increasing t so that the loop g(s,t), $0 \leq s \leq 1$ varies continuously. So g(s,t) is determined by going round loops of fixed t on the sphere. Then g(s,0) = g(0,t) = g(1,t) = 1 but in general g(s,1) is not unity. If it is we have succeeded in lifting φ_∞ to a map $S^2 \to G$ and it must be trivial. In general, however,

$$\varphi_0 = \varphi_\infty(\hat{\underline{z}}(s,1)) = D(g(s,1))\varphi_0 \tag{3.68}$$

so that h(s) = D(g(s,1)) $\in$ H. Thus we have associated with the map $\varphi_\infty : S^2 \to H$ a map $[0,1] \to H$ such that h(0) = h(1) = 1. We can regard h as a map of the circle $S^1 \to H$ and it can be shown that the homotopy class of h as map $S^1 \to H$ (with h(0) = 1) depends only on the homotopy class of φ_∞. Thus we have map $\Pi_2(G/H)$ to $\Pi_1(H)$ the (Abelian) group of classes of maps $S^1 \to H$ (with a fixed point). This map may be shown to be a group isomorphism provided that G is simply connected,

$$\Pi_2(G/H) \cong \Pi_1(H) \tag{3.69}$$

For the 't Hooft-Polyakov case we have $\Pi_2(S^2) \cong \Pi_1(S^1)$, the homotopy classes of maps of the circle into itself. These classes are just labelled by the winding number the number of times, the map encircles the origin and so there is just one for each integer: $\Pi_2(S^2) = \Pi_1(S^1) = \mathbb{Z}$. This corresponds to the possible magnetic charges

N/e, $N \in \mathbb{Z}$.

To make this correspondence between the homotopy class of Φ_N and a closed path in H more concrete we derive a useful expression for h(s), used in this context by Goldstone. It will enable us to find h(s) explicitly in certain contexts, and to deduce quantization conditions.[6,27] Consider a sphere $\underline{x}(s,t) = r\hat{\underline{x}}(s,t)$ in a region in which

$$\mathscr{D}^r \phi = \partial^r \phi + ie\, D(\underset{\sim}{W}^r)\phi = 0 \tag{3.70}$$

Then,

$$\mathscr{D}_t \phi = \frac{\partial \phi}{\partial t} + ie\, D\left(\underset{\sim}{W}^r \frac{\partial x_r}{\partial t}\right)\phi = 0 \tag{3.71}$$

together with

$$\phi(s,0) = \phi_0 \tag{3.72}$$

is sufficient to determine ϕ over the sphere for given r. We construct a solution to these equations we consider $g(s,t) \in G$ defined by

$$\frac{\partial g}{\partial t} + ie\, \underset{\sim}{W}^r \frac{\partial x_r}{\partial t} g = 0, \tag{3.73}$$

subject to $g(s,0) = 1$.

$$\phi(x(s,t)) = D(g(s,t))\,\phi_0 \tag{3.74}$$

then solves eq. (3.71). Further, since $\frac{\partial x}{\partial t} = 0$, for $s = 0,1$, we have $g(0,t) = g(1,t) = 1$ and $g(s,t)$ defines a closed loop in G for each fixed t. In particular, we see from eq. (3.74) that $h(s) = g(s,1)$ is the closed path in H described above.

Now for any $f(s,t) \in G$,

$$\mathscr{D}_t(gf) = \mathscr{D}_t(g)f + g\frac{\partial f}{\partial t} = g\frac{\partial f}{\partial t} \tag{3.75}$$

Consequently,

$$\frac{\partial}{\partial t}(g^{-1}f) = g^{-1}\mathscr{D}_t f \tag{3.76}$$

In particular,

$$\begin{aligned} \frac{\partial}{\partial t}(g^{-1}\mathscr{D}_s g) &= g^{-1}\mathscr{D}_t(\mathscr{D}_s g) \\ &= g^{-1}[\mathscr{D}, \mathscr{D}_s]g \\ &= ie\, g^{-1} \underset{\sim}{G}_{ij}\, g \frac{\partial x^i}{\partial t}\frac{\partial x^j}{\partial s} \quad \text{by eq. (3.73)} \end{aligned} \tag{3.77}$$

Integrating with respect to t from t = 0 to 1,

$$h^{-1}\frac{dh}{ds} = ie\int_0^1 dt\; g^{-1} G_{ij}\, g\,\frac{\partial x^i}{\partial t}\frac{\partial x^j}{\partial s} \tag{3.78}$$

which is the formula we have been seeking.

In the specific case of the 't Hooft-Polyakov model, where G = SU(2), the little group H of Φ_0 consists of rotations about $\Phi_0 = a\hat{z}$.
Thus

$$h(s) = \exp\left(iA(s)\sigma_3/2\right) \tag{3.79}$$

$$\tfrac{1}{2}\sigma_3\frac{dA}{ds} = e\int_0^1 dt\; g^{-1} G_{ij}\, g\,\frac{\partial x^i}{\partial t}\frac{\partial x^j}{\partial s} \tag{3.80}$$

and

$$\frac{dA}{ds} = 2e\int_0^1 \mathrm{Tr}\left(G_{ij}\, g\hat{\Phi}_0 g^{-1}\right)\frac{\partial x^i}{\partial t}\frac{\partial x^j}{\partial s}\,dt \quad \text{for} \quad \hat{\Phi}_0 = \sigma_3/2$$

But

$$\hat{\Phi}(x(s,t)) = g(s,t)\,\hat{\Phi}_0\, g(s,t)^{-1} \tag{3.81}$$

and

$$F^{\mu\nu} = 2\,\mathrm{Tr}\left(G^{\mu\nu}\hat{\Phi}\right) \tag{3.82}$$

So

$$A(s) = e\int_0^s ds\int_0^1 dt\; F_{ij}\frac{\partial x^i}{\partial t}\frac{\partial x^j}{\partial s} \tag{3.83}$$

Asymptotically,

$$F^{ij} = -\frac{g}{4\pi}\,\frac{\epsilon_{ijk}x^k}{r^3} \tag{3.84}$$

giving

$$A(s) = -\frac{eg}{4\pi}\,\Omega(s) \tag{3.85}$$

where $\Omega(s)$ is the area of the surface $\{\hat{x}(s',t) : 0 \leq s' \leq s,\ 0 \leq t \leq 1\}$. So

$$h(s) = \exp\left\{-i\frac{eg}{4\pi}\Omega(s)\sigma_3/2\right\} \tag{3.86}$$

and h(1) = 1 giving the quantization condition

$$e^{ieg\sigma_3/2} = 1, \qquad eg/4\pi \in \mathbb{Z}, \tag{3.87}$$

which we obtained from different considerations in section 3.4.

Thus we have seen that the solution with charge $-4\pi N/e$ is associated with a mapping $S^2 \to G/H \cong S^2$ which is an N-fold covering of the sphere and this is associated with an N-fold map of the circle into itself, h(s), under the isomorphism of $\pi_2(G/H)$ onto $\pi_1(H)$.

If more generally we consider a theory with compact simply connected gauge group G and assume the field tensor has the asymptotic form,

$$\underset{\sim}{G}_{ij} \sim \frac{\epsilon_{ijk} x^k}{r^3} \underset{\sim}{G}(\hat{\underset{\sim}{x}}) \tag{3.88}$$

with

$$\mathcal{D}^i \underset{\sim}{G} = 0 \tag{3.89}$$

Then

$$\underset{\sim}{G}(\hat{\underset{\sim}{x}}(s,t)) = g(s,t)\, \underset{\sim}{G}_0\, g(s,t)^{-1} \tag{3.90}$$

yielding

$$h(s) = \exp(ie \underset{\sim}{G}_0 \Omega(s)) \tag{3.91}$$

and hence the quantization condition

$$\exp(ie \underset{\sim}{G}_0 4\pi) = 1, \tag{3.92}$$

which is further analysed in Ref. 27.

References.

1. G. 't Hooft, Nuclear Physics B79, 276 (1974).

2. A. M. Polyakov, JETP letters 20, 194 (1974).

3. S. Coleman, Phys. Rev. D11, 2088 (1975).

4. P. A. M. Dirac, Proc. Roy. Soc. A133, 60 (1931).

5. S. Coleman 'Classical lumps and their quantum descendants' to be published in the proceedings of the 1975 International School of Subnuclear Physics 'Ettore Majorana'.

6. P. Goddard and D. I. Olive, in preparation; to be submitted in Reports of Progress in Physics.

7. A. C. Scott, F. Y. F. Chu and D. W. McLaughlin, Proc. IEEE 61, 1443 (1973).

8. T. H. R. Skyrme, Proc. Roy. Soc. A247, 260 (1958); Proc. Roy. Soc. A262, 237 (1961).

9. J. Frohlich, Phys. Rev. Letters 34, 833 (1975).

10. G. H. Derrick, J. Math. Phys. 5, 1252 (1964).

11. P. A. M. Dirac, Phys. Rev. 74, 817 (1948).

12. E. Amaddi and N. Cabibbo, 'On the Dirac Magnetic Poles' in 'Aspects of Quantum Theory' edited by Abdus Salam and E. P. Wigner (Cambridge, 1972), 183.

13. B. Zumino, 'Recent developments in the theory of magnetically charged particles' in the proceedings of the 1966 International School of Subnuclear Physics 'Ettore Majorana'.

14. Y. Aharonov and D. Bohm, Phys. Rev. 115, 489 (1959); Phys. Rev. 123, 1511 (1961).

15. D. I. Olive, Nucl. Phys. B113, 413 (1976).

16. S. Weinberg, Phys. Rev. 138B, 988 (1965).

17. J. Schwinger, Phys. Rev. 144, 1087 (1966).

18. M. K. Prasad and C. M. Sommerfield, Phys. Rev. Lett. 35, 760 (1975).

19. B. Julia and A. Zee, Phys. Rev. D11, 2227 (1975).

20. J. C. Taylor, Gauge theories of weak interactions (Cambridge, 1976).

21. E. Corrigan, D. I. Olive, D. B. Fairlie and J. Nuyts, Nucl. Phys. B106, 475 (1976).

22. E. Corrigan and D. I. Olive Nucl. Phys. B110, 237 (1976).

23. E. B. Bogomolny, Soviet J. Nucl. Phys. 24, 801 (1976).

24. L. D. Faddeev, Letters Math. Phys. 1, 289 (1976).

25. S. Coleman, S. Parke, A. Neveu and C. M. Sommerfield, Phys. Rev. D15, 544 (1977).

26. D. G. Boulware, R. N. Cahn, S. D. Ellis and C. Lee, Phys. Rev. D14, 2708 (1976).

27. P. Goddard, J. Nuyts and D. I. Olive, CERN preprint TH 2255 (1976), to be published in Nuclear Physics B.

'MODERN' DIFFERENTIAL GEOMETRY IN ELEMENTARY PARTICLE PHYSICS

(Gauge fields, solitons, superspaces, quantum differential geometry, etc.)

Robert Hermann*

Department of Physics
Lyman Laboratory
Harvard University
Cambridge, Massachusetts

TABLE OF CONTENTS

*Supported in part by the National Science Foundation, Grant NSF-MCS76-06358.

1. INTRODUCTION

Some of the most exciting and fruitful periods in the history of science occur when new mathematical methods are successfully introduced into the study of fundamental physical phenomena. To illustrate, think of the great supernovae of mathematical physics--Newtonian mechanics, relativity, and quantum mechanics.

The theory of "elementary particles" has been under intensive development for the last thirty years. Despite the strong probability (through historical evidence) that new mathematical paradigms would be needed to achieve a fundamental understanding, physicists have been very conservative (and "pragmatic") about the sort of mathematics they consider admissable. To my eye, it has been evident since the introduction of sophisticated Lie group theory in the early 1960's that fundamental new mathematics, particularly concerning Lie groups and nonlinear differential equations, would be required, but I have not been particularly successful in convincing anyone else--mathematician or physicist--that this work is worth doing. In frustration, I decided to develop these ideas in a series of books [1-16], waiting for the day when more specific paths would open up for the development of the ideas. That day is almost here. In the last five years there have been incremental development of a succession of ideas (gauge-Yang-Mills fields, solitons, instantons, "super" spaces,...) that I believe add up to a new way of applying geometry to physics, and that give some clues as to what a *mathematically* unified physical theory will look like.

If it is so self-evident that the answer to the fundamental problems of physics will come from geometry, why haven't physicists learned it? I am afraid that a good deal of the answer must be ascribed to the perversity of the physicists dominant "pragmatic" mathematical ideology. Pragmatists notoriously have no esthetic sense, and "modern" differential geometry (as opposed, say, to tensor analysis, which it replaced and which physicists do know) requires, above all, some aesthetic feeling for the structure of modern mathematics. It does not help either that (at least in the U.S.) the formal mathematical education of most physicists stops at the advanced calculus level. Of course, another perverse feature has been the complete inversion and isolation of the small community of differential geometers. Their fathers and grandfathers took great delight and inspiration in mixing it up with the physicists, but the current generation has completely lost that taste and talent.

The mathematicians who have been most interested in "fundamental physics" have been trained in functional analysis, which is, in my view, not necessarily the most creative source of mathematical ideas for physics. These mathematicians often think their highest goal is to add "mathematical rigor" to physics. However, this is usually not the goal of geometers--good conceptual ideas usually comes first and the "rigor" can be put in later. This distinction goes back to Poincaré; he wrote of two types of mathematician--the "analyste" and the "géomètre". He did not hide his opinion of which type he thought was more creative! (Keep in mind that this does not excuse sloppy mathematical thinking, notation, formulation, etc. In fact, for the "géomètre" these are usually more important than for the "analyste".) Another characteristic of the "modern" geometric point of view is a strong interplay between *algebra* and *geometry*. Ideally, someone wanting to apply modern geometry should have at least a smattering of knowledge of all three main "technical" branches of modern mathematics--analysis, algebra, and topology. This is, of course, a great barrier for the physicist or engineer, who at best will have only a strongly focussed and selected knowledge of mathematics. (Many physicists have told me that they find these geometric matters very interesting, but they do not feel that they can afford to invest the time to acquire the varied mathematical background they would need.)

In these Notes I will try to give the reader some feeling for why I believe a unified geometric theory is in sight. It is unrealistic to hope to start from zero knowledge of either the mathematics or the physics. I hope that the reader is simultaneously reading more basic material in geometry. Among those books written by a mathematician, Boothby [17] is probably best. The geometric parts of Misner, Thorne and Wheeler [18] are the best I know of in the physics literature. Most of the ideas discussed here are to be found in my own books (some are even new); I apologize that I have not made a serious attempt to systematically give other references. (Some standard mathematical references are [2], [19], [20], [21].) It is my contention that three geometric objects--*fiber spaces*, *fiber space connections* (particularly those that I define in [14] to be of *Cartan-Ehresmann type*) and *differential form algebras*--play the key role in these physical applications. Accordingly, I will devote most of these Notes to describing their most relevant properties; unfortunately, I cannot do the physics justice here in the space available. See [27] for a brief survey of the physics of "gauge fields".

I will emphasize the theory of *solitons*. I believe this is the aspect of recent work about which the general theoretical physics community knows the least. (This is evident in the many foolish papers and statements that have appeared in recent years.) To the mathematicians who developed the theory, the theory of "solitons" is <u>not</u> simply the study of special solutions of broad classes of nonlinear field equations. They are mathematical beasts which go along with more fundamental objects--the inverse scattering structure and the Bäcklund structure. (The "prolongation" concepts of Wahlquist and Estabrook are a brilliant "geometrization" and unification of these ideas.) Only "exceptional" field theories seem to have these structures and no one has yet understood their "quantum" analogue. I believe this "exceptionalness" is a very positive feature of the theory; Nature usually chooses exceptionally beautiful mathematical structures to express her laws!

(Think of Newton's inverse square law--it is the only radially symmetric potential which leads to closed orbits!)

Virtually the only physicists I have met who understand the full ramifications of the application of modern differential geometry to physics are Frank Estabrook and Hugo Wahlquist of Jet Propulsion Laboratory. Their general "prolongation" idea [28] is the most striking flash of insight into how differential geometric ideas will mesh with physics. I have greatly benefited from talking to them, and many of the ideas presented here are collaborative intellectual projects. I have also profited from conversations with J. Corones, H. Morris, F. Pirani, P. McCarthy and D. Robinson. This work was done while I was a guest in the Physics Department of Harvard University; I would like to thank Professor S. Glashow for his hospitality.

2. MANIFOLDS AND DIFFERENTIAL FORM ALGEBRAS

The basic geometric object of "modern" differential geometry is a *differentiable manifold.* If X is a set of points, to endow it with a manifold structure means to give the following data:

a) A Hausdorff topological space structure,

b) A countable number of open subsets $U_1, U_2, \ldots$ of X whose union is all of X,

c) A homeomorphism ϕ_i of each U_i, $i = 1,2,\ldots$, with an open subset of R^n. This attaches n numbers to each point of U_i, thought of as its *coordinates.*

d) Whenever U_i and U_j intersect, the function $f_{ij}(\ )$ determining one set of coordinates in terms of the other,

$$\phi_i = f_{ij}(\phi_j)$$

are C^∞ ("infinitely differentiable") functions.

More intuitively, a "manifold" (which we take, for simplicity, as of "differentiability class C^∞") is a space which is *locally* coordinatized by R^n, but globally may not be; however, *do* require that the *different local coordinate systems fit together in a "smooth" way.* Think of the sphere

$$S^2 : |\vec{x}|^2 = x_1^2 + x_2^2 + x_3^2 = 1$$

in R^3. S^2-(north pole), S^2-(south pole) are open subsets which can be coordinatized in this way. For example, "stereographic projection" would be a convenient and natural way to do this. "Spherical coordinates" of the usual type would be another.

This "smoothness" guarantees that differentiability *makes sense independent of coordinates,* and enables techniques of calculus to be developed in a *coordinate-free way.* (From this point of view, "differential calculus on manifolds" is just a way of doing *tensor analysis*. See the book Tensor Analysis on Manifolds by Bishop and Goldberg [21]. In [28a-c] I have tried to describe the relations between classical differential geometry (e.g., "tensor analysis") and the "modern" appraoch

The way that mathematicians like to exhibit this independence of coordinates is to redo calculus, replacing the standard material with more algebraic ideas. Here is one way to do this. (See my book, Differential Geometry and the Calculus of Variations [2] for more detail.) Let $F(X)$ be the set of C^∞ real-valued functions on X. (This means that an $f \in F(X)$ is C^∞ when related back to R^n via coordinate map $\phi_j: U_j \to R^n$.) Elements of $F(X)$ can be added and multiplied, i.e., $F(X)$ forms a *ring*. A *vector field* is a *derivation* of this ring. Each such derivation can be multiplied by functions, i.e., the vector fields form a *module* over the ring $F(X)$. $\mathscr{D}^1(X)$, the *one-differential* forms, are *defined* as the *dual* module. $\mathscr{D}^r(X)$, the n-differential forms for $r \geq 2$, are the r-fold skew-symmetric tensor product of $\mathscr{D}^1(X)$ with itself. $\mathscr{D}^0(X)$, the *zero-forms*, are defined as $F(X)$. $\mathscr{D}(X)$ is the direct sum $\mathscr{D}^0(X) \oplus \mathscr{D}^1(X) \oplus \cdots$ of all these modules. It has two algebraic operators:

$$\phi: \mathscr{D}^r(X) \to \mathscr{D}^{r+1}(X) \quad ,$$

called *exterior derivative*, and *exterior multiplication*

$$\wedge : \mathscr{D}^r(X) \times \mathscr{D}^s(X) \to \mathscr{D}^{r+s}(X) \quad .$$

Now, all of the *differential geometric* information about X (e.g., properties of Riemannian matric, affine connections, calculus of variation, etc.) can be expressed in terms of differential forms, particularly if one follows ideas of Elie Cartan. Many ideas of physics (Maxwell fields, Yang-Mills fields, etc.) can also be beautifully stated in terms of differential forms.

This assignment

$$X \to \mathscr{D}(X)$$

is a typical example of what mathematicians call a *functor*. If $\phi: X \to Y$ is a C^∞ map between manifolds, then it induces a "pull-back" map

$$\phi^*: \mathscr{D}(Y) \to \mathscr{D}(X)$$

which goes in the *opposite* direction to ϕ, i.e., if

$$\phi = \phi_1\phi_2$$

the composition of two such maps, then

$$\phi^* = \phi_2^*\phi_1^* \quad .$$

A *coordinate system* for a manifold X is a set

$$(x^1, \ldots, x^n)$$

of elements of $F(X) \equiv \mathscr{D}^0(X)$ such that the C^∞ map $x \to (x^1(x), \ldots, x^n(x))$ of $X \to R^n$ is onto, and C^∞ *invertible*. Given such a coordinate system, $\mathscr{D}(X)$ can be described more explicitly in a theory that ties up with tensor analysis (i.e., elements of $\mathscr{D}^r(X)$ as r-fold, skew-symmetric tensors.) Choose the range of indices

$$1 \leq i, i_1, \ldots \leq n \quad ,$$

and the summation convention on these indices. The x^i are elements of $\mathscr{D}^0(X)$. Hence, their "differentials" dx^i are elements of $\mathscr{D}^1(X)$. We can then use the exterior multiplication to build up higher order elements

$$dx^{i_1} \wedge \cdots \wedge dx^{i_m} \quad .$$

The dx^i *anti-commute:*

$$dx^i \wedge dx^j = - dx^j \wedge dx^i$$

x^i and dx^j *commute:*

$$x^i \, dx^j = dx^j \, x^i \quad .$$

One proves now that the elements $\omega \in \mathscr{D}^r(X)$, $r = 1, 2, \ldots$, can be written uniquely in the form:

$$\omega = a_{i_1 \cdots i_r} \, dx^{i_1} \wedge \cdots \wedge dx^{i_r} \quad ,$$

with coefficients $(a_{i_1 \cdots i_r})$ which are elements of $F(X) \equiv \mathscr{D}^0(X)$, and which depend *skew-symmetrically* on their indices. This representation, and the following algebraic rules,

$$d(d\omega) = 0 \quad , \qquad \text{for } \omega \in \mathscr{D}(X)$$

$$d(\omega_1 \wedge \omega_2) = d\omega_1 \wedge \omega_2 + (-1)^{\text{degree } \omega_1} \omega_1 \wedge d\omega_2$$

determines the action of d completely:

$$d\omega = \frac{\partial a_{i_1 \cdots i_r}}{\partial x^i} \, dx^i \wedge dx^{i_1} \wedge \cdots \wedge dx^{i_r}$$

$$\equiv da_{i_1 \cdots i_r} \wedge dx^{i_1} \wedge \cdots \wedge dx^{i_r}$$

Thus, we see that in principle "differential geometry" can be "algebracized". The prototype for this is work by H. Cartan [29] that was enormously influential for mathematicians of my generation, but is not well-known nowadays. The "superspace-supersymmetry" game recently put together by various physicists is, from the mathematician's point of view, an intriguing variant of H. Cartan's ideas. I will now elaborate a bit with some formal definitions.

2. AN ALGEBRAIC GENERALIZATION OF DIFFERENTIAL GEOMETRY. SUPERSPACE AND "QUANTUM GEOMETRY" AS SPECIALIZATIONS

Definition. A set $\mathcal{D}$ will be called a *differential form algebra* if it has the following properties:

a) $\mathcal{D}$ is a vector space, say, over the real numbers as field of scalars,

b) $\mathcal{D}$ is *integer-graded*, i.e., as a direct sum

$$\mathcal{D}^0(X) \oplus \mathcal{D}^1(X) \oplus \cdots$$

of subspaces labelled by a nonnegative integer.

c) $\mathcal{D}$ is an algebra, i.e., a bilinear map $\mathcal{D} \times \mathcal{D} \to \mathcal{D}$ is given. Denote this product structure by

$$(\omega_1, \omega_2) \to \omega_1 \wedge \omega_2 \quad .$$

It satisfies:

d) $\mathcal{D}^r \wedge \mathcal{D}^s \subset \mathcal{D}^{r+s}$,

e) It is *associative*, i.e.,

$$\omega_1 \wedge (\omega_2 \wedge \omega_3) = (\omega_1 \wedge \omega_2) \wedge \omega_3$$

for $\omega_1, \omega_2, \omega_3 \in \mathcal{D}$

For $r = s = 0$ this means that $\mathcal{D}^0$ is a *subalgebra*. Denote the multiplication in $\mathcal{D}^0$ simply as juxtaposition: $(f,g) \to fg$.

f) There is given a linear map $d\colon \mathcal{D} \to \mathcal{D}$ such that $d(\mathcal{D}^r) \subset \mathcal{D}^{r+1}$.

$$d(fg) = (df)g + f(dg) \tag{3.1}$$

for $f,g \in \mathcal{D}^0$

$$d^2 = 0 \tag{3.2}$$

i.e.,

$$d(d\omega) = 0, \qquad \text{for all } \omega \in \mathcal{D}$$

$$d(f\omega) = df \wedge \omega \tag{3.3}$$

if $d\omega = 0$, $f \in \mathcal{D}^0$

g) $$d(\omega f) = (-1)^p\, d\omega \wedge df \tag{3.4}$$

if $\omega \in \mathcal{D}^p$, $f \in \mathcal{D}^0$ and if $d\omega = 0$

h) $\mathcal{D}$ is generated by $\mathcal{D}^0$, and the operations $d, \wedge$ plus the vector space structure, i.e., every element $\omega \in \mathcal{D}^r$ can be written as a linear combination of elements of the form

$$f(df_1) \wedge g_1(df_2)g_2 \cdots g_r \wedge df_r \tag{3.5}$$

for $f_1, \ldots, f_r;\ g_1, \ldots, g_r \in \mathcal{D}$

Note especially that we do <u>not</u> assume that multiplication in $\mathcal{D}^0$ --which determines everything else--is commutative, as it was for standard "differential geometry". This gives us an additional degree of freedom, which can, in fact, be exploited to fit in (at least) two additional cases of interest for physics-- "superspace" and what I call [30] "quantum" (boson) differential geometry. To see such possibilities, suppose f, g are two elements of $\mathcal{D}^0$, with:

$$fg \pm gf = c \quad , \tag{3.6}$$

and $c \in R$. (Suppose the algebra structure on $\mathcal{D}^0$ has a unit element "1", so that c is just $c1$.) Apply (3.1)

$$(df)g + fdg \pm (dg)f \pm gdf = 0 \tag{3.7}$$

<u>Suppose</u> (as an Ansatz) that we want to require that df, dg be "independent". (3.5) then requires that

$$(df)g \pm gdf = 0 \quad . \tag{3.8}$$

Apply d to both sides of (3.8) using (3.1) - (3.4):

$$- df \wedge dg \pm dg \wedge df = 0 \tag{3.9}$$

Hence:

df, dg anticommute (commute) if $+1$ (-1) occurs in (3.6).

Thus, we see that $\mathcal{D}$, d, r will be determined if we *assume* that $\mathcal{D}^0$ is generated (as an algebra) by pairs of elements (f,g) satisfying commutation relations like (3.6).

Example 1) "Superspace":

$\mathcal{D}^0$ is generated by elements (x^i, y^a), $1 \leq i,j \leq n$; $1 \leq a,b \leq m$, satisfying

$$x^i x^j = x^j x^i \; ; \quad y^a y^b = - y^b y^a \; ; \quad x^i y^a = y^a x^y \quad . \tag{3.10}$$

Their differentials must then satisfy the following commutation relations [31,36]:

$$\begin{aligned} dx^i \wedge dx^j &= - dx^j \wedge dx^i \\ dx^i \wedge dy^a &= - dy^a \wedge dx^i \\ dy^a \wedge dy^b &= +1\, dy^b \wedge dy^a \quad . \end{aligned} \tag{3.11}$$

This enables us to write an element $\omega \in \mathcal{D}^p$ in the following form:

$$f_{i_1 \cdots i_r a_1 \cdots a_{i_s}} dx^{i_1} \wedge \cdots \wedge dx^{i_r} \wedge dy^{a_1} \wedge \cdots \wedge dy^{a_s} \tag{3.12}$$

with $r + s = p$, and the f's are elements of $\mathcal{D}^0$ which depend skew-symmetrically on the indices i, symmetrically on the indices a.

Now, one can use the commutation relations (3.10) to write an $f \in \mathcal{D}^0$ in the form

$$f = g_a(x) y^a + g_{a_1 a_2}(x) y^{a_1} y^{a_2} + \cdots \tag{3.13}$$

Using this in (3.12) shows that ω can be written as sums of the form

$$\theta_{a,a_1 \cdots a_r} y^a \wedge dy^{a_1} \wedge \cdots \wedge dy^{a_r} + \theta_{ab,a_1 \cdots a_r} y^a y^b \wedge dy^{a_1} \wedge \cdots \wedge dy^{a_r} + \cdots \tag{3.14}$$

where the θ's are *differential forms of the usual type in terms of the variables* x. This is the algebraic prototype of the physicists talking about using "superspace" to put together in a "unified" way "fields" of the usual type of various spins.

Example 2) Quantum differential geometry:

Suppose that $\mathcal{D}^0$ is generated (as an associative algebra) by elements that we label

$$(x^i, p_j) \quad , \qquad 1 \leq i,j \leq n \quad ,$$

which satisfy the following *commutation* relations:

$$\begin{aligned} x^i x^j &= x^j x^i \; ; \quad p_i p_j = p_j p_i \; ; \\ p_i x^j - x^j p_i &= \hbar \delta_i^j \end{aligned} \tag{3.15}$$

$\hbar$ is a constant, which is, of course, to be identified (up to 2π) with Planck's constant. Thus relations (3.15) are essentially the *Heisenberg commutation relations*. (However, it is important to note that they are not to be considered as *operator* relations, but as commutation relations for an "abstract" associative algebra.)

We can now assume that $\mathcal{D}^0$ is part of a differential form algebra $\mathcal{D}^n$, $n \geq 0$. As an Ansatz, let us say that the dp_i, dx^j are linearly independent (in the $\mathcal{D}^0$-module sense). As we have seen, this condition, plus the commutation relations (3.15), determines the commutation relations between their differentials. In fact, they take the following form:

$$dx^i \wedge dx^j = - dx^j \wedge dx^i$$

$$dx^i \wedge dp_j = - dp_j \wedge dx^i$$

$$dp_i \wedge dp_j = - dp_j \wedge dp_i$$

$$x^i dp_j = dp_j x^i \quad ,$$

etc.

In words, the differentials dx^i, dp_j satisfy exactly the *same commutation relations* as they do "classically", i.e., when $\hbar = 0$.

The usual quantum-mechanical story may now be obtained by asking for irreducible representations of $\mathcal{D}^0$ by Hermitian operators. However, it is interesting to note that it is not necessary to do so, even to form the "dynamics". Thus, it is possible to completely decouple (as far as differential-geometric properties go) the structure of quantum mechanics from the need to use a Hilbert space (which always seemed to me to be the weakest link in the mathematical structure of quantum mechanics--and which becomes especially acute when field theories enter the picture). To see this, it is most convenient to follow Cartan's brilliant (but difficult and eccentric) approach to classical mechanics presented in his book , Lecons sur les Invariants Intégraux [39]. Introduce another variable "t" which commutes with everything. Let h be an element of $\mathcal{D}^0$. Set

$$\omega = dp_i \wedge dx^i - dh \wedge dt \tag{3.16}$$

Write:

$$dh = h_i dx^i + h^i dp_i \quad . \tag{3.17}$$

The h_i, h^i are *defined* by this relation. (In case $\hbar = 0$, they are, as usual, the partial derivatives

$$h_i = \frac{\partial h}{\partial x^i} ; \qquad h^i = \frac{\partial h}{\partial p_i} .$$

However, to take care of this *quantum mechanically*, usually involves some sort of *Wick ordering*. The proscription (3.17) for defining these objects thus contains within itself a *differential-geometric* version of this.

Now, substitute (3.17) into (3.16). Notice that:

$$\omega = (dp_i + h_i dt) \wedge (dx^i - h^i dt) \qquad (3.18)$$

The "Cauchy characteristic" equations [2,39] can now be read off:

$$\frac{dx^i}{dt} = h^i ; \qquad \frac{dp_i}{dt} = - h_i .$$

They are the "Hamilton-Heisenberg" equations of motion, but in their quantum mechanical form. These differential-geometric techniques can now be generalized to handle field theories [4,5,8]. Perhaps most interesting and important from the foundational point of view, replacing the $\mathcal{D}^0$ defined above by one appropriate to "superspace" leads in a very natural way to a procedure for quantization which yields the *Dirac equation*, as shown by Berezin and Marinov [38]. I regard this as the most convincing piece of evidence that these algebraic concepts are important for physics!

There is, of course, much more material involved in the foundations of manifold theory. I wanted to put down just enough to give the reader the flavor--assigning an "algebraic" object--the differential forms--to each manifold. (This is a typical example of a "functor".) Differential geometry can thus be viewed as a "game", recapturing the "geometric" properties in terms of this algebra. The physicists who are developing the theory of "superspaces" are then engaged in the very interesting project of slightly modifying the rules of this game.

The main point to this is that *much of physics can be rewritten in terms of differential forms*, which, in turn, exhibits the differential-geometric unity and interconnections of much of physics. This is especially powerful when combined with the second major "geometric object", the *fiber spaces*. Let us now turn to these objects.

4. FIBER SPACES

Let X,Y be manifolds, and let $\pi: Y \to X$ be a (smooth) map. I want to say what it means for π to define a "fiber space". There are various technical ways of doing this. I will choose the one which appears closest to the way the theory of fiber spaces is involved in the theory of "gauge fields".

First of all, let us suppose that π is what topologists call an *open map.* This means that there are coordinate open sets $U_Y \subset Y$, $U_X \subset X$ such that

$$\pi(U_Y) = U_X \quad .$$

Let (x^i), $1 \leq i,j \leq n$, be a coordinate system of functions on U_X. Let

$$\pi^*(x^i)$$

be the pull-back into functions on Y, i.e.,

$$\pi^*(x^i)(y) = x^i(\pi y) \quad .$$

We now <u>suppose</u> that the following condition is satisfied:

> There are an additional set (y^a), $1 \leq a,b \leq m$, of smooth functions on U_X such that the functions
>
> $$(\pi^*(x^i), y^a)$$
>
> form a coordinate system for U_Y.

For notational convenience, we leave off the π^*, and just label this coordinate system (which is "specially adapted" to the fiber space) as (x^i, y^a).

These coordinates are not, of course, uniquely defined. They can be changed to

$$x_1^i = f^i(x) \tag{4.1}$$

$$y_1^a = g^a(x,y) \tag{4.2}$$

The special "triangular" nature of the allowed coordinate changes (4.1) is what is typical of the "fiber space" situation.

However, it is our aim to think as much as possible in coordinate-free terms, and use such coordinates as little as possible. Given a point $x_0 \in X$, the set $Y(x_0) \equiv \pi^{-1}(x_0)$ of all points of X which map into x_0 is called the *fiber of the fiber space above the point* x_s.

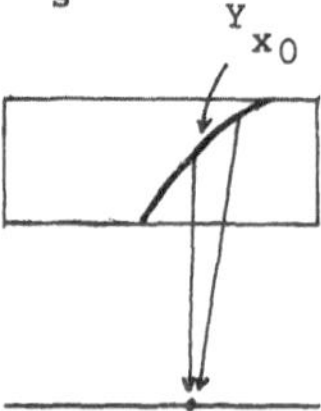

The fibers are *submanifolds* of Y; the y^a are then *coordinates* of these fibers. It is this geometric picture--the manifold Y "fibered" by giving a submanifold at each point--which is the most important. (In fact, there is a more general differential geometric concept--a *foliation*--which has this property. They are also important, but at the moment lie outside our domain.)

We want to endow these fibers with *geometric* structures. One convenient way is to prescribe a group acting on each fiber. One does this by postulating a group G^x assigned to each $x \in X$, and an action

$$(x,y) \to gy$$

of G^x by transformation on Y_x. (One must also prescribe G^x as "varying smoothly with x" in an appropriate way. We ignore such technical details.) Call this structure $\{G^x\}$ a *bundle of groups*. Giving this action on Y is called *giving the fiber space a bundle of structure groups*. Often, each group G_x is isomorphic to a single group G. One then speaks of the fiber space as having a *structure group* G.

In physics, one most often (but not always!) encounters fiber spaces which are product spaces

$$Y = X \times F$$

$\pi: Y \to X$ is the Cartesian product map.

Giving the group bundle

$$\{G^x\} \quad ,$$

a map $x \to \underline{g}(x) \in G^x$ determines a transformation on Y.

$$\underline{g}(x,f) = (x,\underline{g}(x)f) \tag{4.3}$$

$$\text{for } x \in X, \quad f \in F \quad .$$

Physicists call this a *gauge transformation*.

5. CONNECTIONS

Let $\pi: Y \to X$ be a fiber space map. Let $\mathcal{D}^r(X)$ and $\mathcal{D}^r(Y)$ denote the r-th degree differential forms of X and Y. Any differential form on Y which is a linear combination (with coefficients in $\mathcal{D}^0(Y)$) of forms in $\pi^*(\mathcal{D}^r(X))$ is said to be *vertical*. $\mathcal{V}^r(\pi)$ denotes the set of vertical r-forms.

Focus on $\mathcal{V}^1(\pi)$. In the "adapted" coordinates (x^i,y^a) described in Section 3, they are those which can be written in the form

$$\theta = f_a(x,y)dx^a \tag{5.1}$$

Notice that there is no *uniquely defined* "complementary" set of one-forms. In these coordinates dy^a forms such a set, but it is not at all unique. For example, the change of variables (4.2) creates a new complementary set dy_1^a, with

$$dy_1^a = \frac{\partial g^a}{\partial x^i} dx^i + \frac{\partial g^a}{\partial y^b} dy^b$$

What we must do is to *prescribe* a complementary set. In fact, this turns out to be precisely the geometric object called a *connection*.

Definition. A *connection* for the fiber space (Y,X,π) is a set $\mathscr{P}$ of one-forms on Y such that:

> $\mathscr{P}$ is a $\mathscr{D}^0(Y)$-module, i.e., if $\theta_1,\theta_2 \in \mathscr{P}$, $f \in \mathscr{D}^0(Y)$, then
>
> $$\theta_1 + \theta_2,\ f\theta,\ \in \mathscr{P} \quad .$$
>
> $\mathscr{D}^1(Y)$ is a direct sum (as an $\mathscr{D}^0(Y)$-module) of the vertical one-form $\mathscr{V}^1(\pi)$ and $\mathscr{P}$.

Here is what this means in the adapted coordinates (x^i,y^a). $\mathscr{P}$ has a basis (as a $\mathscr{D}^0(Y)$-module) consisting of a set of one-forms θ^a of the form:

$$\theta^a = dy^a - \omega^a \quad , \tag{5.1}$$

where the ω^a are *vertical* one-forms. They are said to be the *components* of the connection in this adapted coordinate system.

In order to see this relevance to physics, let us see how the components change when the adapted coordinates are changed. For example, suppose that

$$y^a = f^a(x,z) \quad ,$$

where z^a are new "fiber" coordinates. Then,

$$dy^a = f_i^a dx^i + f_b^a dz^b \tag{5.2}$$

$$\left(f_i^a = \frac{\partial f}{\partial x^i} ; \qquad f_b^a = \frac{\partial f^a}{\partial z^b} \right) \quad .$$

Insert (5.2) into (5.1):

$$\theta^a = f_i^a dx^i + f_b^a dz^b - \omega^a = f_b^a(dz^b - (f^{-1})^b_c(\omega^c - f_i^c dx^i)) \quad .$$

Set:

$$\theta'^a = dz^a - \omega'^a \tag{5.3}$$

with

$$\omega'^a = (f^{-1})^a_b(\omega^b - f^b_i dx^i) \tag{5.4}$$

(5.4) exhibits the change of the components when the fiber coordinates are changed (in an x-dependent way, i.é., as what physicists traditionally call a "gauge transformation).

Linear Connections and Gauge Transformations

The ideas developed above are valid for both linear and nonlinear situations. As usual, they are simplest to understand in the linear situations.

Consider a connection described by formula (5.1). It is linear (in these coordinates) if the θ^a are of the following form:

$$\theta^a = dy^a - M^a_{bi}(x)y^b dx^i \tag{5.5}$$

Denoting y^a as a vector $\underline{y}$, (6.1) can be written more conveniently in vector-matrix form:

$$\underline{\theta} = d\underline{y} - \underline{M}_i(x)\underline{y}\, dx^i \quad . \tag{5.6}$$

(The $\underline{M}_i(x)$ are n matrix $(m \times m)$-valued functions of x.)

We can now consider the changes in the "adapted" coordinates (x,z) which preserve the linear property of the vectorial connection form $\underline{\theta}$. Among these are the following transformation, sometimes called "linear, local gauge transformations".

$$\underline{y} = \underline{G}(x)\underline{z} \quad . \tag{5.7}$$

We can readily find the expression for the connection in the new coordinates:

$$d\underline{y} = d\underline{G}\underline{z} + \underline{G}d\underline{z}$$

$$\underline{\theta} = d\underline{G}\underline{z} + \underline{G}d\underline{z} - \underline{M}_i\underline{y}\, dx^i$$

$$= \underline{G}(d\underline{z} - \underline{G}^{-1}\underline{M}_i\underline{G}dx^i + \underline{G}^{-1}d\underline{G}) \quad .$$

Set

$$\underline{\theta}' = d\underline{z} - \underline{G}^{-1}\underline{M}_i\underline{G}dx^i + \underline{G}^{-1}d\underline{G} \tag{5.8}$$

The $\underline{\theta}'$ are the bases for the connection forms in the "new" coordinates $(x,\underline{z})$.

Notice the typical "non-tensorial", "gauge-like" transformation law:

$$\underline{M}_i \to \underline{G}^{-1}\underline{M}_i\underline{G} + \underline{G}^{-1}dG \tag{5.9}$$

It is the second term on the right hand side of (5.9) which is "non-tensorial".

6. PARALLEL TRANSPORT AND CURVATURE

Let $\pi: Y \to X$ continue to be a fiber space. Assume also that Y has a global adapted coordinate system (y^a, x^i). Suppose a connection is given, defined by a basis of one-forms

$$\theta^a = dy^a - \Gamma^a_i(y,x)\, dx^i \quad . \tag{6.1}$$

The Γ^a_i are arbitrary functions of these variables.

Suppose that we give a curve $t \to x(t)$, $a \le t \le b$, in the base X. Let $x^i(t)$ be its components in these coordinates. Substitute these functions $x^i(t)$ into (6.1), obtaining the following one-forms:

$$\theta^a = dy^a - \Gamma^a_i(y,x(t))\, \frac{dx^i}{dt}\, dt \quad . \tag{6.2}$$

The variables y^a are still to be thought of as functions of the parameter variable t given along with the curve.

Now, let us ask for y as a function of t, so that the forms θ^a become zero. Thus $y^a(t)$ must then satisfy the following differential equations:

$$\frac{dy^a}{dt} = \Gamma^a_i(y(t),x(t))\, \frac{dx^i}{dt} \tag{6.3}$$

The usual existence and uniqueness theorem for ordinary differential equations (assuming, for simplicity, that it can be applied over all of the interval $a \le t \le b$) implies that there is a unique solution of $t \to y^a(t)$ (given $x(t)$) beginning at a given point $y^a(0)$. Let us say that a curve satisfying (6.2) is *horizontal*.

We can sum up as follows:

THEOREM 6.1. Let $\pi: Y \to X$ be a fiber space map, and let $\mathcal{P}$ be a collection of one-forms on Y which defines a connection. Given a curve $t \to x(t)$, $0 \le t \le b$, in X, and a point $y_0 \in Y$ such that $\pi(y_0) = x(a)$, there is a unique horizontal curve in Y, $t \to y(t)$ (found locally by solving a time-dependent system of ordinary differential equations) such that:

$$\pi(y(t)) = x(t) \quad , \qquad \text{for all } t.$$

This curve is called a *horizontal lift* of the base curve $t \to x(t)$

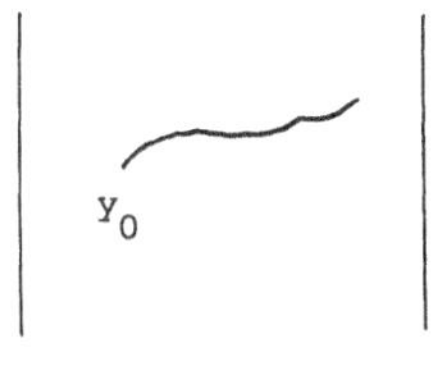

Now, fix the curve $t \to x(t)$ in the base space X. For $x \in X$, let

$$F(X) = \pi^{-1}(X) \equiv \textit{fiber above } x \quad .$$

For each point y of $F(x(0))$, there is then a unique horizontal lift beginning at y at $t = a$. Fell this lifted curve, and we obtain a point y' of $F(x(b))$ which is a function solely of y and the base curve. This defines a map

$$F(x(a)) \to f(x(b))$$

between fibers, called *parallel transport* along the curve. It is the basic geometric operation of connection theory.

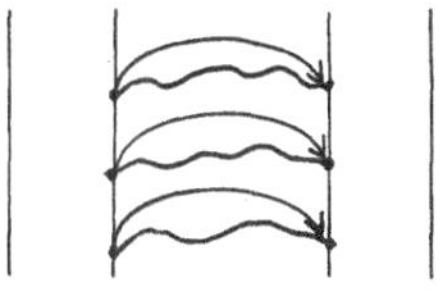

In general, this transformation of fibers will be path dependent. However, let us suppose that it is independent of path. Returning to the local coordinates (x^i, y^a) used above, we see that there will be functions $y^a(x)$ such that if $t \to x(t)$ is a curve in X, its parallel transport is the curve

$$t \to y^a(x(t)) \quad .$$

In particular, $y(x)$ will satisfy the Pfaffian differential equations

$$dy^a = \Gamma^a_i(y(x),x)\, dx^i \quad . \tag{6.4}$$

In the classical literature these are called *Mayer equations*. The condition that they can be solved in this way is traditionally known as "complete integrability", i.e., that the relations obtained by applying exterior derivative d to both sides, and the relation $d^2 = 0$, are satisfied *identically*. In modern differential geometry, this is done (following Cartan [39,40]) by saying that the forms θ^a *define a completely integrable Pfaffian system* in the sense that there are one-forms ω^a_b such that

$$d\theta^a = \omega^a_b \wedge \theta^b \quad . \tag{6.5}$$

In other words, it can be shown that the existence of one-forms ω^a_b satisfying (6.5) is *equivalent* to parallel transport in the fiber space being independent of path.

Let us now work out the conditions for (6.4) using (6.1).

$$\begin{aligned} d\theta^a &= -d\Gamma^a_i \wedge dx^i \\ &= -\Gamma^a_{i,b} dy^b \wedge dx^i - \Gamma^a_{i,j}\, dx^j \wedge dx^i \ , \quad \left(\Gamma^a_{i,b} = \frac{\partial \Gamma^a_i}{\partial y^b} \ ; \quad \Gamma^a_{i,j} = \frac{\partial \Gamma^a_i}{\partial x^j} \right) \\ &= -\Gamma^a_{i,b}\, (\theta^b + \Gamma^b_j dx^j) \wedge dx^i - \Gamma^a_{i,j} dx^j \wedge dx^i \end{aligned}$$

Now, the forms

$$\Omega^a_{ij} = (-\Gamma^a_{i,b}\Gamma^b_j + \Gamma^a_{j,b}\Gamma^b_i - \Gamma^a_{i,j} + \Gamma^a_{j,i})\, dx^j \wedge dx^i \tag{6.6}$$

can only be written in form (6.5) if they vanish identically, since the one-forms θ^a, dx^i are linearly independent. These two forms are called the *curvature forms*. It can be verified (a tedious calculation) that the curvature transforms *tensorially* whereas the θ^a do not. In fact, the curvature can be defined in a completely coordinate-free way. It then turns out to be a more exotic object that we have considered up to now--a *two-differential form on* X *with values in the vector bundle over* X *whose fibers are the vertical vector fields*, i.e., the vector fields on Y which are tangent to the fibers of π.

The curvature becomes a less formidable object in case the connection is linear. To work out this case--which is the basic one needed for physics--revert to the vector-matrix notation used earlier:

$$\underline{\theta} = d\underline{y} - \underline{M}_i(x)\underline{y}\,dx^i$$

Then

$$d\underline{\theta} = -d\underline{M}_i\underline{y}dx^i - \underline{M}_i d\underline{y} \wedge dx^i$$

$$= -d\underline{M}_i\underline{y}dx^i - \underline{M}_i(\underline{\theta} + \underline{M}_j\underline{y}dx^j) \wedge dx^i$$

Set:

$$\underline{\Omega} = d\underline{M}_i \wedge dx^i + \underline{M}_i\underline{M}_j\,dx^j \wedge dx^i \tag{6.7}$$

$$= \left(\frac{\partial \underline{M}_i}{\partial x^j} - \frac{\partial \underline{M}_j}{\partial x^i} + \frac{1}{2}[\underline{M}_i, \underline{M}_j]\right) dx^j \wedge dx^i \tag{6.8}$$

This is now the object whose vanishing ensures complete integrability, i.e., path-independence (locally) of parallel transport. Of course, the object occurring in (6.7)

$$\underline{F}_{ij} = \frac{\partial \underline{M}_i}{\partial x^j} - \frac{\partial \underline{M}_j}{\partial x^i} + \frac{1}{2}[\underline{M}_i, \underline{M}_j] \tag{6.9}$$

is very familiar to "gauge" physicists--the *curvature tensor of the Yang-Mills field.*

The fact that it is the *commutator* that appears in (6.9) is significant--it indicates that Lie algebras should get into the game. We will consider the appropriate way of doing this after a short diversion to consider the topic of "Lie algebra-valued one-forms", which is of interest in its own right.

7. LIE ALGEBRA-VALUED ONE-FORMS AND THE CURVATURE OPERATOR

The analytical part of the theory of connections "with structure group" involves the topic described by the title of this section.

Forget about fiber spaces for the moment, and consider a manifold X. $\mathcal{D}^r(X)$, $r = 0,1,\ldots,$ denotes the r-th degree differential forms. $\mathcal{D}^0(X)$ itself forms a (commutative, associative) algebra.

Let $\mathcal{G}$ be a real Lie algebra. The tensor product

$$\mathcal{D}^r(X) \otimes \mathcal{G} \stackrel{\text{def}}{\equiv} \mathcal{D}^r(X, \mathcal{G})$$

is defined as a $\mathcal{G}$-valued differential form on X. Explicitly, an element $\underset{\sim}{\omega} \in \mathcal{D}^r(X, \mathcal{G})$ is a linear combination

$$\underset{\sim}{\omega} = \underset{\sim}{\omega}^1 \otimes A_1 + \cdots + \omega^p \otimes A_p \tag{7.1}$$

with

$$A_1, \ldots, A_p \in \mathscr{G}, \qquad \omega^1, \ldots, \omega^p \in \mathscr{D}^r(X)$$

Define an operation

$$D: \mathscr{D}^1(X, \mathscr{G}) \to \mathscr{D}^2(X, \mathscr{G}) \tag{7.2}$$

by means of the following formula:

$$D\underset{\sim}{\omega} = d\omega^1 \otimes A_1 + \cdots + d\omega^p \otimes A_p + \frac{1}{2} \sum_{\alpha,\beta=1}^{p} (\omega^\alpha \wedge \omega^\beta) \otimes [A_\alpha, A_\beta] \tag{7.3}$$

Note that D is a nonlinear differential operator (if $\mathscr{G}$ is a nonabelian Lie algebra) called the *curvature operator*. (We will see why it is given this "geometric" name in the next section.)

D can be put into a more familiar form (to physicists) if the A are chosen as a basis for the Lie algebra $\mathscr{G}$. Let $C^\gamma_{\alpha\beta}$ be the structure constants of the Lie algebra, i.e.,

$$[A_\alpha, A_\beta] = C^\gamma_{\alpha\beta} A_\gamma \quad . \tag{7.4}$$

Suppose that

$$\omega^\alpha = \Gamma^\alpha_i \, dx^i \quad . \tag{7.5}$$

Then,

$$D\underset{\sim}{\omega} = d\Gamma^\alpha_i \wedge dx \otimes A_\alpha + \frac{1}{2} \Gamma^\alpha_i \, dx^i \wedge \Gamma^\beta_j \, dx^j \, C^\gamma_{\alpha\beta} \otimes A_\gamma$$

$$= \frac{1}{2} F^\gamma_{ij} \, dx^i \wedge dx^j \otimes A_\gamma \quad , \tag{7.6}$$

with

$$F^\gamma_{ij} = \left(\frac{\partial \Gamma^\alpha_1}{\partial x^j} - \frac{\partial \Gamma^\alpha_j}{\partial x^i} \right) + \frac{1}{2} (\Gamma^\alpha_i \Gamma^\beta_j - \Gamma^\alpha_j \Gamma^\beta_i) \, C^\gamma_{\alpha\beta} \tag{7.7}$$

which is again the familiar "Yang-Mills field" formula (reducing to one half of "Maxwell's equations" if $\mathscr{G}$ is abelian).

The vanishing of $D\underset{\sim}{\theta}$ has a special geometric significance. Let G be a Lie group whose Lie algebra is $\mathscr{G}$. Let η^α be the left-invariant one-forms on G which are "dual" to the basis A_α of $\mathscr{G}$. (The η^α are called the *Cartan-Maurer forms.*) Then, $D\underset{\sim}{\theta} = 0$, i.e.,

$$d\omega^\alpha + \frac{1}{2} C^\gamma_{\alpha\beta}\omega^\alpha \wedge \alpha^\beta = 0$$

implies that there is a map

$$\phi: X \to G$$

such that

$$\phi^*(\eta^\alpha) = \omega^\alpha \quad .$$

We shall see that this has a special importance in the theory of solitons.

8. CONNECTIONS WITH STRUCTURE GROUPS

Return to a fiber space $\pi: Y \to X$, with adapted coordinates (x^i, y^a). Let G be a Lie group of transformations on the variables y.

Definition. A connection for the fiber space has G for *structure group* if it is defined in these coordinates by one-forms θ^a of the type

$$\theta^a = dy^a - \underset{\sim}{\omega}(y^a) \tag{8.1}$$

with $\underset{\sim}{\omega} \in \mathscr{D}^1(X, \mathscr{G})$.

Let us see what this means analytically and geometrically. Suppose that A_α is a basis for $\mathscr{G}$, with

$$\underset{\sim}{\omega} = \Gamma^\alpha_i(x)\ dx^i \otimes A_\alpha \quad , \tag{8.2}$$

$$A_\alpha = A^a_\alpha(y) \frac{\partial}{\partial y^a} \quad . \tag{8.3}$$

Then,

$$\theta^a = dy^a - \Gamma^\alpha_i(x) A^a_\alpha(y)\ dx^i \quad . \tag{8.4}$$

Suppose that $t \to x(t)$ is a curve in the base space X. Looking at (8.4), we see that the differential equations for the parallel-transported curve are

$$\frac{dy^a}{dt} = \Gamma^a_i(x(t)) A^a_\alpha(y) \quad . \tag{8.5}$$

There are equations which are associated with the action of the group G --parallel transport is essentially defined by the action of a curve $t \to g(t)$ in G in the space of y, i.e.,

$$y(t) = g(t)(y(0)) \quad . \tag{8.6}$$

(Technically, the time-dependent differential equations (8.5) are called a *Lie system* associated with the Lie group G. They were studied extensively by Lie himself and then by Vessiot. They appear in many places in physics and engineering [13].)

Here is a main result which may be proved by a straightforward computation.

THEOREM 8.1. The connection defined--via formula (8.1)--by the Lie algebra-valued one-form $\underset{\sim}{\theta}$ is flat, i.e., parallel transport is (locally) independent of the path, if and only if the curvature two-form $D\underset{\sim}{\theta}$ is identically zero.

In Part III of E. Cartan's Collected Works [22] many examples are described of geometrically interesting connections with various structure groups. For example, $X = R^n$ if $Y = R^n \times R^n = T(X)$ ($\equiv$ the tangent bundle to X) then there are four choices of G of prime geometric interest:

a) G = group of affine transformations in R^n;

b) G = group of orthogonal transformations on R^n;

c) G = group of (nonlinear) projective transformations on R^n, i.e., the space of maps $R^n \to R^n$ which map hyperplanes into hyperplanes;

d) G = group of (nonlinear) conformal maps, i.e., the maps $R^n \to R^n$ which preserve angles.

The corresponding connections are associated with *affine*, *Riemannian*, *projective*, and *conformal* geometry. Geometrically, this whole business is a marvelous realization of Klein's "Erlanger program" for studying "geometry" with a group-theoretic motivation and unification. Up to now, only the Riemannian connections of general relativity have been of interest for physics--one must now ask what the connections that appear in Yang-Mills and soliton theory have to do with geometry! However, I will not pursue these general geometric ideas further in this review--keep in mind that they are sitting in the background while we develop the analytical machinery needed to understand the physics of "gauge fields" and "solitons".

9. GAUGE TRANSFORMATIONS OF LIE-ALGEBRA VALUED ONE-FORMS

Let X be a manifold, G a Lie group, $\mathcal{G}$ its Lie algebra. Let

$$M(X,G)$$

be the space of maps $X \to G$. If X is fixed, denote M(X,G) as $\underset{\sim}{G}$. (In physics, three choices of X are used in various contexts: $X = R$, R^3 ($\equiv$ space), R^4 ($\equiv$ space-time).) Denote an element of $\underset{\sim}{G}$ by $\underset{\sim}{g}$.

We can introduce multiplication into $\underset{\sim}{G}$:

$$\underset{\sim}{g}_1\underset{\sim}{g}_2(x) = \underset{\sim}{g}_1(x)\underset{\sim}{g}_2(x)$$

for $x \in X$.

With this multiplication, $\underset{\sim}{G}$ is a group called the *gauge group*. It can be made to act naturally as a transformation group in the following way:

Choose Y as any space on which G acts. Let $\underset{\sim}{G}$ act in $X \times Y$ as follows:

$$\underset{\sim}{g}(x,y) = (x,\underset{\sim}{g}(x)(y)) \tag{9.1}$$

for $(x,y) \in X \times Y$.

For example, X would be R^4 ($\equiv$ Minkowski space-time), $G = SU(2)$, $Y = \mathbb{C}^2$; $SU(2)$ = natural linear action on $\mathbb{C}^2$. Physicists might write on element of $X \times Y$ as

$$\begin{pmatrix} p(x) \\ n(x) \end{pmatrix} ,$$

thinking of $p(x)$ as the "proton field" at the point x, $n(x)$ as the "neutron field" at point x.

$\underset{\sim}{g}$ they would write as

$$\begin{pmatrix} g_1^1(x), & g_1^2(x) \\ g_2^1(x), & g_2^2(x) \end{pmatrix} ,$$

i.e., a "local" SU(2)-gauge transformation. The action (2.1) is then

$$\begin{pmatrix} p(x) \\ n(x) \end{pmatrix} \to \begin{pmatrix} g_1^1(x), & g_1^2(x) \\ g_2^1(x), & q_2^2(x) \end{pmatrix} \begin{pmatrix} p(x) \\ n(x) \end{pmatrix}$$

The next step should be to work out how connections transform under "gauge transformations". Working this out in full detail is rather tedious, so I will take a short cut, working it out in the special case of a linear connection and gauge transformation and guessing the complete answer.

Let $\underline{y}$ denote an m-vector so that the fiber space $Y = X \times R^m$. Let

$$\underline{\theta} = d\underline{y} - \underline{M}_j(x)\underline{y}\, dx^i \tag{9.2}$$

be a linear connection.

Let G be a group of $m \times m$ matrices. Let $x \to g(x)$ be a map of $X \to G$. Set:

$$\underline{y}' = g(x)^{-1}\underline{y}$$

or

$$\underline{y} = g(x)\underline{y}' \quad .$$

Then,

$$d\underline{y} = (dg)\underline{y}' + gd\underline{y}' \quad .$$

Hence,

$$\underline{\theta} = dg\underline{y}' + gd\underline{y}' - \underline{M}_i(x)g\underline{y}'\,dx^i \quad ,$$

$$g^{-1}\underline{\theta} = d\underline{y}' + g^{-1}dgy' - g^{-1}\underline{M}_i(x)gy'\,dx^i \quad .$$

Set:

$$\underline{\omega} = \underline{M}_i(x)\underline{y}\,dx^i \tag{9.3}$$

$$\underline{\omega}' = g(x)^{-1}M_i(x)g(x)dx^i - g^{-1}(x)dg(x) \tag{9.4}$$

$$\underline{\theta} = d\underline{y} - \underline{\omega}\underline{y} \tag{9.5}$$

$$\underline{\theta}' = d\underline{y}' - \underline{\omega}'\underline{y}' \tag{9.6}$$

Then,

$$\underline{\theta}' = g^{-1}\underline{\theta} \quad . \tag{9.7}$$

$\underline{\theta}'$ represents the *transformed connection* under the gauge transformation. (9.4) show how the Lie algebra-valued one-forms on the base transform under this gauge transformation. Now, (9.4) can be written in a matrix-independent way as:

$$\underline{\omega}' = \mathrm{Ad}\, g(x)(\underline{\omega}) - g^*(\underline{\eta}) \quad , \tag{9.8}$$

where η is the left Lie algebra-valued (Maurer-Cartan) one-form on G.

We can now abstract this material out of the connection-theoretic context in which it was derived. We can say that two Lie-algebra $\mathcal{G}$ valued one-forms $\underline{\omega},\underline{\omega}'$ on a manifold X are *gauge-equivalent* if there exists a map $g\colon X \to G$ such that (9.8) is satisfied. At the "Yang-Mills field" level, this means "gauge covariance" in the usual way.

Particularly notice how $\underline{\omega}$ and $\underline{\omega}'$ can have quite different *singularity structure*; the inhomogeneous term, i.e., the second term on the right hand side of (9.8), can possibly cancel a singularity present in $\underline{\omega}$. This possibility is particularly important in the study of "instantons" and "monopoles".

We now want to construct "gauge invariants". The obvious one is the curvature two-form described earlier.

$$D_{\underline{\omega}}\underline{\omega} = d\underline{\omega} + \frac{1}{2}[\underline{\omega},\underline{\omega}] \quad . \tag{9.9}$$

It is readily seen that it transforms *tensorially* under a gauge transformation, i.e.,

$$D\underline{\omega}' = \text{Ad } g(x)(D\underline{\omega}) \quad . \tag{9.10}$$

In particular, the *Yang-Mills field equations* can be constructed from these invariants. If $\mathcal{G}$ is semi-simple and if X has a Riemannian metric, there is a "duality" operator

$$\Omega \to *\Omega$$

on $\mathcal{G}$-valued two-forms. ($*\Omega$ is an (n-r)-form, where $n = \dim X$.) This is an obvious generalization of the "Hodge" duality operator [41,42]. Then, the covariant derivative with respect to ω can be applied to $*\Omega$

$$D_{\underline{\omega}}*\Omega \quad . \tag{9.11}$$

It is an (n-1)-form, with values in $\mathcal{G}$. The vanishing of (9.11) then expresses the "Yang-Mills" equations; they are, by their very construction, "gauge invariant", i.e., if (ω,Ω) satisfies

$$\begin{aligned} \Omega &= D_{\omega}\omega \\ D_{\omega}*\Omega &= 0 \quad , \end{aligned} \tag{9.12}$$

and if ω' is gauge-related to ω, then it too satisfies these equations. This invariance of the system of nonlinear differential equations (9.12) under the "infinite parameter" group of all gauge transformations is what makes them so important physically, but it also creates substantial mathematical difficulties. (A general theory of higher spin "gauge-invariant" equations is presented briefly in [14].)

Now, it is possible to construct invariants of gauge transformations whose regularity give a *necessary* condition that singularity may be "gauged" away. These invariants are called *characteristic classes* by mathematicians. We will now briefly indicate how they may be defined.

10. CHARACTERISTIC CLASSES

Suppose $\mathcal{G}$ is a Lie algebra, X is a manifold, and ω is a Lie-algebra valued one-form on X. Its curvature

$$\Omega = d\omega + \frac{1}{2}[\omega,\omega] \equiv D_{\omega}\omega$$

defines, for each $x \in X$, a bilinear mapping

$$\Omega(x): M_x \times M_x \to \mathcal{G}$$

which is skew-symmetric. We can then construct a quadralinear mapping

$$\Omega(x) \times \Omega(x): M_x \times M_x \times M_x \times M_x \to \mathcal{G} \otimes \mathcal{G}$$

by the following formula:

$$\Omega(x) \times \Omega(x)(v_1, v_2, v_3, v_x) = \Omega(x)(x_1, v_2) \otimes \Omega(x)(v_3; v_4) \qquad (10.1)$$

We can then follow this with any linear mapping

$$\alpha: \mathcal{G} \otimes \mathcal{G} \to R$$

to obtain a mapping

$$\Omega_\alpha \equiv \alpha(\Omega(x) \otimes \Omega(x)): M_x \times M_x \times M_x \times M_x \to R \quad ,$$

i.e.,

$$\Omega_\alpha = \alpha(\Omega(x)(v_1, v_2) \otimes \Omega(x)(v_3, v_4)) \qquad (10.2)$$

This formula will define a differential form on X if α is *symmetric*, i.e.,

$$\alpha(A_1 \otimes A_2) = \alpha(A_2 \otimes A_1) \qquad (10.3)$$

for $A_1, A_2 \in \mathcal{G}$

Suppose in addition that α satisfies the following condition:

$$\alpha(\text{Ad } g(A_1) \otimes \text{Ad } g(A_2)) = \alpha(A_1 \otimes A_2) \qquad (10.4)$$

for $A_1, A_2 \in \mathcal{G}$

(One says that α is *invariant* under the *adjoint representation* of $\mathcal{G}$.) One can then prove that Ω_α is a *closed* differential form, i.e.,

$$d\Omega_\alpha = 0 \quad , \qquad (10.5)$$

(this is a consequence of the *Bianchi identities*) and:

Ω_α *is invariant under gauge transformations applied to* ω . (10.6)

One can, in fact, apply this construction with α to any element of the tensor algebra of $\mathcal{G}$ which is invariant under permutations of Ad G, obtaining a closed differential form of degree $2r$ (r = degree α), which is called the *characteristic class* generated by α and the connection. These classes are important for topological purposes and in studies of the behavior of connections near "singularities". These objects have played a prominent role in recent work by physicists on "monopoles" and "instantons".

Luckily the algebraic problem of finding all such α's is solved, at least for $\mathcal{G}$ = "classical" semisimple Lie algebra. They are related to objects which are also very familiar to physicists, the *Casimir operators* of the *universal enveloping algebra of* $\mathcal{G}$ [1,7].

11. DIFFERENTIAL EQUATIONS DEFINED BY VANISHING OF THE CURVATURE. ESTABROOK-WAHLQUIST PROLONGATIONS

Continue with X as a manifold, $\mathcal{G}$ as a Lie algebra,

$$\omega: T(X) \to \mathcal{G}$$

as a Lie-algebra valued one-form on X. The curvature form

$$\Omega = d\omega + \frac{1}{2}[\omega,\omega]$$

is then a $\mathcal{G}$-valued two-form on X. If (A_a), $1 \leq a,b \leq m$, is a basis for $\mathcal{G}$, then

$$\Omega = \Omega^a \otimes A_a^i \quad ,$$

where Ω^a are scalar valued two-forms on X.

Following E. Cartan [40] one can now define the *exterior differential system* on X generated by the Ω^a. Submanifolds

$$\phi: Z \to X$$

are said to be *integral submanifolds* of this system if the forms Ω^a are zero when restricted to the submanifolds, i.e., if

$$\phi^*(\Omega^a) = 0 \quad . \tag{11.1}$$

For such a submanifold, the *Cartan-Maurer* operations hold, i.e.,

$$d\phi^*(\omega) + \frac{1}{2}[\phi^*(\omega),\phi^*(\omega)] = 0$$

Then there is a map

$$\sigma_\phi: X \to G$$

(where G = Lie group whose Lie algebra is $\mathcal{G}$) such that:

$$\phi^*(\omega) \;=\; \text{pull back under } \sigma_\phi \text{ of the left-invariant form on } G.$$

We then see the possibility of defining differential equations (for the maps ϕ) in this way. The differential equations which appear prominently in soliton-inverse scattering theory seem to be all defined in this way.

Here is a more geometric way of thinking about this in terms of the theory of connections. Let

$$\pi\colon Y \to X$$

be a fiber space, with X a base space. Suppose a connection is given for this fiber space, defined by a set $\mathscr{P}$ of one-forms on X. One asks for the submanifolds

$$Z \subset X \tag{11.2}$$

which have the property that parallel transport along curves in Z *is independent of the path.* These are the submanifolds on which the curvature vanishes, i.e., which satisfies (11.1).

Alternately, one can look for the mappings

$$\beta\colon Z \to Y$$

such that

$$\beta^*(\mathscr{P}) \;=\; 0 \quad . \tag{11.3}$$

The projection in X

$$\pi\beta\colon Z \to X$$

will then (at least locally) be the submanifolds (11.2).

(11.3) define a set of differential equations which Estabrook and Wahlquist call [28] the *prolongation* of the differential equation (11.2). The foundation of their method for studying nonlinear differential equations is this process of constructing such prolongations. For two differential equations--the Korteweg-de Vries and nonlinear Schrödinger (in one space variable)--they have, in a sense, determined <u>all</u> such prolongations. J. Corones, R. Dodd and J. Gibbon and H. Morris have partially extended their work to other sets of equations. One important feature of this is that once prolongations of this type are found, there are "algorithms" for finding Bäcklund transformations, the soliton structures and heirarchy, and the "superposition formulas". (From the physical point of view, this is probably the most significant and remarkable property of these equations, particularly from the point of view of *elementary particles*; although they are *nonlinear*, these are formulas (which are nonlinear) which enable "simple" solutions to be combined to make "more complicated" ones. These formulas were first found by Bianchi for the Sine-Gordon equation and by Wahlquist and Estabrook for Korteweg-de Vries [44].)

Finding the prolongations is closely related to the "inverse scattering" structure. Work by Wadati, Sanuki and Konno [67,68] gives the most convenient (so far) way to write down the prolongations. A glance at these papers should show the reader the unified role that the Lie group SL(2,R) plays in the study of the various equations (Sine-Gordon, Korteweg-de Vries, nonlinear Schrödinger,...) which *so far* are the unique examples with the same properties. (I believe there are other "nice" examples yet to be found associated with other groups--even infinite dimensional ones!)

12. SL(2,R) PROLONGATIONS, PROLONGATION AND BÄCKLUND TRANSFORMATIONS

Let G be the Lie group SL(2,R) of 2×2 real matrices. Its Lie algebra is the set of 2×2 real matrices of trace zero:

$$A = \begin{pmatrix} A_{11} & A_{12} \\ A_{21} & -A_{11} \end{pmatrix} \tag{12.1}$$

Let X be a manifold. A $\mathcal{G}$-valued one-form is then a matrix

$$\omega = \begin{pmatrix} \omega_{11} & \omega_{12} \\ \omega_{21} & -\omega_{11} \end{pmatrix} \tag{12.2}$$

of scalar one forms. Suppose X is the space of variables $(x,t,u,u_x,u_t,\ldots)$. (Let us now work in the context of the "theory of nonlinear waves"; (x,t) are independent variables $u_x,u_t,\ldots$ then stand for derivatives.) Consider ω of the form

$$\omega = A\,dx + B\,dt \tag{12.3}$$

where A and B are 2×2 matrices of trace zero consisting of functions of the variables $u,u_x,u_t,\ldots$. Set

$$\begin{aligned}\Omega &\equiv d\omega + \frac{1}{2}[\omega,\omega]\\ &= dA\wedge dx + dB\wedge dt + \frac{1}{2}[Adx+Bdt,\ Adx+Bdt]\\ &= dA\wedge dx + dB\wedge dt + [A,B]\,dx\wedge dt\quad .\end{aligned}$$

Finding a submanifold on which $\Omega = 0$ means finding u as function of u(x,t) such that:

$$A(u(x,t),\ldots)_t - B(u(x,t),\ldots)_x = [A,B](u(x,t),\ldots) \tag{12.4}$$

(12.4) then defines a system of partial differential equations which we call "PDE" for u(x,t). Requiring that these equations coincide with a *given* system of equations (e.g., Korteweg-de Vries, Sine-Gordon) then determines differential equations for A and B. These equations (which sometimes can be solved) play the role in soliton theory analogous to the "Yang-Mills" equations in the theory of gauge fields.

Once such A,B are found, *Bäcklund transformations* can be constructed. Here is the method of Estabrook and Wahlquist [28,44] for doing this.

Let

$$y = \begin{pmatrix} y_1 \\ y_2 \end{pmatrix}$$

be a two-vector. Set:

$$\theta = dy - \omega y \quad .$$

Let Y be the space of variables y_1, y_2. θ are then one-forms on $X \times Y$, which define a *linear connection*. Ω are the curvature forms. Now, look for a map

$$\beta: X \times Y \to X$$

such that

> $\beta^*(\Omega)$ lies in the Grassmann algebra ideal generated by Ω, θ.

β is called a *Bäcklund map*. It enables one to generate new solutions of the underlying partial differential equations for u(x,t). Namely, if u(x,t) is one solution of PDE, if y(x,t) is defined as follows

$$\begin{aligned} y_x &= A(u(x,t),\ldots)y \\ y_t &= B(u(x,t),\ldots)y \quad , \end{aligned} \tag{12.5}$$

then

$$\beta(x,t,u(x,t),u_x(x,t),\ldots,y(x,t)) = (x,t,u'(x,t),\ldots) \quad ,$$

((12.5) are called the *prolongation equations*) β then maps each solution of the component system (consisting of (2.5) and the PDE's for u) into a solution u' of PDE. Convenient formulas for the Bäcklund maps are given in a paper by Konno and Wadati [68]. In these formulas β is determined by a simple natural map

$$(u,y) \rightarrow u' \quad .$$

We shall (engaging in a bit of "abus de langage") call this the "Bäcklund map", and denote it by β.

13. SOLITONS

As in Section 12, suppose a $\mathcal{G}$-valued one-form ω is given such that a given set of equations for $u(x,t)$ are the conditions for

$$d\omega + \frac{1}{2}[\omega,\omega] \equiv \Omega = 0 \quad .$$

Suppose also that

$$\beta\colon X \times Y \rightarrow X$$

is given as a Bäcklund map.

Start off with the solution

$$u(x,t) \equiv 0 \quad .$$

The integrability conditions (12.4) are then

$$[A(0),B(0)] = 0 \quad . \tag{13.1}$$

The prolongation equations (12.5) take the following form:

$$\begin{aligned} y_x &= A(0)y \\ y_t &= B(0)y \quad . \end{aligned} \tag{13.2}$$

Equations (13.2) can be solved explicitly:

$$\begin{aligned} y(x,t) &= \exp(xA(0))(y(0,t)) \\ &= \exp(xA(0) + tB(0))(y(0,0)) \quad . \end{aligned} \tag{13.3}$$

Thus the map $Z\ (\equiv R^2) \rightarrow G$ determined by the "rest" solution of the PDE is the map

$$(x,t) \rightarrow \exp(x(A(0) + tB(0)) \quad . \tag{13.4}$$

Notice that its image in G is a *one-parameter subgroup*. Then we can set

$$u' = \beta(y,u)$$

where $u \equiv 0$, y is given by (13.3), u' is called the *one-soliton solution* of the PDE.

The next prolongation equations are:

$$\begin{aligned} y'_x &= A(u',\ldots)y' \\ y'_t &= B(u',\ldots)y \quad . \end{aligned} \tag{13.5}$$

The

$$u'' = \beta(y',u')$$

are called the *two-soliton* solutions.

One continues in this way to define sequences $(0;y;u',y';u'',y'',\ldots)$, called the *soliton ladder*. Now, it looks like these equations for these functions become progressively harder to solve. In fact, in all the cases we know about, there is a "miracle", and they are all solvable in terms of explicit *rational functions* of the functions of the form (13.3). This is a consequence of the *superposition formulas of Bianchi type*.

14. THE SOLITON LADDER FOR KORTEWEG-DE VRIES

To illustrate the generalities described in Sections 12 and 13, let us turn to the Korteweg-de Vries (abbreviated K-dV) equation, which is the main example which has motivated this line of research in nonlinear waves and solitons. I will follow the formalism developed by Konno and Wadati [68] which is very convenient. Set:

$$A = \begin{pmatrix} \lambda & u \\ -1 & -\lambda \end{pmatrix}$$

$$B = \begin{pmatrix} -4\lambda^2 - 2\lambda u - u_x, & -u_{xx} - 2\lambda u_x - 4\lambda^2 u - 2u^2 \\ 4\lambda^2 + 2u, & 4\lambda^2 + 2\lambda u + u_x \end{pmatrix}$$

λ is a parameter (essentially the "eigenvalue parameter" of the "inverse scattering problem"). The PDE satisfied by u is K-dV:

$$\boxed{u_t + 6uu_x + u_{xxx} = 0} \tag{14.1}$$

The equation (14.1), together with the prolongation equations

$$y_x = Ay \quad ; \quad y_t = By \quad ;$$
$$y = \begin{pmatrix} y_1 \\ y_2 \end{pmatrix} \qquad (14.2)$$

form a composite PDE system for (u,y) which is "completely integrable" in *the classical sense* used, e.g., by E. Cartan [39,40]. (Workers in "nonlinear waves" often use the term "complete integrability" for another property; it is often confusing because the K-dV equation is also "completely integrable" in their sense; and there is no obvious relation between the two concepts. I think it is best to reserve the term for its *classical* version, especially because the newer use of the term is rather confusing and ill-defined.)

The Bäcklund map is given by the following formula:

$$\beta(u,y) = -\left(u + 2\left(\frac{y_1}{y_2}\right)^2 + 4\lambda \frac{y_1}{y_2}\right) \qquad (14.3)$$

We can start the calculation of the soliton ladder off with the choice $u = 0$. Then

$$A(0) = \begin{pmatrix} \lambda & 0 \\ -1 & -\lambda \end{pmatrix}$$

$$B(0) = \begin{pmatrix} -4\lambda^2, & 0 \\ 4\lambda^2, & 4\lambda^2 \end{pmatrix} = -4\lambda^2 A(0) \qquad (14.4)$$

$$y_x = A(0)y$$
$$y_t = -4\lambda^2 A(0)y \qquad (14.5)$$

$$u' = -2\left(\frac{y_1}{y_2}\right)^2 - 4\lambda \frac{y_1}{y_2} \qquad (14.6)$$

Now, equations (14.5) take the form

$$y_{1,x} = \lambda y_1$$
$$y_{2,x} = -y_1 - \lambda y_2 \quad ,$$

hence:

$$\left(\frac{y_1}{y_2}\right)_x = \frac{y_2(\lambda y_1) - y_1(-y_1 - \lambda y_2)}{y_2^2}$$

$$= 2\lambda\left(\frac{y_2}{y_2}\right) + \left(\frac{y_1}{y_2}\right)^2 = -\frac{1}{2}u' \tag{14.7}$$

Hence, the second prolongation equations are given as:

$$y'_x = \begin{pmatrix} \lambda , & -2\left(\frac{y_1}{y_2}\right)_x \\ -1 , & 0 \end{pmatrix} y' \tag{14.8}$$

$$y'_t = B(u')y \tag{14.9}$$

(The time-derivatives eventually becomes so complicated in the K-dV theory that it is difficult to write them out.)

$$u'' = -\left(u' + 2\left(\frac{y'_1}{y'_2}\right)^2 + 4\lambda\left(\frac{y'_1}{y'_2}\right)\right) \tag{14.10}$$

is then the *two-soliton* solution of *K-dV*. Approached directly, the equations (14-8) and (14.9) are messy. However, the superposition formula (proved in this case by Wahlquist and Estabrook [44]) enables one to express the general solution of (14.8)-(14.9) in terms of explicit natural functions of the general solution of (14.5) for λ and *another* value of λ, say λ'. These formulas then show that the two-solitons "decouple" as $x \to \pm\infty$ into a sum of "one-solitons" with *shifted phase*. This leads to the remarkable and basic physical property of solitons--*they interact in a very "particle" like way, preserving their shape.*

No one knows if such properties can persist for nonlinear PDE's in more than one space variable, but I am confident that the basic "Bäcklund-prolongation" formalism must generalize in some way; if so, it would be very useful in a wide spectrum of nonlinear physics and engineering problems.

15. EXTERIOR DIFFERENTIAL SYSTEMS, GENERALIZED CONSERVATION LAWS, AND KINKS

Up to now, I have emphasized the more traditional approach to prolongations--Bäcklund transformations, solitons, etc. It is important to realize that it can be described beautifully also in terms of E. Cartan's theory of exterior differential systems [40, 28, 16], which is, in a sense, the "pure" geometric theory of differential equations. If there is a reasonable extension of the theory to nonlinear PDE's in more than one space variable, I believe that it will turn up most readily in the context of Cartan's framework. (See the papers by H. Morris [54,66] as a first step.)

I cannot go into the full details. I want to point out here a sequence of concepts which leads to a systematic geometric way of thinking about the *global, topological solutions* of nonlinear PDE's, which in a sense is a realization of the ideas of D. Finkelstein and Rubinstein [69,70].

Let X be a manifold. An *exterior differential system* is a collection of differential forms, denoted by ED, with the following properties:

$$ED + ED \subset ED$$

$$\theta_1 \wedge \theta_2 \in ED$$

for $\theta_2 \in ED$, θ_1 an arbitrary differential form on X

$$d\theta \in ED \quad \text{if} \quad \theta \in ED \quad .$$

Let ED be such a system. An *integral submanifold* is a map between manifolds

$$\phi: Z \to X$$

such that:

a) The indirect map $\phi_*: T(Z) \to T(X)$ on tangent bundles is one-to-one (this is the "submanifold" condition);

b) $\phi^*(\theta) = 0$, for all $\theta \in ED$.

Partial differential equations of any type with n independent variables, give rise to exterior differential systems with the property that there is a one-one correspondence between their solutions and certain integral submanifolds with $Z = R^n$. (Typically, there might be certain limiting and degenerate integral submanifolds which do not arise from solutions. In [16] I have explained how these limiting cases can be used to geometrize certain ideas of "singular perturbation theory".) The most extensive and useful discussion of this is in Cartan's book [40]. Goursat's book [71] is also very useful as a guide to the classical literature. Cartan usually worked with ED's generated by one-forms. (These are also called *Pfaffian systems*.) Harrison, Estabrook and Wahlquist have shown [43,28] that it is also very useful and important to work with ED's generated by two-forms. In fact, the first step in constructing "prolongations" and "Bäcklund transformations" for a given system of PDE's is to write it in this way.

Let ED be a given exterior differential system. A *conservation law* is a differential form θ such that

$$d\theta \in ED \tag{15.1}$$

See [4,8,11] for a description of the relation between this concept and the physicists' notion of "conserved currents".

In general, θ is a form of arbitrary degree r. (For field theories with n independent variables, typically $r = n-1$.) Here is a way to usefully generalize:

Definition. A *generalized conservation law* for ED of degree r is a set θ^i, $1 \leq i,j \leq m$, of r-forms and a set ω^i_j of *one*-forms such that:

$$d\theta^i - \omega^i_j \wedge \theta^j \in ED \tag{15.2}$$

Remark. Here is a geometric meaning of this condition in the case $r = 1$. Let

$$\phi: Z \to X$$

be an integral submanifold. (15.2) says that the "pulled-back" Pfaffian system

$$\phi^*(\theta') = 0 \tag{15.3}$$

is *completely integrable*, i.e., there are functions (z^i, f^i_j) on Z such that:

$$dz^i = f^i_j \phi^*(\theta^j) \quad . \tag{15.4}$$

In the Estabrook-Wahlquist theory [28] a simplified version of these relations is encountered. Again working with the case $r = 1$ (and we do not really know any interesting examples in higher degrees), suppose that the ω^i_j are of the following form

$$\omega^i_j = -\frac{1}{2} c^i_{jk} \theta^k \tag{15.5}$$

where (c^i_{jk}) are the *structure constants* of a Lie algebra $\mathcal{G}$. This mans that (if G is a Lie group whose Lie algebra is $\mathcal{G}$) there is a basis η^i for the left-invariant ("Cartan-Maurer") one-forms on G such that

$$d\eta^i + \frac{1}{2} c^i_{jk} \eta^j \wedge \eta^k = 0 \quad .$$

Then (15.) means that

$$d\theta^i + \frac{1}{2} c^i_{jk} \theta^j \wedge \theta^k \in ED \quad . \tag{15.6}$$

If (15.4) is satisfied, the θ^i are said to be a *generalized conservation law for* ED *with* G *as structure group.*

Let

$$\phi: Z \to X$$

be an integral submanifold of ED. Then (15.6) means that

$$d\phi^*(\theta^i) + \frac{1}{2} C^i_{jk} \phi^*(\theta^j) \wedge \phi^*(\theta^k) = 0 \tag{15.7}$$

This implies (if Z is simply connected) that there is a map

$$\alpha_\phi: Z \to G$$

such that

$$\alpha^*_\phi(\eta^i) = \phi^*(\theta^i) \quad . \tag{15.8}$$

Thus, we can assign (with the help of the generalized conservation law) to each integral submanifold ϕ (i.e., to each solution of the underlying PDE) a map $Z \to G$.

Let H be a closed subgroup of G. The coset space G/H is then a *homogeneous space* of G. Consider the map $\alpha_\phi: Z \to G$ followed by the map $\pi G \to G/H$. We obtain a map $\pi\alpha_\phi: Z \to G/H$. The "homotopy" properties of this map are of interest--as the "kink" associated with the solution of the PDE.

In the cases with which we are familiar, Z is R^2, parametrized, say, by variables x and t. Denote a point of G/H as p. Thus, $\pi\alpha_\phi$ is a map

$$(x,t) \to p(x,t) \quad .$$

Suppose that:

$$\lim_{x\to\pm\infty} p(x,t) = P \quad . \tag{15.9}$$

Then, for fixed t, the curve

$$x \to p(x,t)$$

goes from p_- to p_+ as $x \to \pm\infty$. This curve defines an element of

$$\pi_1(G/H) \quad ,$$

the *fundamental group* of G/H.

Now relation (15.9) is more likely if G/H is *compact*. It is easy to guess what H must be for the usual equations (K-dV, Sine-Gordon, etc.). There are generalized conservation laws associated with SL(2,R). Let H = set of matrices of the form

$$\begin{pmatrix} a & b \\ 0 & a^{-1} \end{pmatrix} \quad .$$

G/H is compact, in fact, it is homeomorphic to the circle S^1 whose fundamental group is the integers. Thus, we can define an integer topological invariant, which is Finkelstein's "kink" number. It would be interesting to get other Lie groups G into the game. (I believe there are "interesting" classes of PDE's which are yet to be discovered associated with at least all the simple Lie groups.) If G is noncompact semisimple, the generalization of what we did for $G = SL(2,R)$ is to use the *Iwasawa decomposition* [19.1]

$$G = KAN \quad ,$$

with K = maximal compact subgroup of G and A,N subgroups. (A is abelian, N nilpotent.) Set:

$$H = MAN \quad ,$$

where M is a subgroup of K. Then G/H is compact; topologically, G is K/M. Thus, knowing the topological properties of the compact group, it can readily be arranged that G/H has topological properties which are potentially useful for generalizations of the "kink" idea. However, these interesting possibilities must await further progress in discovering new sorts of generalized conservation laws.

16. SUMMARY AND FURTHER COMMENTS

"Modern" differential geometry has a characteristic flavor. (It is really not all that "modern", but has its roots solidly in the 19-th century, which was really much more of a Golden Age of Geometry than today. The work of Sophus Lie and Elie Cartan is the basis for most of what we do.) The basic objects are manifolds, vector fields, differential forms, fiber spaces, and connections. The mathematical formalism underlies much of science and engineering. (There are many unexploited possibilities of applying these ideas in areas outside of physics--systems and control theory, continuum mechanics, chemistry, biology, even economics!)

In this article I have concentrated on "soliton" theory as that area in which to illustrate the influence of these geometric ideas. (This is also the area in which I am working and which I know best.) However, it should be clear that the *mathematics* is closely related to that involved in the study of "gauge fields", particularly the topics of "instantons" and "monopoles". Now, I have encountered among physicists an attitude that the theory of solitons is a curiosity of little significance for elementary particle physics. This may, in fact, turn out to be so (self-fulfilling prophecy?), but I am somewhat disturbed that people who express this put-down usually do not understand the full mathematical ramifications and structure of soliton theory, particularly the marvelous techniques (inverse scattering, Bäcklund transformation, Bianchi-style superposition, Estabrook-Wahlquist prolongations, etc.) that go along with it. What we are doing is developing new mathematics especially adapted to the geometric structure of certain types of non-linear partial differential equations--and that mathematics is closely linked to Lie theory. I believe these techniques will turn out to be classics of 20-th century mathematics--comparable historically to elliptic functions, say, in the 19-th century--and I want to see as wide as possible an interaction with physics and engineering. I hope more physicists will invest the time and effort needed to understand the mathematical foundation of these ideas, which are so full of promise!

BIBLIOGRAPHY

1. R. Hermann, Lie Groups for Physicists, W.A. Benjamin, Inc., New York, 1966.

2. R. Hermann, Differential Geometry and the Calculus of Variations, Academic Press, New York, 1969. Second edition to be published by Math Sci Press, Brookline, Mass.

3. R. Hermann, Fourier Analysis on Groups and Partial Wave Analysis, W.A. Benjamin, New York, 1969.

4. R. Hermann, Lie Algebras and Quantum Mechanics, W.A. Benjamin, New York, 1970.

5. R. Hermann, Vector Bundles in Mathematical Physics, Parts I and II, W.A. Benjamin, New York, 1970.

6. R. Hermann, Lectures on Mathematical Physics, Vol. 1, W.A. Benjamin, New York, 1970.

7. R. Hermann, Lectures on Mathematical Physics, Vol. II, W.A. Benjamin, Reading, Mass, 1972.

8. R. Hermann, Geometry, Physics and Systems, Marcel Dekker, New York, 1973.

9. R. Hermann, Physical Aspects of Lie Group Theory, University of Montreal Press, Montreal, 1974.

10. R. Hermann, Energy-Momentum Tensors, Vol. IV of Interdisciplinary Mathematics, Math Sci Press, Brookline, Mass., 1973.

11. R. Hermann, Topics in General Relativity, Vol. V of Interdisciplinary Mathematics, Math Sci Press, Brookline, Mass., 1973.

12. R. Hermann, Topics in the Mathematics of Quantum Mechanics, Vol. VI of Interdisciplinary Mathematics, Math Sci Press, Brookline, Mass., 1973.

13. R. Hermann, Geometric Structure Theory of Systems-Control Theory and Physics, Part A, Vol. IX of Interdisciplinary Mathematics, Math Sci Press, Brookline, Mass.

14. R. Hermann, Gauge Fields and Cartan-Ehresmann Connections, Part A, Vol. X of Interdisciplinary Mathematics, Math Sci Press, Brookline, Mass., 1975.

15. R. Hermann, Geometric Structure of Systems-Control Theory and Physics, Part B, Vol. XI of Interdisciplinary Mathematics, Math Sci Press, Brookline, Mass., 1976.

16. R. Hermann, Geometry of Non-Linear Differential Equations, Bäcklund Transformations and Solitons, Parts A and B, Vols. XII and XIV of Interdisciplinary Mathematics, Math Sci Press, Brookline, Mass., 1976, 1977.

16a. R. Hermann, Toda Lattices, Cosymplectic Manifolds, Bäcklund Transformations and Kinks, Part A, Interdisciplinary Mathematics, Vol. XV, Math Sci Press, Brookline, Mass., 1977.

17. W. Boothby, An Introduction to Differentiable Manifolds and Riemannian Geometry, Academic Press, New York, 1975.

18. C. Misner, K. Thorne and J. Wheeler, Gravitation, W.H. Freeman, San Francisco, 1973.

19. A. Helgason, Differential Geometry and Symmetric Spaces, Academic Press, 1962.

20. S. Kubayashi and K. Nomizu, Foundations of Differential Geometry, Interscience, New York, 1973.

21. R. Bishop and S. Goldberg, Tensor Analysis on Manifolds, Macmillan, New York, 1968.

22. E. Cartan, Oeuvres completes, Gauthier-Villar, Paris, 1952.

23. C. Chevalley, Lie Groups, Princeton Univ. Press, 1946.

24. G. Darboux, Théorie genérale des surfaces, Chelsea, New York.

25. A. Forsyth, Theory of Differential Equations, Dover, New York.

26. A. Kumpera and D.C. Spencer, Lie Equations, Princeton Univ. Press, 1972.

27. J.C. Taylor, Gauge Invariance of Weak Interaction, Cambridge Univ. Press, 1976.

28. H.D. Wahlquist and F.B. Estabrook, Prolongation structures of nonlinear evolution equations, J. Math. Phys. 16 (1975), 1-7.

28a. Sophus Lie's 1880 Transformation Group Paper, comments and additional material by R. Hermann (Lie Groups: History, Frontiers and Applications, Vol. 1) Math Sci Press, Brookline, Mass., 1975.

28b Ricci and Levi-Civita's Tensor Analysis Paper, translation, comments and additional material by R. Hermann (Lie Groups: History, Frontiers and Applications, Vol. 2) Math Sci Press, 1975.

28c Sophus Lie's 1884 Differential Invariant Paper, comments and additional material by R. Hermann (Lie Groups: History, Frontiers and Applications, Vol. 3) Math Sci Press, 1975.

29. H. Cartan, Notions d'algebre, différentielles, Colloque de topologie de Bruxelles, Masson and Cie, Paris, 1950.

30. R. Hermann, Quantum and Fermion Differential Geometry, Part A, Math Sci Press, Brookline, Mass.

31. F. Mansouri, Differential geometry in graded manifolds, J. Math. Phys. 18 (1977) 52.

32. R. Casabuoni, Nuovo Cim 33A (1976), 389.

33. F. Berezin and G.I. Kac , Mat. Sb. USSR, 82 (1970), 124; Eng. translation 11 (1970), 311.

34. J. Wess and B. Zumino, Nucl. Phys. B70 (1974), 39.

35. L. Corwin, Y. Ne'eman and S. Sternberg, Reviews of Modern Physics 47 (1975), 573.

36. B. Zumino, in Gauge Theories and Modern Field Theory, ed. by R. Arnowitt and P. Nath, MIT Press, 1976.

37. B. Kostant, Graded manifolds, graded Lie theory and prequantization, preprint, MIT Math. Dept.

38. F. Berezin and M. Marinov, Particle spin dynamics as the Grassmann variant of classical mechanics, preprint, Moscow, 1976.

39. E. Cartan, Lecons sur les invariants integraux, Hermann et Cie, Paris, 1971.

40. E. Cartan, Les systèmes différentielles exterieures et leurs applications géométriques,

41. W.V.D. Hodge, The Theory and Applications of Harmonic Integrals, Cambridge Press, 1941.

42. G. de Rham, Varietés différentiables, Hermann, Paris, 1955.

43. B.K. Harrison and F.B. Estabrook, Geometric appraoch to invariance groups and solution of partial differential systems, J. Math. Phys. 12 (1971) 653-666.

44. H.D. Wahlquist and F.B. Estabrook, Bäcklund transformation for solution of the Korteweg-de Vries equation, Phys. Rev. Lett. 31 (1973), 1386-1390.

45. F. Estabrook, Comments on generalized Hamiltonian dynamics, Phys. Rev. D8 (1973), 2740-2743.

46. F.B. Estabrook, Some old and new techniques for the practical use of exterior differential forms, in Robert M. Miura, ed., Bäcklund Transformation, the Inverse Scattering Method. Solitons and Their Application, Lecture Notes in Mathematics, No. 515, Springer-Verlag, Berlin, New York, 1976.

47. H.D. Wahlquist, Bäcklund transformation of potentials of the Korteweg-de Vries equation and the interaction of solitons with cnoidal waves, in Robert M. Miura, ed., Bäcklund Transformation, the Inverse Scattering Method. Solitons and Their Application, Lecture Notes in Mathematics, No. 515, Springer-Verlag, Berlin, New York, 1976.

48. J. Corones and F.J. Testa, Pseudopotentials and their applications, in Robert M. Miura, ed., Bäcklund Transformation, the Inverse Scattering Method. Solitons and Their Application, Lecture Notes in Mathematics, No. 515, Springer-Verlag, Berlin, New York, 1976.

49. F.B. Estabrook and H.D. Wahlquist, The geometric approach to sets of ordinary differential equations and Hamiltonian Dynamics, SIAM Rev. 17 (1975), 201-220.

50. R. Hermann, The pseudopotentials of Estabrook and Wahlquist, the geometry of solitons, and the theory of connections, Phys. Rev. Lett. 36 (1976), 835.

51. H.C. Morris, Prolongation structures and a generalized inverse scattering problem, J. Math. Phys. 17 (1976), 1867-1869.

52. J. Corones, Solitons and simple pseudopotentials, J. Math. Phys. 17 (1976), 756-759.

53. F.B. Estabrook and H.D. Wahlquist, Prolongation structures of nonlinear evolution equations, II, J. Math. Phys. 17 (1976) 1293-1297.

54. C. Morris, Prolongation structures and nonlinear evolution equations in two spatial dimensions, J. Math. Phys. 17 (1976), 1870-1872.

55. H.C. Morris, Prolongation structures and nonlinear evolution equations in two spatial dimensions, II: A generalized nonlinear Schrödinger equation, J. Math. Phys. 18 (1977), 285-288.

56. B.K. Harrison, Remarks on the problem of two neighboring black holes, Utah Academy Proceedings 53 (1976), 67-74.

57. H.C. Morris, A prolongation structure for the AKNS system and its generalization, J. Math. Phys. 18 (1977), 533-536.

58. J.C. Corones, Solitons, pseudopotentials and certain Lie algebras, J. Math. Phys. 18 (1977), 163-164.

59. H.C. Morris, Prolongation structures and nonlinear evolution equations in two spatial dimensions, III: A general class of equations, TCD 1976-7 (submitted to J. Math. Phys., 1976).

60. R.K. Dodd and J.D. Gibbon, The prolongation structure of some higher order Korteweg-de Vries equations (preprint, 1977).

61. H.C. Morris, A generalized prolongation structure and the Bäcklund transformation of the anticommuting massive thirring model, TCD-1977-2 (preprint, 1977).

62. R.K. Dodd and J.D. Gibbon, The prolongation structure of a class of nonlinear evolution equations (preprint, 1977).

63. M. Crampin, F.A.E. Pirani and D.C. Robinson, The soliton connection (preprint, 1977).

64. H.C. Morris , Prolongation structure of nonlinear evolution equations in two and three dimensiona, Seminar/Institute on Differential and Algebraic Geometry for Control Engineers, NASA 1976.

65. H.C. Morris, Soliton solutions and the higher order Korteweg-de Vries equations, J. Math. Phys. 18 (1977), 530-532.

66. R. Hermann, ed., Proceedings of the Ames (NASA) 1976 Conference on Geometric Non-Linear Waves. Articles by F. Estabrook, R. Hermann, H. Wahlquist, J. Corones, H. Morris, R. Gardner, A. Scott; MATH SCI PRESS, Brookline, Mass., 1977.

67. M. Wadati, H. Sanuki and K. Konno, Prog. Theor. Phys. 53 (1975), 419.

68. K. Konno and M. Wadati, Prog. Theor. Phys. 53 (1975), 1652.

69. D. Finkelstein, Kinks, J. Math. Phys. 7 (1966), 1218-1228.

70. D. Finkelstein and D. Rubenstein, J. Math. Phys. 9 (1968), 1762.

71. E. Goursat, Lecons sur le problème de Pfaff, Hermann, Paris, 1972.

SUPERSYMMETRY- SUPERGRAVITY

J. Wess
University of Karlsruhe

Table of contents

SUPERSYMMETRY - SUPERGRAVITY

Supersymmetry has been a successful concept in the framework of renormalizable Lagrangian field theories [1]. Theories, on which a supersymmetric structure is imposed have the tendency that divergencies cancel.

The gauging of supersymmetry leads to a theory which is invariant under general coordinate transformations - the graviton couples to matter fields covariantly. Such couplings had been known to be highly unrenormalizable. Supersymmetry improves the situation - at least up to the two-loop order renormalizability can be proved for special cases [2]. These models tell us that the last word on quantum field theory and the theory of gravitation has not been spoken yet. It is, from what we have learned up to now, conceivable that the framework of renormalizable quantum field theory might be large enough to comprise the theory of gravitation.

To construct models of supergravity is still an art - and hard work as well. In these lectures I intend to follow the geometrical approaches because, at the moment, it seems to me to be the most systematic one - though due to technicalities not the easiest or fastest one. [3]

The lectures are organized as follows: In the first lecture the concept of supersymmetry is introduced. We discuss the algebra and its representations. A field theoretical model which is supersymmetric and gauge invariant under an additional SU(N) group will be discussed.

In the second lecture we shall develop methods of differential geometry which allow us to construct covariants under supersymmetry, under a SU(N)-gauge group or under supergauge transformations. These covariants can be used to write down covariant field equations and Lagrangians.

In the third lecture we shall study a SU(N)-gauge invariant supersymmetric theory using these methods. This might serve as an example to learn about the technicalities mentioned before.

In the last lecture we shall apply the methods developed to construct a theory of supergravity.

FIRST LECTURE: SUPERSYMMETRY

Algebra:

The algebra is defined as follows:

$$[P_m, P_n] = 0 \quad , \quad [P_m, Q_\alpha] = [P_m, \bar{Q}_{\dot\alpha}] = 0$$

$$\{Q_\alpha, \bar{Q}_{\dot\beta}\} = 2\sigma^m{}_{\alpha\dot\beta} P_m$$

$$\{Q_\alpha, Q_\beta\} = \{\bar{Q}_{\dot\alpha}, \bar{Q}_{\dot\beta}\} = 0$$

P_m is the energymomentum operator, Q_α is a Weyl spinor, $\bar{Q}_{\dot\alpha}$ is its complex conjugate.

Superspace:

We want to find a representation of this algebra in terms of differential operators. For this purpose we introduce the concept of superspace. It consists of the common 4-dimensional space and a set of anticommuting variables $\theta^\alpha, \bar{\theta}^{\dot\alpha}$. An element of this space will be denoted by $z^M \sim \{x^m, \theta^\mu, \bar{\theta}^{\dot\mu}\}$.

The index notation is the following: Latin letters (m) denote a Lorentz four vector, Greek letters ($\mu, \dot\mu$) a Lorentz-spinor, a capital Latin letter (M) denotes a superspace index.

Generators:

A "finite group"-element can be defined as follows:

$$G(x, \theta, \bar\theta) = \exp i \{\theta^\alpha Q_\alpha + \bar\theta_{\dot\alpha} \bar{Q}^{\dot\alpha} - x_m P^m\}$$

Using Hausdorffs formula we find:

$$G(y, \xi, \bar\xi)\, G(x, \theta, \bar\theta) = G(x + y - i\xi\sigma\bar\theta + i\theta\sigma\bar\xi, \theta + \xi, \bar\theta + \bar\xi)$$

The "Group element" $G(y, \xi, \bar\xi)$ induces a motion in superspace:

$$G(y, \xi, \bar\xi) : \{x, \theta, \bar\theta\} \longrightarrow \{x + y - i\xi\sigma\bar\theta + i\theta\sigma\bar\xi, \theta + \xi, \bar\theta + \bar\xi\}$$

The infinitesimal generators of this motion are:

$$P_m = i \frac{\partial}{\partial x^m}$$

$$Q_\alpha = \frac{\partial}{\partial \theta^\alpha} - i \sigma^m_{\alpha\dot\beta} \bar\theta^{\dot\beta} \frac{\partial}{\partial x^m}$$

$$\bar Q_{\dot\alpha} = - \frac{\partial}{\partial \bar\theta^{\dot\alpha}} + i \theta^\beta \sigma^m_{\beta\dot\alpha} \frac{\partial}{\partial x^m}$$

They represent the algebra.

<u>Covariant derivatives:</u>

It is of interest to study the tangents to the curves generated by group elements in superspace. They span the so-called covariant "vector fields" and they have the property that they are carried into themselves by group transformations. This means that the generators of the group commute (anticommute) with the vector fields. This property leads to an easy construction. Because of the associative law

$$\Big(G(y,\zeta,\bar\zeta)\, G(x,\theta,\bar\theta) \Big)\, G(z,\Delta,\bar\Delta) =$$

$$= G(y,\zeta,\bar\zeta) \Big(G(x,\theta,\bar\theta)\, G(z,\Delta,\bar\Delta) \Big)$$

the vector field can be obtained as the infinitesimal of the right multiplication of the group.

$$D_a = \frac{\partial}{\partial x^a}$$

$$D_\alpha = \frac{\partial}{\partial \theta^\alpha} + i \sigma^m_{\alpha\dot\beta} \bar\theta^{\dot\beta} \frac{\partial}{\partial x^m}$$

$$\bar D_{\dot\alpha} = - \frac{\partial}{\partial \bar\theta^{\dot\alpha}} - i \theta^\beta \sigma^m_{\beta\dot\alpha} \frac{\partial}{\partial x^m}$$

These "covariant" derivatives have the properties:

$$\{D_\alpha, Q_\beta\} = \{\bar{D}_{\dot\alpha}, Q_\beta\} = \{D_\alpha, \bar{Q}_{\dot\beta}\} = \{\bar{D}_{\dot\alpha}, \bar{Q}_{\dot\beta}\} = 0$$

$$\{D_\alpha, \bar{D}_{\dot\beta}\} = -2i\,\sigma^m{}_{\alpha\dot\beta}\,\frac{\partial}{\partial x^m}$$

A convenient notation will be:

$$D_A = \tilde{e}_A{}^M \frac{\partial}{\partial z^M}$$

Vierbein:

From above we find:

$$\tilde{e}_A{}^M = \begin{pmatrix} \delta_a^m & 0 & 0 \\ i\sigma^m{}_{\alpha\dot\beta}\bar\theta^{\dot\beta} & \delta_\alpha^\mu & 0 \\ i\theta^\beta\sigma^m{}_{\beta\dot\gamma}\epsilon^{\dot\gamma\dot\alpha} & 0 & \delta^{\dot\alpha}_{\dot\mu} \end{pmatrix}$$

We also introduce the inverse matrix:

$$e_N{}^A\,\tilde{e}_A{}^M = \delta_N{}^M \qquad \tilde{e}_A{}^M\,e_M{}^B = \delta_A{}^B$$

$$e_M{}^A = \begin{pmatrix} \delta_m^a & 0 & 0 \\ -i\sigma^a_{\mu\dot\beta}\bar\theta^{\dot\beta} & \delta_\mu^\alpha & 0 \\ -i\theta^\beta\sigma^a{}_{\beta\dot\gamma}\epsilon^{\dot\gamma\dot\mu} & 0 & \delta^{\dot\mu}_{\dot\alpha} \end{pmatrix}$$

$e_M{}^A$ will be called the generalized flat-space vierbein.

Superfields:

Functions of z^M are called superfields. As an example we consider the real vector superfield, it is defined through:

$$V(x,\theta,\bar\theta) = V^+(x,\theta,\bar\theta)$$

Each superfield should be understood as a power series in the variables $\theta, \bar\theta$, it will always be a finite polynomial in $\theta, \bar\theta$.

$$\begin{aligned} V(x,\theta,\bar\theta) = {} & C(x) + i\theta\chi(x) - i\bar\theta\bar\chi(x) + \\ & + \tfrac{i}{2}\theta\theta\,(M(x)+iN(x)) - \tfrac{i}{2}\bar\theta\bar\theta\,(M(x)-iN(x)) - \\ & - \theta\sigma^m\bar\theta\, v_m(x) + i\theta\theta\,\bar\theta\bar\lambda(x) - \\ & - \tfrac{1}{2}\theta\theta\,\partial_m\chi(x)\sigma^m\bar\theta - \\ & - i\bar\theta\bar\theta\,\theta\lambda(x) + \tfrac{1}{2}\bar\theta\bar\theta\,\theta\sigma^m\partial_m\bar\chi(x) + \\ & + \theta\theta\bar\theta\bar\theta\left(\tfrac{1}{2}D(x) + \tfrac{1}{4}\Box C(x)\right) \end{aligned}$$

The transformation property of the common fields can be easily obtained from the definition:

$$\begin{aligned} \delta V & = \delta C + i\theta\,\delta\chi + \ldots + \theta\theta\bar\theta\bar\theta\left(\tfrac{1}{2}\delta D + \tfrac{1}{4}\Box\delta C\right) \\ & = (\xi Q + \bar\xi\bar Q)\,V \end{aligned}$$

$Q, \bar{Q}$ are here the differential operators from above. A comparison of the individual powers yields:

$$\delta C = i(\xi\chi - \bar{\xi}\bar{\chi})$$

$$\delta\chi = \xi(M+iN) + \sigma^m\bar{\xi}(\partial_m C + i v_m)$$

$$\delta\bar{\chi} = \bar{\xi}(M-iN) + \xi\sigma^m(\partial_m C - i v_m)$$

$$\delta N = \xi(i\lambda - \sigma^m\partial_m\bar{\chi}) + \bar{\xi}(-i\bar{\lambda} + \bar{\sigma}_m\partial^m\chi)$$

$$\delta M = \xi(\lambda + i\sigma_m\partial^m\bar{\chi}) + \bar{\xi}(\bar{\lambda} + i\bar{\sigma}_m\partial^m\chi)$$

$$\delta v_m = \xi\partial_m\chi + \bar{\xi}\partial_m\bar{\chi} + i\xi\sigma_m\bar{\lambda} + i\bar{\xi}\bar{\sigma}_m\lambda$$

$$\delta\lambda = \xi\sigma^{mn}(\partial_m v_n - \partial_n v_m) + \xi D$$

$$\delta\bar{\lambda} = \bar{\xi}\bar{\sigma}^{mn}(\partial_m v_n - \partial_n v_m) + \bar{\xi} D$$

$$\delta D = -\xi\sigma^m\partial_m\bar{\lambda} + \bar{\xi}\bar{\sigma}^m\partial_m\lambda$$

Observe that the component of the highest power in $\theta, \bar{\theta}$ transforms into a total space-time derivative. The integrated quantity will, therefore, be invariant under supersymmetric transformations. It is easy to see from the definition of Q and $\bar{Q}$ that this statement is true in general. By the multiplication of superfields we obtain a superfield again. This allows us to construct supersymmetric Lagrangians.

The covariant derivatives can be used to restrict a superfield and it is possible to do it in such a way as to avoid equations in the x^m space. As an example, the scalar superfield can be defined as a superfield S (complex) which satisfies the equation $D_\alpha S = 0$.

The components of $S + S^{+}$ can be identified with the components of V and we obtain:

$$C = A + A^{*}, \quad \chi = -i\Psi, \quad \bar{\chi} = i\bar{\Psi}, \quad v^{m} = -i\partial^{m}(A - A^{*})$$

$$N = -(F + F^{*}), \quad M = -i(F - F^{*}), \quad \lambda = 0, \quad D = 0$$

There are no restrictions on the fields A, χ and F, which are the components of a scalar superfield. We are going to use a scalar superfield to define a gauge transformation.

Gauge Theories:

Abelian Case:

We define a gauge transformation:

$$V \rightarrow V + i(\Lambda - \Lambda^{+}), \quad \bar{D}\Lambda = 0, \quad D\Lambda^{+} = 0$$

The component fields v_m, λ and D transform as follows:

$$v_m \rightarrow v_m + \partial_m (A + A^{*}), \quad \lambda \rightarrow \lambda, \quad D \rightarrow D$$

The transformation law of the vector field v_m indicates that the above transformation might have something to do with the supersymmetric generalization of a gauge transformation.

The superfield $W_\alpha = \bar{D}\bar{D} D_\alpha V$ is invariant under the gauge transformation:

$$\bar{D}\bar{D} D_\alpha(\Lambda - \Lambda^{+}) = \bar{D}\bar{D} D_\alpha \Lambda = \bar{D}\{\bar{D}, D_\alpha\}\Lambda \sim \bar{D}\partial\Lambda = 0$$

Due to its definition W_α satisfies the covariant equations:

$$\bar{D}_{\dot{\beta}} W_\alpha = 0, \quad D^{\alpha} W_\alpha - \bar{D}_{\dot{\beta}} \bar{W}^{\dot{\beta}} = 0$$

This reduces the independent fields in

W_α to $v_{mn} = \partial_m v_n - \partial_n v_m$, λ and D

Because $\bar{D}_{\dot\beta} W_\alpha = 0$ we find that $W^\alpha W_\alpha + \bar{W}_{\dot\beta} \bar{W}^{\dot\beta}$ gives rise to the following Lagrangian:

$$L = -\frac{1}{2} v_{mn}^{\ 2} - \frac{i}{2} \bar\lambda \sigma \partial \lambda + \frac{1}{2} D^2$$

This Lagrangian is invariant under the gauge transformation

$$v_m \to v_m + \partial_m a \ , \quad \lambda \to \lambda \ , \quad D \to D$$

and under the supersymmetry transformation

$$\delta v_{mn} = i \xi (\partial_m \sigma_n - \partial_n \sigma_m) \bar\lambda + i \bar\xi (\partial_m \sigma_n - \partial_n \sigma_m) \lambda$$

$$\delta \lambda = \xi \sigma^{mn} v_{mn} + \xi D$$
$$\delta D = - \xi \sigma^m \partial_m \bar\lambda + \bar\xi \bar\sigma^m \partial_m \lambda$$

as well.

<u>Nonabelian Case:</u>

T^ℓ_{ij} are the generators of a compact Lie algebra , $\ell = 1, \ldots, L$.

$V_\ell , \Lambda_\ell , \ell = 1, \ldots, L$ are superfields.

We define the matrix field:

$$V = T^\ell_{ij} V_\ell \qquad V_\ell^+ = V^\ell$$

$$\Lambda = T^\ell_{ij} \Lambda_\ell \qquad \bar{D} \Lambda_\ell = 0$$

A gauge transformation can be defined as follows:

$$e^V \to e^{-i\Lambda^+} e^V e^{i\Lambda} ,$$

$$V \to V + i (\Lambda - \Lambda^+) + \ldots$$

A gauge covariant expression can be defined as before:

$$W_\alpha = \bar{D} \bar{D} e^{-V} D_\alpha e^V$$

The gauge invariant expression

$$\mathrm{Tr}\,(W_\alpha W^\alpha + \overline{W}^{\dot\alpha}\overline{W}_{\dot\alpha})$$ gives rise to

the following Lagrangian:

$$L = \mathrm{Tr}\left\{ -\frac{1}{2} v_{mn}^2 - \frac{i}{2}\bar\lambda\sigma^m \mathcal{D}_m\lambda + \frac{1}{2}D^2 \right\}$$

where
$$v_{mn} = \partial_m v_n - \partial_n v_m + i\,[v_m, v_n]$$
$$\mathcal{D}_m\lambda = \partial_m\lambda + i\,[v_m, \lambda]$$

This Lagrangian is gauge invariant and supersymmetric as well. The way it was constructed was not very systematic, actually I presented it in a way which should make it obvious that it was guessed. This should make it easier to accept the formalism which I am going to present in the next lecture and which will allow a more systematic treatment.

SECOND LECTURE: DIFFERENTIAL FORMS, EXTERIOR DERIVATIVES AND STRUCTURE EQUATIONS

It is within the framework of differential geometry that General Relativity and Yang Mill's theories can be treated by the same formalism.[4] Therefore, it seems to be the right concept for gauging supersymmetry. For this purpose, the formalism of differential geometry has to be extended to be applicable to Grassmann spaces (superspace).
We start with a few definitions:

Differential Forms:

Elements of the space are denoted by

$z^M = (x^m, \theta^\mu, \bar{\theta}^{\dot{\mu}})$, they have the multiplication properties:

$$[x^m, x^n] = [x^m, \theta^\mu] = [x^m, \bar{\theta}^{\dot{\mu}}] = 0 .$$

$$\{\theta^\mu, \theta^\nu\} = \{\theta^\mu, \bar{\theta}^{\dot{\nu}}\} = \{\bar{\theta}^{\dot{\mu}}, \bar{\theta}^{\dot{\nu}}\} = 0 .$$

or, in short $z^M z^N = (-1)^{m(M)\, n(N)} z^N z^M$ where $m(M) = 0$ for $M = m$, a vector index and $m(M) = 1$ for $M = \mu, \dot{\mu}$, a spinor index. For the differentials dz^M we define: $dz^M dz^N = -(-1)^{m(M)\, n(N)} dz^N dz^M$
in more detail

$$dx^m dx^n = - dx^n dx^m$$
$$dx^m d\theta^\mu = - d\theta^\mu dx^m$$
$$dx^m d\bar{\theta}^{\dot{\mu}} = - d\bar{\theta}^{\dot{\mu}} dx^m$$

$$d\theta^\mu d\theta^\nu = d\theta^\nu d\theta^\mu$$
$$d\theta^\mu d\bar{\theta}^{\dot{\nu}} = d\bar{\theta}^{\dot{\nu}} d\theta^\mu$$
$$d\bar{\theta}^{\dot{\mu}} d\bar{\theta}^{\dot{\nu}} = d\bar{\theta}^{\dot{\nu}} d\bar{\theta}^{\dot{\mu}}$$

For any integer p we define a linear space which is spanned by the basis elements

$$dz^{M_1} \dots dz^{M_p} ,$$

the products are restricted by the multiplication rule from above. A p-form is now a linear combination of these basis elements with z-dependent coefficients, certain differentiation properties being understood. For example: A function $f(z)$ is called a zero form, $dz^M f_M(z)$ is a one form, $dz^M dz^N f_{NM}$ is a two form, and so on. Forms can be multiplied, the product of a p-form and a q-form gives a $p+q$ form. In this process we have to keep in mind that

$$z^M dz^N = (-)^{n(N)m(M)} dz^N z^M$$

<u>Exterior Derivatives:</u>

There is another process to obtain a $p+1$ form from a p-form, this is the exterior differentiation. σ is supposed to be a p-form:

$$\sigma = dz^{M_1} \dots dz^{M_p} f_{M_p \dots M_1}(z)$$

$$d\sigma = dz^{M_1} \dots dz^{M_p} dz^N \frac{\partial}{\partial z^N} f_{M_p \dots M_1}(z)$$

is a $p+1$ form. The exterior differentiation has the property that

$$dd = 0 \qquad \text{(Poincaré Lemma)} .$$

Also the converse Poincaré Lemma holds within topological restrictions. From $d\omega = 0$ follows $\omega = d\sigma$.

<u>Vierbein:</u>

In order to choose an arbitrary basis for our forms, and to adjust them to our supersymmetric space, we introduce the vierbein matrix

$$dz^M E_M{}^A(x, \theta, \bar{\theta}) = E^A$$

and its inverse

$$\tilde{E}_A{}^M E_M{}^B = \delta_A^B \qquad E_M{}^B \tilde{E}_B{}^N = \delta_M^N$$

<u>Structure Group:</u>

σ^t, $t = 1, \ldots, N$ is a set of p-forms which transform under a Lie group:

$\sigma'^t = \sigma^r X_r{}^t$, $\sigma' = \sigma X$ in matrix notation

$X_r{}^t = X_r{}^t(x, \theta, \bar{\theta})$ is an element of the Lie group.

<u>First Structure Equation:</u>

We differentiate the p-form σ and we obtain a $p+1$ form:

$$d\sigma^r = \sigma^t \varphi_t{}^r + \Omega^r \quad , \quad d\sigma = \sigma\varphi + \Omega$$

$\varphi_t{}^r$ is a 1-form $\varphi_t{}^r = E^A \varphi_{A\,t}{}^r$ and is supposed to be Lie-algebra valued in its matrix structure. We assign the following transformation properties to φ :

$$\varphi' = X^{-1}\varphi X + X^{-1}dX$$

$$\varphi'_\ell{}^r = X^{-1}{}_\ell{}^t \varphi_t{}^s X_s{}^r + X^{-1}{}_\ell{}^t dX_t{}^r$$

Due to this transformation property the $p+1$ form Ω transforms as follows:

$$\Omega' = \Omega X \qquad \Omega'^r = \Omega^s X_s{}^r$$

Proof:

$$\begin{aligned}
\Omega' &= d\sigma' - \sigma'\varphi' = \\
&= d(\sigma X) - \sigma X(X^{-1}\varphi X + X^{-1}dX) = \\
&= \sigma\, dX + d\sigma\, X - \sigma\varphi X - \sigma\, dX = \\
&= (\sigma\varphi + \Omega)X - \sigma\varphi X = \Omega X
\end{aligned}$$

In the context of General Relativity φ is called the connection form and Ω the torsion, in the context of a Yang Mill's theory φ will be the Yang Mill's potential and Ω a covariant derivative.

Second Structure Equation:

We differentiate the 1-form φ and define:

$$d\varphi = \varphi\varphi + F \quad , \quad d\varphi_t{}^r = \varphi_t{}^\ell \varphi_\ell{}^r + F_t{}^r$$

F is a Lie algebra valued 2-form and the meaning of this equation is that F transforms like a tensor:

$$F' = X^{-1} F X \quad , \quad F'_t{}^r = X^{-1}{}_t{}^s F_s{}^\ell X_\ell{}^r$$

Proof:

$$\begin{aligned} F' &= d\varphi' - \varphi'\varphi' = \\ &= d\left(X^{-1}\varphi X + X^{-1}dX\right) - \left(X^{-1}\varphi X + X^{-1}dX\right)\left(X^{-1}\varphi X + X^{-1}dX\right) = \\ &= X^{-1}\varphi\, dX + X^{-1} d\varphi\, X - dX^{-1}\varphi X - dX^{-1} dX - \\ &\quad - X^{-1}\varphi\varphi X - X^{-1}dX\, X^{-1}\varphi X - X^{-1}\varphi\, dX - \\ &\quad - X^{-1}dX\, X^{-1}dX \end{aligned}$$

Because $X^{-1}dX = -dX^{-1}X$ we find

$$F' = X^{-1} d\varphi\, X - X^{-1}\varphi\varphi X = X^{-1} F X$$

General Relativity: F is the curvature tensor,

Yang Mill's Theory: F is the Yang Mill's field.

Bianchi Identities:

It could seem to be natural to continue the process of exterior differentiation on Ω and F to obtain new tensor quantities. But due to Poincaré's lemma $dd = 0$ we do not find new quantities but merely identities among the old ones.

From $dd\sigma = 0$ follows:

$$d\Omega + \sigma d\varphi - d\sigma\varphi = 0$$

or using the second structure equation:

$$d\Omega = \Omega\varphi - \sigma F$$

and analogously from $dd\varphi = 0$

$$dF = F\varphi - \varphi F$$

Yang Mills:

First we are going to apply the formalism to the well known Yang Mill's case:

$$z^M \sim (x^m) \quad , \quad E_M{}^A = \delta_M{}^A$$

The space is the flat Minkowski space. The structure group is a Lie-group which acts on a set of p - $(0-)$ forms which span a representation of the Lie group for fixed x .

$$X = e^{i\Lambda(x)} \quad , \quad \Lambda(x) = \sum_\ell \lambda_\ell(x) T^\ell$$

$$\sigma' = \sigma X$$

The matrices T_ℓ are the generators of the group. For the connection form to be "Lie Algebra"-valued means that it can be written as

$$\varphi = dx^n \varphi_n \qquad \text{and} \qquad \varphi_n = \sum_\ell V_n^\ell T_\ell$$

The fields $V_n^\ell(x)$ are the well known Yang Mill's potentials.

The transformation law

$$\varphi' = X^{-1}\varphi X + X^{-1} dX$$

reduces, infinitesimally, to the well known transformation law of Yang Mill's potentials:

$$\varphi_a' = i[\varphi_a, \Lambda] + i\partial_a \Lambda$$

The first structure equation $\Omega = d\sigma - \sigma\varphi$ defines a tensor quantity in terms of derivatives and the Yang Mill's potentials. It is the well known covariant derivative:

$$\Omega^t = dx^a \left(\frac{\partial}{\partial x^a} \sigma^t - \sigma^s T^\ell{}_s \, {}^t V_a^\ell \right) = dx^a \left(\mathcal{D}_a \sigma \right)^t$$

The second structure equation

$$d\varphi = \varphi\varphi + F$$

defines a tensor field F in terms of the Yang Mill's potential φ and its derivatives. It is the well known Yang Mill's Field:

$$F = \frac{1}{2} dx^a dx^b F_{ab} = dx^a dx^b \left(\frac{\partial}{\partial x^a} \varphi_b + \varphi_a \varphi_b \right)$$

$$F_{ab} = \frac{\partial}{\partial x^a} \varphi_b - \frac{\partial}{\partial x^b} \varphi_a + [\varphi_a, \varphi_b]$$

The first identity $\sigma F = - d\Omega + \Omega\varphi$ tells us that there is a relation between the derivative of a covariant derivative and the Yang Mill's Field. It is left to the reader to work it out in detail, the answer is:

$$(\mathcal{D}_m \mathcal{D}_n - \mathcal{D}_n \mathcal{D}_m)\, \sigma = - \sigma F_{mn}$$

The second identity $dF = \varphi F - F\varphi$ tells us that there holds a cyclic identity for the covariant derivatives of the Yang Mill's Fields:

$$\mathcal{D}_l F_{nm} + \mathcal{D}_n F_{ml} + \mathcal{D}_m F_{ln} = 0 ,$$

General Relativity:

In this case the vierbein $E_M{}^A$ is supposed to be an independent field variable $e_m{}^a(x)$. Under general coordinate transformations (Einstein transformations)

$$x'^m = x'^m(x)$$

it transforms like

$$e'_m{}^a = \frac{\partial x^n}{\partial x'^m} e_n{}^a$$

such that the form

$$e^a = dx^m e_m{}^a$$

is a scalar density under Einstein transformations.

$$e'^a(x') = e^a(x) .$$

Note that the form, which is obtained by exterior differentiation from a scalar-density form will again be a scalar-density form.
The structure group is the local Lorentz transformation and the vierbein is supposed to transform like:

$$e'_m{}^b(x) = e_m{}^a(x) L_a{}^b(x)$$

The indices (m) which transform under Einstein transformations are called world indices, the indices (a) which transform under local Lorentz transformations are called Lorentz indices. Lorentz indices can be raised and lowered by η^{ab} . With the help of the vierbein and its inverse world indices can be transformed into Lorentz indices

and vice versa. This way any Einstein tensor density can be transformed into a Einstein scalar density-Lorentz tensor. Theseare the components of a tensor with respect to a preferred coordinate basis described by the vierbein.
For the form σ we take the vierbein form itself and we use the first structure equation to define the torsion Ω^a :

$$de^a = e^b \varphi_b{}^a + \Omega^a$$

$$dx^m dx^n \frac{\partial}{\partial x^n} e_m{}^a = dx^m dx^n \left(e_m{}^b \varphi_{n\,b}{}^a + \frac{1}{2} \Omega_{nm}{}^a \right)$$

$\Omega^a = \frac{1}{2} dx^m dx^n \Omega_{nm}{}^a$ is a two form which transforms like a Lorentz vector-Einstein scalar density. In a torsion-free theory, $\Omega_{nm}{}^a = 0$, the first structure equation can be used to express the connection $\varphi_{n\,b}{}^a$ in terms of the vierbein. The connection form is Lie algebra valued, this means that

$\varphi_{\ell\,mn} = e_m{}^a e_n{}^b \varphi_{\ell\,ba}$ is antisymmetric in n and m:
$\varphi_{\ell\,mn} = -\varphi_{\ell\,nm}$. This allows us to solve the equation:

$$e_\ell{}^a \left(\frac{\partial}{\partial x^n} e_{ma} - \frac{\partial}{\partial x^m} e_{na} \right) = \varphi_{mn\ell} - \varphi_{nm\ell}$$

$$\begin{aligned} \varphi_{nm,\ell} = -\frac{1}{2} \Big\{ & e_\ell{}^a \partial_n e_{ma} - e_n{}^a \partial_m e_\ell{}^a + \\ & + e_m{}^a \partial_\ell e_{n\,a} - e_m{}^a \partial_n e_\ell{}^a + \\ & + e_n{}^a \partial_\ell e_{m\,a} - e_\ell{}^a \partial_m e_n{}^a \Big\} \end{aligned}$$

The second structure equation defines the curvature tensor in terms of the connection.

$$d\varphi = \varphi\varphi + F \qquad F = \frac{1}{2} dx^m dx^n R_{nm}$$

$$dx^m dx^n \left(\frac{\partial}{\partial x^n} \varphi_{m\,a}{}^b - \varphi_{m\,a}{}^c \varphi_{n\,c}{}^b \right) = \frac{1}{2} dx^m dx^n R_{mn\,a}{}^b$$

or

$$R_{mn\,a}{}^b = \frac{\partial}{\partial x^n} \varphi_{m\,a}{}^b - \frac{\partial}{\partial x^m} \varphi_{n\,a}{}^b - \varphi_{m\,a}{}^c \varphi_{n\,c}{}^b + \varphi_{n\,a}{}^c \varphi_{m\,c}{}^b$$

As R_{mn} is the coefficient of a two-form it follows that $R_{mn\,a}{}^b = -R_{nm\,a}{}^b$ Again $R_{mn\,a}{}^b$ is Lie algebra valued in ab. This means for

$$R_{mn\,k\ell} = e_k{}^b e_\ell{}^a R_{mn\,ba}$$

$$R_{mn\,k\ell} = - R_{mn\,\ell k}$$

For a torsion free theory, the curvature tensor can be expressed in terms of the vierbein. The connection with the Riemannian theory is established by defining the metric tensor

$$g_{mn} = e_m{}^a e_{na}$$

The first identity:

$$dz^\ell dz^m dz^n \frac{\partial}{\partial x^n} \Omega_{m\ell}{}^a =$$

$$= dz^\ell dz^m dz^n \left(\Omega_{m\ell}{}^b \varphi_{n\,b}{}^a - e_\ell{}^b R_{nm\,b}{}^a \right)$$

tells us the symmetry relation:

$$R_{nm\ell}{}^{a} + R_{m\ell n}{}^{a} + R_{\ell nm}{}^{a} =$$

$$= \Omega_{m\ell}{}^{b}\varphi_{nb}{}^{a} + \Omega_{\ell n}{}^{b}\varphi_{mb}{}^{a} + \Omega_{nm}{}^{b}\varphi_{\ell b}{}^{a} +$$

$$+ \frac{\partial}{\partial x^{n}}\Omega_{m\ell}{}^{a} + \frac{\partial}{\partial x^{m}}\Omega_{\ell n}{}^{a} + \frac{\partial}{\partial x^{\ell}}\Omega_{nm}{}^{a}$$

In the torsion-free case it is just the cyclic condition on $R_{\ell mn}{}^{a}$.
The second identity ' $dF = F\varphi - \varphi F$,

$$dx^{\ell}dx^{m}dx^{n}\Big(\frac{\partial}{\partial x^{n}}R_{m\ell a}{}^{b} +$$

$$+ \varphi_{na}{}^{c}R_{m\ell c}{}^{b} - R_{nma}{}^{c}\varphi_{\ell c}{}^{b}\Big) = 0 .$$

In the torsion-free case this is just the well known Bianchi identity.

THIRD LECTURE: SUPERSYMMETRIC GAUGE THEORY

Now we are going to apply the methods of differential geometry to superspace. In order to have supersymmetry explicit we use the generalized flat-space vierbein $e_M{}^A$ from our first lecture to define the flat space. The structure group is again a Lie group that acts on a set of p forms

$$X = e^{i\Lambda(x,\theta,\bar\theta)}$$

$$\Lambda(x,\theta,\bar\theta) = \sum_\ell \lambda_\ell(x,\theta,\bar\theta)\, T^\ell$$

The matrices T^ℓ are generators of the group ,

$$\sigma' = \sigma X$$

The connection form

$$\varphi = dz^M \varphi_M(x,\theta,\bar\theta) = e^A \varphi_A(x,\theta,\bar\theta)$$

is Lie Algebra valued, this means

$$\varphi_A(x,\theta,\bar\theta) = \sum_\ell V_A^\ell(x,\theta,\bar\theta)\, T_\ell$$

These superfields are the generalization of the Yang Mills potentials. The transformation law of these potentials is, according to our general rules:

$$\varphi' = X^{-1}\varphi X + X^{-1} dX$$

$$\varphi_A' = X^{-1}\varphi_A X + X^{-1} D_A X$$

$$D_A = \tilde{e}_A{}^N \frac{\partial}{\partial z^N}$$

the first structure equation defines the covariant derivative:

$$\Omega = d\sigma - \sigma\varphi \quad , \quad \text{for a 0-form e.g:}$$

$$e^A \Omega_A{}^r = e^A D_A \sigma^r - e^A \sigma^t \varphi_{At}{}^r$$

It means that

$$D_A \sigma^r - \sigma^t \varphi_{At}{}^r = (\mathcal{D}_A \sigma)^r$$

transforms like a linear representation again. Moreover, due to our choice of the vierbein, $\mathcal{D}_A \sigma$ is a superfield if σ is a superfield.

The second structure equation $F = d\varphi - \varphi\varphi$ tells us how to construct tensor quantities in terms of the Yang Mills potentials. We compute:

$$d\varphi = e^A d\varphi_A + de^A \varphi_A$$

$$d\varphi_A = e^B D_B \varphi_A \quad , \quad de^A = e^B D_B e^A$$

$$\varphi \cdot \varphi = e^A e^B \varphi_B \varphi_A$$

and we obtain for $F = \frac{1}{2} e^A e^B F_{BA}$, $F_{BA} = -(-)^{ab} F_{AB}$

$$\begin{aligned} F_{BA} = {} & D_B \varphi_A - (-)^{ab} D_A \varphi_B + \\ & + \tilde{e}_A{}^M (D_B e_M{}^C) \varphi_C (-)^{b(a+m)} - \tilde{e}_B{}^M (D_A e_M{}^C) \varphi_C (-)^{bm} \\ & - \varphi_B \varphi_A + \varphi_A \varphi_B (-)^{ab} \end{aligned}$$

In more detail, this contains the following tensor quantities that can be used to formulate covariant (Lorentz covariant, supersymmetric, gauge covariant) equations:

$$F_{ab} = \frac{\partial}{\partial x^a}\varphi_b - \frac{\partial}{\partial x^b}\varphi_a + [\varphi_a, \varphi_b]$$

$$F_{a\alpha} = \frac{\partial}{\partial x^a}\varphi_\alpha - D_\alpha \varphi_a + [\varphi_a, \varphi_\alpha]$$

$$F_{a\dot\alpha} = \frac{\partial}{\partial x^a}\varphi_{\dot\alpha} - \bar{D}_{\dot\alpha}\varphi_a + [\varphi_a, \varphi_{\dot\alpha}]$$

$$F_{\alpha\beta} = D_\alpha \varphi_\beta + D_\beta \varphi_\alpha + \{\varphi_\alpha, \varphi_\beta\}$$

$$F_{\dot\alpha\dot\beta} = \bar{D}_{\dot\alpha}\varphi_{\dot\beta} + \bar{D}_{\dot\beta}\varphi_{\dot\alpha} + \{\varphi_{\dot\alpha}, \varphi_{\dot\beta}\}$$

$$F_{\alpha\dot\alpha} = D_\alpha \varphi_{\dot\alpha} + \bar{D}_{\dot\alpha}\varphi_\alpha + 2i\sigma^a{}_{\alpha\dot\alpha}\varphi_a + \{\varphi_\alpha, \varphi_{\dot\alpha}\}$$

We would like to impose a set of covariant equations such that the number of independent component fields, contained in the superfield φ_A is largely reduced without restricting the x_n dependence of the independent fields.

Abelian case:

Let me first discuss the Abelian case as an example. If we impose the equation

$$F_{\alpha\beta} = D_\alpha \varphi_\beta + D_\beta \varphi_\alpha = 0$$

$$F_{\dot\alpha\dot\beta} = \bar{D}_{\dot\alpha}\varphi_{\dot\beta} + \bar{D}_{\dot\beta}\varphi_{\dot\alpha} = 0$$

we learn that there exist superfields V and U such that:

$$\varphi_\alpha = -i D_\alpha V$$

$$\varphi_{\dot\alpha} = i \bar{D}_{\dot\alpha} U$$

These potentials U and V are only determined up to scalar superfields, $V \rightarrow V + S^{+}$, where $\bar{D}S = 0$, does not change φ ; the same is true for

$$U \rightarrow U + T \quad , \quad \bar{D}T = 0$$

Therefore, under gauge transformations the potentials U and V transform like:

$$V \rightarrow V + \Lambda + S^{+} \qquad \bar{D}S = 0$$

$$U \rightarrow U - \Lambda + T \qquad \bar{D}T = 0$$

This yields the desired transformations:

$$\varphi_{\alpha} \rightarrow \varphi_{\alpha} - i D_{\alpha} \Lambda$$

$$\varphi_{\dot{\alpha}} \rightarrow \varphi_{\dot{\alpha}} - i \bar{D}_{\dot{\alpha}} \Lambda$$

If we impose the additional equation

$$F_{\alpha\dot{\beta}} = D_{\alpha}\varphi_{\dot{\beta}} + \bar{D}_{\dot{\beta}}\varphi_{\alpha} + 2i\sigma^{a}{}_{\alpha\dot{\beta}}\varphi_{a} = 0$$

we can express φ_a in terms of U and V :

$$\varphi_{a} = \frac{1}{4}\bar{\sigma}_{a}{}^{\dot{\beta}\alpha}\left\{ D_{\alpha}\bar{D}_{\dot{\beta}}U - \bar{D}_{\dot{\beta}}D_{\alpha}V \right\}$$

Under gauge transformations:

$$\varphi_{a} \rightarrow \varphi_{a} - i\partial_{a}\Lambda$$

We have now achieved that the superfield φ_A can be expressed in terms of the fields U and V . The quantities F_{ab} , $F_{a\alpha}$ and $F_{a\dot{\alpha}}$ are also functions of U and V and no further equation can be imposed without restricting the x dependence of the component fields of U and V (field equations). All invariants can be expressed in terms of the invariant:

$$W_\alpha = \bar{D}\bar{D} D_\alpha (U+V) \quad , \quad \bar{W}_{\dot\beta} = DD\bar{D}_{\dot\beta}(U+V)$$

$$F_{a\alpha} = \frac{i}{8}\,\bar{\sigma}_a{}^{\dot\beta\delta}\,\epsilon_{\delta\alpha}\,\bar{W}_{\dot\beta} \quad , \quad F_{a\dot\alpha} = -\frac{i}{8}\,\epsilon_{\dot\alpha\dot\beta}\,\bar{\sigma}_a{}^{\dot\beta\alpha}\,W_\alpha$$

$$F_{ab} = \frac{i}{64}\left\{ (\epsilon\sigma^a\bar{\sigma}^b)^{\alpha\beta}\,(D_\beta W_\alpha + D_\alpha W_\beta) + (\bar{\sigma}^a\sigma^b\epsilon)^{\dot\alpha\dot\beta}\,(\bar{D}_{\dot\alpha}\bar{W}_{\dot\beta} + \bar{D}_{\dot\beta}\bar{W}_{\dot\alpha})\right\}$$

From the definition of W_α , $\bar{W}_{\dot\alpha}$ follows:

$$\bar{D}_{\dot\beta} W_\alpha = 0 \quad , \quad D_\beta \bar{W}_{\dot\alpha} = 0 \quad , \quad D^\alpha W_\alpha - \bar{D}_{\dot\beta}\bar{W}^{\dot\beta} = 0$$

This quantity W_α is very much alike to the gauge invariant superfield W_α of our first lecture, except that $U+V$ do not satisfy any reality condition. Only $U+V$ enter the gauge invariant expressions, this is because $U-V$ is gauge dependent and can be transformed away completely.

$$U-V \to U-V-2\Lambda+T-S^+$$

In order to impose a reality condition it would seem to be natural to demand $\varphi_a^+ = \varphi_a$ (one real vector field). This leads to the condition $U = V^+$ and restricts the gauge transformation to the "subgroup" $\Lambda = -\Lambda^+$, $S = T$. This shows that the condition $\varphi_a = \varphi_a^+$ is too strong if we want gauge invariance under a gauge group such that $\bar{D}\Lambda = 0$. We can relax this condition by demanding $\varphi_a = \varphi_a^+$ up to gauge transformations. This yields: $U+V = (U+V)^+$ and restricts the gauge group only by $S = T$. Now we can choose a gauge where $U = 0$, this leaves the subgroup $\Lambda = S$ as a gauge group.

The potential $V = V^{+}$, which is left as independent field transforms like $V \to V + S + S^{+}$. We also find $\overline{W}_{\dot\alpha} = W_{\dot\alpha}^{+}$

Thus we have reduced our gauge theory to exactly the theory which we have treated in our last lecture. The Lagrangian is:

$$L \sim W^{\alpha} W_{\alpha} + W_{\dot\alpha}^{+} W^{+\dot\alpha}$$

Before I am going to treat the nonabelian case I would like to demonstrate how we could have arrived at the same result by using the Bianchi identities. In this case the Bianchi identity

$$dF + \varphi F - F\varphi = 0 \qquad \text{means:}$$

$$e^{A} e^{B} e^{C} \{ D_C F_{BA} + \varphi_C F_{BA} - F_{BA} \varphi_C (-)^{c(a+b)} \} = 0$$

spelt out in its components:

$$1) \quad \mathcal{D}_d F_{cb} + \mathcal{D}_c F_{bd} + \mathcal{D}_b F_{dc} = 0$$

$$2) \quad \mathcal{D}_\delta F_{cb} + \mathcal{D}_c F_{b\delta} + \mathcal{D}_b F_{\delta c} = 0$$

$$3) \quad \mathcal{D}_{\dot\delta} F_{cb} + \mathcal{D}_c F_{b\dot\delta} + \mathcal{D}_b F_{\dot\delta c} = 0$$

$$4) \quad \mathcal{D}_\delta F_{\gamma b} - \mathcal{D}_\gamma F_{b\delta} + \mathcal{D}_b F_{\gamma\delta} = 0$$

$$5) \quad \overline{\mathcal{D}}_{\dot\delta} F_{\dot\gamma b} - \mathcal{D}_{\dot\gamma} F_{b\dot\delta} + \mathcal{D}_b F_{\dot\gamma\dot\delta} = 0$$

$$6) \quad \overline{\mathcal{D}}_{\dot\delta} F_{\gamma b} - \mathcal{D}_\gamma F_{b\dot\delta} + \mathcal{D}_b F_{\gamma\dot\delta} + 2i\,\sigma^{r}{}_{\gamma\dot\delta} F_{rb} = 0$$

$$7) \quad \mathcal{D}_\delta F_{\gamma\beta} + \mathcal{D}_\gamma F_{\beta\delta} + \mathcal{D}_\beta F_{\delta\gamma} = 0$$

$$8) \quad \overline{\mathcal{D}}_{\dot\delta} F_{\gamma\beta} + \mathcal{D}_\gamma F_{\beta\dot\delta} + \mathcal{D}_\beta F_{\dot\delta\gamma} + 2i\,\sigma^{r}{}_{\gamma\dot\delta} F_{r\beta} + 2i\,\sigma^{r}{}_{\beta\dot\delta} F_{r\gamma} = 0$$

$$9) \quad \overline{\mathcal{D}}_{\dot\delta} F_{\dot\gamma\beta} + \overline{\mathcal{D}}_{\dot\gamma} F_{\beta\dot\delta} + \mathcal{D}_\beta F_{\dot\delta\dot\gamma} + 2i\,\sigma^{r}{}_{\beta\dot\gamma} F_{r\dot\delta} + 2i\,\sigma^{r}{}_{\beta\dot\delta} F_{r\dot\gamma} = 0$$

$$10) \quad \overline{\mathcal{D}}_{\dot\delta} F_{\dot\gamma\dot\beta} + \overline{\mathcal{D}}_{\dot\gamma} F_{\dot\beta\dot\delta} + \overline{\mathcal{D}}_{\dot\beta} F_{\dot\delta\dot\gamma} = 0$$

If we demand $F_{\alpha\beta} = F_{\dot\alpha\dot\beta} = F_{\alpha\dot\beta} = 0$ we learn from (8) and (9)

$$F^{\alpha\beta} = \frac{1}{8} \bar{W}_{\dot\delta} \bar\sigma^{a\,\dot\delta\beta} \quad , \quad F^{a\dot\beta} = \frac{1}{8} \bar\sigma^{a\,\dot\beta\gamma} W_\gamma$$

W_γ and $\bar{W}_{\dot\gamma}$ are gauge invariant. From (6) follows:

$$D^\delta W_\delta = \bar{D}_{\dot\delta} \bar{W}^{\dot\delta}$$

and

$$F_{ab} = \frac{i}{64} \{ (\epsilon\sigma_a\bar\sigma_b)^{\alpha\beta} (D_\beta W_\alpha + D_\alpha W_\beta) + + (\bar\sigma_a \sigma_b \epsilon)^{\dot\alpha\dot\beta} (\bar{D}_{\dot\alpha}\bar{W}_{\dot\beta} + \bar{D}_{\dot\beta} \bar{W}_{\dot\alpha}) \}$$

All gauge invariants can be expressed in terms of W_α and $\bar{W}_{\dot\alpha}$, as we know already. From (4) and (5) follows

$$\bar{D}_{\dot\beta} W_\alpha = D_\alpha \bar{W}_{\dot\beta} = 0$$

Therefore, it follows that W_α is restricted in exactly the same way as we have found from the explicit solution above.

Non-Abelian Case:

We impose the same equations as in the abelian case:

$$F_{\alpha\beta} = F_{\dot\alpha\dot\beta} = F_{\alpha\dot\beta} = 0$$

From $F_{\alpha\beta} = 0$:

$$D_\alpha \varphi_\beta + D_\beta \varphi_\alpha - \{ \varphi_\alpha , \varphi_\beta \} = 0$$

we learn

$$\varphi_\alpha = - e^{-V} D_\alpha e^{V}$$

from $F_{\alpha\dot{\beta}} = 0$

$$\varphi_{\dot{\alpha}} = - e^{U} \bar{D}_{\dot{\alpha}} e^{-U}$$

The transformation properties under gauge transformations which yield the right transformation law for φ_A are

$$e^{V} \rightarrow e^{S^+} e^{V} e^{\Lambda}$$

$$e^{-U} \rightarrow e^{-T} e^{-U} e^{\Lambda}$$

These equations can be solved for the transformation properties of V and U

$$V \rightarrow V + \Lambda + S^+ + \ldots$$

$$U \rightarrow U - \Lambda + T + \ldots$$

From $F_{\alpha\dot{\beta}} = 0$ follows that also the field φ_a can be expressed in terms of U and V.
The tensor quantities F_{ab}, $F_{a\alpha}$ and $F_{a\dot{\alpha}}$ are functions of U and V. They can be expressed in terms of the two invariants W_α, $\bar{W}_{\dot{\alpha}}$. This follows easily from the identities. It also follows from the identities that

$$\bar{\mathcal{D}}_{\dot{\beta}} W_\alpha = \mathcal{D}_\alpha \bar{W}_{\dot{\beta}} = 0$$

$$\mathcal{D}^\alpha W_\alpha = \bar{\mathcal{D}}_{\dot{\alpha}} \bar{W}^{\dot{\alpha}}$$

The derivatives $\mathcal{D}$ are the fully covariant derivatives, defined through the first structure equation.

Because $\mathrm{Tr}\,(W^\alpha W_\alpha)$

is gauge invariant, it follows that

$$\bar{D}_{\dot{\beta}} \operatorname{Tr} (W^{\alpha} W_{\alpha}) \equiv \bar{\mathcal{D}}_{\dot{\beta}} \operatorname{Tr} (W^{\alpha} W_{\alpha}) = 0$$

and, therefore, $\operatorname{Tr} W^{\alpha} W_{\alpha}$ and $\operatorname{Tr} \bar{W}_{\dot{\alpha}} \bar{W}^{\dot{\alpha}}$

can be used to construct a supersymmetric, gauge invariant Lagrangian. It remains to impose the reality condition which relates $\bar{W}_{\dot{\alpha}}$ to $W_{\dot{\alpha}}^{+}$ and which restricts the gauge group to the subgroup of scalar superfields $\Lambda = S$. This can again be done by requiring that $i\varphi_a$ should be hermitean up to a gauge transformation and by fixing the gauge such that $U = 0$, then $V^{+} = V$ In this gauge the invariant W_{α} takes the value:

$$W_{\alpha} = \bar{D}\bar{D}\, e^{-V} D_{\alpha}\, e^{V}$$

Thus we have derived the non-abelian supersymmetric gauge theory from our general formalism. This method can be applied to extended supersymmetry and an application will be given in a forthcoming paper by R. Grimm, M. Sohnius and the author.

FOURTH LECTURE: SUPERGRAVITY

In a gauged supersymmetric theory we can expect that the currents of supersymmetry act as sources of fields: the energy momentum tensor as source of the graviton, the supercurrent as source of a spin 3/2 particle. To formulate supergravity will, therefore, require at least one spin 2 and one spin 3/2 field and it is indeed possible to construct such a theory. We shall derive such a theory with the help of our general methods of differential geometry.

The generalized vierbein $E_M{}^A$ is introduced as an independent variable as well as the connection form $\phi_{M,A}{}^B$. Our main task will be to find covariant equations which will reduce the number of independent component fields of $E_M{}^A$ and $\phi_{M,A}{}^B$ to just one spin 2 and one spin 3/2 field. Under general coordinate transformations

$$dz'^M = dz^N \frac{\partial z'^M}{\partial z^N}$$

E and ϕ transform as follows:

$$E'_M{}^A = \frac{\partial z^N}{\partial z'^M} E_N{}^A$$

$$\phi'_{M,A}{}^B = \frac{\partial z^N}{\partial z'^M} \phi_{N,A}{}^B$$

Next we have to choose the structure group, under which the vierbein form $dz^M E_M{}^A = E^A$ and the connection form

$\phi_A{}^B = dz^M \phi_{MA}{}^B$ transform as follows:

$$E^{A'} = E^B X_B{}^A$$

$$\phi'_A{}^B = X^{-1}{}_A{}^C \phi_C{}^D X_D{}^B + X^{-1}{}_A{}^D dX_D{}^B$$

$\phi_A{}^B$ is, as we know, Lie algebra valued. In order to reduce the number of independent fields as much as possible, we shall choose the smallest possible structure group - this is the Lorentz group itself. The parameters of the transformation are arbitrary functions of $z^M \sim \{x^m, \theta^\mu, \bar{\theta}^{\dot{\mu}}\}$ From this choice it follows that the 8 by 8 matrix $\phi_A{}^B$ contains only 6 independent superfields. The first and second structure equations define torsion and curvature.

$$\Omega^A = dE^A - E^B \phi_B{}^A$$

$$R_A{}^B = d\phi_A{}^B - \phi_A{}^C \phi_C{}^B$$

The curvature form $R_A{}^B$ is also Lie algebra valued.
An explicit form of these equations is:

$$\Omega_{BC}{}^A = (-)^{b(c+m)} \tilde{E}_C{}^M \tilde{E}_B{}^N \partial_N E_M{}^A -$$

$$- (-)^{cm} \tilde{E}_B{}^M \tilde{E}_C{}^N \partial_N E_M{}^A - \phi_{BC}{}^A + (-)^{bc} \phi_{CB}{}^A$$

$$R_{DEA}{}^{B} = (-)^{d(e+m)} \tilde{E}_E{}^M \tilde{E}_D{}^N \partial_N \Phi_{MA}{}^B$$

$$- (-)^{d(e+a+c)} \Phi_{EA}{}^C \Phi_{DC}{}^B -$$

$$- (-)^{em} \tilde{E}_D{}^M \tilde{E}_E{}^N \partial_N \Phi_{MA}{}^B +$$

$$+ (-)^{e(a+c)} \Phi_{DA}{}^C \Phi_{EC}{}^B$$

To get some hint for the equation we are looking for, we compute the torsion Ω^o for the flat space: $E^A = e^A$, $\Phi_A{}^B = 0$. We obtain:

$$\Omega^o{}_{\alpha\dot{\beta}}{}^a = \Omega^o{}_{\dot{\beta}\alpha}{}^a = 2i\,\sigma^a_{\alpha\dot{\beta}}$$

all other components of the torsion are zero. We might try to postulate these equations also for the general case:

$$\Omega_{\alpha\dot{\beta}}{}^a = \Omega_{\dot{\beta}\alpha}{}^a = 2i\,\sigma^a{}_{\alpha\dot{\beta}} \quad , \text{ all other components zero.}$$

For our choice of structure group, these would be covariant equations. It can, however, be shown that the only solution is the flat space, $E^A = e^A$. Therefore, we have to relax this set of equations. By linearizing the field equations:

$$E_M{}^A = e_M{}^A + \kappa H_M{}^A$$

and by solving them explicitly one can show that the equations:

$$\Omega_{ab}{}^{c} = \Omega_{\beta\gamma}{}^{\alpha} = \Omega_{\beta\gamma}{}^{\dot\alpha} = \Omega_{\dot\beta\alpha}{}^{\gamma} = \Omega_{\beta\gamma}{}^{a} = \Omega_{\beta c}{}^{a} = 0$$

$$\Omega_{\alpha\dot\beta}{}^{a} = \Omega_{\dot\beta\alpha}{}^{a} = 2i\,\sigma^{a}{}_{\alpha\dot\beta}$$

imply algebraic relations only. They reduce the number of dynamical independent fields to exactly one spin 2 and one spin 3/2 field.

This set of equations can also be solved completely - without linearization. One would proceed as follows: Choose a special gauge such that for $\theta = \bar\theta = 0$ the vierbein and the connection take the form:

$\theta = \bar\theta = 0$:

$$E_m{}^a = e_m{}^a(x) \,, \quad E_m{}^\alpha = \tfrac{1}{2}\psi_m{}^\alpha(x) \,, \quad E_{m\dot\alpha} = \tfrac{1}{2}\psi_{m\dot\alpha}(x)$$

$$E_\mu{}^a = 0 \,, \quad E_\mu{}^\alpha = \delta_\mu^\alpha \,, \quad E_{\mu\dot\alpha} = 0$$

$$E_a{}^{\dot\mu} = 0 \,, \quad E_\alpha{}^{\dot\mu} = 0 \,, \quad E_{\dot\alpha}{}^{\dot\mu} = \delta_{\dot\alpha}^{\dot\mu}$$

$$\phi_{m,\alpha}{}^{\beta} = \varphi_{m,\alpha}{}^{\beta}(x)$$

The fields $e_m{}^\alpha$, $\psi_m{}^\alpha$ and $\varphi_{m\,\alpha}{}^\beta$ are to be identified with the usual four-space vierbein, Rarita-Schwinger-field, connection respectively. The field equations determine the higher powers in $\theta, \bar\theta$ of E and ϕ in terms of the fields just introduced - the physical fields - and in terms of arbitrary gauge fields which have no dynamical meaning. Because there is only a finite power of $\theta, \bar\theta$ different from zero, these methods allow us to compute E and ϕ in a finite number of steps. However, even this is beyond our ambition here. We shall rather manipulate the equations and use the identities to illuminate the physical content of the theory. The equation

$$\Omega_{\beta c}{}^{\dot\alpha} = \Omega_{\beta c}{}^{\alpha} = 0$$

contains the x - dependent dynamics of the theory.
Note that $\Omega_{ab}{}^\gamma$, $\Omega_{ab}{}^{\dot\gamma}$ are the only components of the torsion which are left undetermined. If we would put them to zero we would end up with the flat space only.
We proceed as follows:

$$\Omega_{MN}{}^C = E_M{}^A E_N{}^B \Phi_{BA}{}^C (-)^{an}$$

is the torsion with world-indices M, N Due to our equations we obtain:

$$\Omega_{MN}{}^c = 2i\, E_M{}^\alpha E_N{}^{\dot\beta} \sigma^c_{\alpha\dot\beta} + 2i\, E_m{}^{\dot\alpha} E_N{}^\beta \sigma^c{}_{\beta\dot\alpha}$$

We learn that in our special gauge

$$\Omega_{mn}{}^c \Big|_{\theta=\bar\theta=0} = \mathcal{T}_{mn}{}^c = \frac{i}{2}\left(\psi_m \sigma^c \bar\psi_n - \psi_n \sigma^c \bar\psi_m\right)$$

This is the connection between the torsion in four-space and the spin density of the Rarita-Schwinger-field.

Furthermore:

$$\Omega_{MN}{}^{\beta} = E_M{}^a E_N{}^b \Omega_{ba}{}^{\beta} \quad , \quad \Omega_{mn}{}^{\beta} = E_m{}^a E_n{}^b \Omega_{ba}{}^{\beta}$$

Here $\Omega_{ba}{}^{\beta}$ has to be expressed in terms of the vierbein and the connection due to the first structure equation.

$$\Omega_{MN}{}^{\beta} = E_M{}^A E_N{}^B \Omega_{BA}{}^{\beta} (-)^{an} =$$

$$= E_M{}^A E_N{}^B \Big\{ (-)^{b(a+m')} \tilde{E}_A{}^{M'} \tilde{E}_B{}^{N'} \partial_{N'} E_{M'}{}^{\beta} -$$

$$- (-)^{am'} \tilde{E}_B{}^{M'} \tilde{E}_A{}^{N'} \partial_{N'} E_{M'}{}^{\beta} -$$

$$- \Phi_{BA}{}^{\beta} + (-)^{ba} \Phi_{AB}{}^{\beta} \Big\} (-)^{an} =$$

$$= (-)^{nm} \partial_N E_M{}^{\beta} - \partial_M E_N{}^{\beta} -$$

$$- \Phi_{MN}{}^{\beta} + (-)^{nm} \Phi_{NM}{}^{\beta}$$

We are interested in $\Omega_{mn}{}^{\beta}$ and obtain:

$$\Omega_{mn}{}^{\beta} = \partial_n E_m{}^{\beta} - \partial_m E_n{}^{\beta} - \Phi_{mn}{}^{\beta} + \Phi_{nm}{}^{\beta}$$

However

$$\phi_{mn}{}^{\beta} = E_n{}^{A}\,\phi_{mA}{}^{\beta} = E_n{}^{\alpha}\,\phi_{m\alpha}{}^{\beta}$$

because $\phi_{mA}{}^{B}$ is Lie algebra valued. For $\theta, \bar\theta = 0$, in our special gauge, this becomes

$$\Omega_{mn}{}^{\beta}\Big|_{\theta=\bar\theta=0} = T_{mn}{}^{\beta} =$$

$$= \tfrac{1}{2}\left(\partial_n \psi_m{}^{\beta} - \partial_n \psi_m{}^{\beta} - \psi_n{}^{\alpha}\varphi_{m\alpha}{}^{\beta} + \psi_m{}^{\alpha}\varphi_{n\alpha}{}^{\beta}\right) =$$

$$= \tfrac{1}{2}\left(D_n \psi_m{}^{\beta} - D_m \psi_n{}^{\beta} \right)$$

where we have introduced the usual four-space covariant derivative D_n.

In order to proceed we are going to use the Bianchi identities:

$$d\Omega^{A} + \Omega^{B}\phi_B{}^{A} - E^{B}R_B{}^{A} = 0$$

$$dR_A{}^{B} + R_A{}^{C}R_C{}^{B} - \phi_A{}^{C}R_C{}^{B} = 0$$

They can be spelled out in more detail:

$$E^{C}E^{B}E^{A}\left\{ E_A{}^{M}\partial_M \Omega_{BC}{}^{D} + T_{AB}{}^{C'}T_{C'C}{}^{D} - R_{AB,C}{}^{D}\right\} = 0$$

$$E^{C}E^{B}E^{A}\left\{ E_A{}^{M}\partial_M R_{BCD}{}^{F} + T_{AB}{}^{C'}R_{C'CD}{}^{F}\right\} = 0$$

where $\mathcal{D}_M$ is the covariant derivative, e.g.

$$\mathcal{D}_M v^A = \partial_M v^A + (-)^{bm} v^B \phi_{M,B}{}^A$$

$$\mathcal{D}_M v_A = \partial_M v_A - \phi_{M,A}{}^B v_B$$

I do not intend to write all the components of these identities - I think the publishers would not like it. But let me list those which I am going to use in this lecture. We put equal zero the respective components of the torsion and obtain:

$$(1) \quad R_{\varepsilon d \dot{\gamma}}{}^{\dot{\alpha}} = 2i\sigma^b{}_{\varepsilon\dot{\gamma}} \Omega_{bd}{}^{\dot{\alpha}}$$

$$(2) \quad R_{\alpha\beta c}{}^d = 0$$

$$(3) \quad \mathcal{D}_{\dot{\alpha}} R_{\beta b,c}{}^d + \mathcal{D}_\beta R_{\dot{\alpha} b,c}{}^d + 2i\sigma^{c'}{}_{\beta\dot{\alpha}} R_{c'bc}{}^d = 0$$

$$(4) \quad R_{\varepsilon d\gamma}{}^\alpha + R_{\gamma d\varepsilon}{}^\alpha = 0$$

$$(5) \quad R_{\dot{\varepsilon} d\dot{\gamma}}{}^{\dot{\alpha}} + R_{\dot{\gamma} d\dot{\varepsilon}}{}^{\dot{\alpha}} = 0$$

$$(6) \quad R_{\varepsilon dc}{}^a - R_{\varepsilon cd}{}^a = 2i\sigma^a{}_{\varepsilon\dot{\delta}} \Omega_{dc}{}^{\dot{\delta}}$$

From (1) follows:

$$(7) \quad \sigma^b{}_{\varepsilon\dot{\gamma}} \Omega_{bd}{}^{\dot{\gamma}} = 0 .$$

because $\mathcal{R}_A{}^B$ is Lie algebra valued and our Lie group is the Lorentz group. This equation is, for $\theta = \bar{\theta} = 0$ in our special gauge the Rarita Schwinger equation. From $\Omega_{mn}{}^{\beta} = E_m{}^a E_n{}^b \Omega_{ba}{}^{\beta}$ follows for $\theta = \bar{\theta} = 0$

$$\Omega_{ba}{}^{\beta}\Big|_{\theta=\bar{\theta}=0} = \tilde{e}_b{}^n \tilde{e}_a{}^m \Omega_{mn}{}^{\beta}\Big|_{\theta=\bar{\theta}=0} = \tilde{e}_b{}^n \tilde{e}_a{}^m T_{mn}{}^{\beta}$$

$$\sigma^b{}_{\varepsilon\dot{\gamma}}\, \Omega_{bd}{}^{\dot{\gamma}} = 0 \qquad \text{becomes}$$

$$\sigma^b{}_{\varepsilon\dot{\gamma}}\, \tilde{e}_b{}^m \tilde{e}_d{}^n \left\{ D_m \bar{\Psi}_n{}^{\dot{\gamma}} - D_n \bar{\Psi}_m{}^{\dot{\gamma}} \right\} = 0$$

This is the Rarita Schwinger equation.

To derive Einstein's equation we shall need the following relations:

$$\text{(a)} \quad R_{\alpha\beta,\,c}{}^{d} = 0$$

$$\text{(b)} \quad R_{\alpha b,\,cd} = 2i\, \sigma_{b\,\alpha\dot{\beta}}\, \Omega_{cd}{}^{\dot{\beta}}$$

$$\text{(c)} \quad R_{ab,\,c}{}^{b} = 0$$

$$\text{(d)} \quad R_{\alpha b,\,c}{}^{b} = 0$$

(a) follows from the identity (2);

(d) follows from (b) and (1) ;

(c) follows from (d) and (3) ;

thus we are left to prove (b). From (6) we learn that $R_{\varepsilon d \gamma}{}^{\alpha}$ is antisymmetric in ε and γ , therefore:

$$R_{\varepsilon d \gamma}{}^{\alpha} = \epsilon_{\varepsilon\gamma} X_d^{\alpha}$$

From $R_{\varepsilon d \gamma}{}^{\gamma} = 0$ follows $X_d^{\alpha} = 0$ and, therefore,

$$\text{(e)} \quad R_{\varepsilon d \gamma}{}^{\alpha} = 0$$

From (6) and from the property

$$R_{\varepsilon\, dca} = - R_{\varepsilon d\, ac} \qquad \text{(Lie algebra)}$$

follows:

$$\text{(f)} \quad R_{\varepsilon dca} = i \left\{ \sigma_{a \varepsilon \dot{\delta}} \Omega_{dc}{}^{\dot{\delta}} + \sigma_{d \varepsilon \dot{\delta}} \Omega_{ac}{}^{\dot{\delta}} + \sigma_{c \varepsilon \dot{\delta}} \Omega_{ad}{}^{\dot{\delta}} \right\},$$

we have to show that this yields (b). We use again that $R_A{}^B$ is Lie algebra valued:

$$R_{\varepsilon d \gamma}{}^{\alpha} = \frac{1}{4} (\sigma^a \bar{\sigma}^b)_{\gamma}{}^{\alpha} R_{\varepsilon d\, ba}$$

and obtain from (e)

$$(\sigma^a \bar{\sigma}^b)_{\gamma}{}^{\alpha} R_{\varepsilon d\, ba} = 0$$

This we apply to (f) and using (7) we obtain

$$(\sigma^a \bar{\sigma}^b)_{\gamma}{}^{\alpha} \Omega_{ab}{}^{\delta} = 0$$

From this relation, together with the property $\Omega_{ab}{}^{\dot\delta} = -\Omega_{ba}{}^{\dot\delta}$ and from (7) follows that

$$\Omega_{ab}{}^{\dot\delta} = (\epsilon\bar\sigma_a\sigma_b)_{\dot\alpha\dot\beta}\, X^{\dot\beta\dot\alpha\dot\delta}$$

where $X^{\dot\beta\dot\alpha\dot\delta}$ is totally symmetric. This, in turn, implies:

$$\sigma_a\Omega_{dc} + \sigma_d\Omega_{ca} + \sigma_c\Omega_{ad} = 0$$

and has (b) as a consequence.

Now let me show how Einstein's equation follows from (a,b,c,d). We compute

$$R_{MN,A}{}^{B} = E_M{}^{D} E_N{}^{C} R_{CD,A}{}^{B} (-)^{nd} \, .$$

From the structure equation:

$$R_{mna}{}^{b} = \partial_n\phi_{ma}{}^{b} - \partial_m\phi_{na}{}^{b} - \\ - \phi_{ma}{}^{\ell}\phi_{n\ell}{}^{b} + \phi_{na}{}^{\ell}\phi_{m\ell}{}^{b}$$

This shows that for $\theta = \bar\theta = 0$ we obtain

$$R_{mna}{}^{b}\Big|_{\theta=\bar\theta=0} = \mathcal{R}_{mna}{}^{b} \, ,$$

the usual four-space curvature tensor. On the other hand we have

$$R_{mna}{}^{b} = E_m{}^{d}E_n{}^{c}R_{cda}{}^{b} + E_m{}^{\delta}E_n{}^{c}R_{c\delta a}{}^{b} + \\ + E_m{}^{d}E_n{}^{\gamma}R_{\gamma da}{}^{b} + E_{m\dot\delta}E_n{}^{c}R_c{}^{\dot\delta}{}_a{}^{b} + E_m{}^{d}E_{n\dot\delta}R^{\dot\delta}{}_{da}{}^{b}$$

because $R_{\gamma\delta a}{}^{b} = 0$ due to (a)

For $\theta = \bar{\theta} = 0$ this relation becomes, using the relation (b) as well:

$$\mathcal{R}_{mn\,ab} = e_m{}^d e_n{}^c R_{cd\,ab} -$$

$$-i\left(e_n{}^c \psi_m{}^\delta - e_m{}^c \psi_n{}^\delta\right) \sigma_{c\delta\dot{\sigma}} \Omega_{ab}{}^{\dot{\sigma}} -$$

$$-i\left(e_n{}^c \bar{\psi}_{m\dot{\delta}} - e_m{}^c \bar{\psi}_{n\dot{\delta}}\right) \bar{\sigma}_c{}^{\dot{\delta}\sigma} \Omega_{ab\,\sigma} \Bigg|_{\theta=\bar{\theta}=0}$$

We invert the vierbein $e_m{}^d$, we use $R_{cd\,a}{}^d = 0$ (d)

and $\sigma^b{}_{\varepsilon\dot{\gamma}} \Omega_{bd}{}^{\dot{\gamma}} = 0$ (7) . With the obvious

four-space notation

$$\mathcal{R}_{ab\,c}{}^d = \tilde{e}_a{}^m \tilde{e}_b{}^n R_{nm\,c}{}^d$$

we obtain Einstein's equation:

$$\mathcal{R}_{ab\,c}{}^b + i\,e^{bm} \psi_m{}^\alpha \sigma_{a\,\alpha\dot{\beta}} \mathcal{T}_{bc}{}^{\dot{\beta}} +$$

$$+ i\,e^{bm} \bar{\psi}_m{}^{\dot{\beta}} \sigma_{a\,\alpha\dot{\beta}} \mathcal{T}_{bc}{}^\alpha = 0$$

Remember that we have already derived:

$$\mathcal{T}_{bc}{}^\alpha = \frac{1}{2} \tilde{e}_b{}^m \tilde{e}_c{}^n \left(D_n \psi_m{}^\alpha - D_m \psi_n{}^\alpha\right)$$

We were able to derive the dynamical equations of supergravity, i.e. Einstein's equation, Rarita Schwinger equation, the connection between torsion and spin density in the form which was first obtained by Deser and Zumino in their paper on supergravity. This we have achieved by starting from purely geometrical considerations in superspace.

I would like to thank Richard Grimm for helpful discussions and for reading the manuscript.

Notation:

$$\eta^{mn} = (-1, 1, 1, 1)$$

$$\varepsilon_{0123} = 1$$

$$\sigma^{m}{}_{\alpha\dot{\beta}} = (1, \vec{\sigma}) \qquad \bar{\sigma}^{m\,\dot{\alpha}\beta} = (1, -\vec{\sigma})$$

$$\sigma^{mn} = \frac{1}{4}\left(\sigma^{m}\bar{\sigma}^{n} - \sigma^{n}\bar{\sigma}^{m}\right)$$

$$\bar{\sigma}^{mn} = \frac{1}{4}\left(\bar{\sigma}^{m}\sigma^{n} - \bar{\sigma}^{n}\sigma^{m}\right)$$

$$\epsilon^{\alpha\beta}, \ \epsilon^{\dot{\alpha}\dot{\beta}} : \qquad \epsilon^{12} = -\epsilon^{21} = 1, \ \epsilon^{11} = \epsilon^{22} = 0$$

$$\epsilon^{\alpha\beta}\epsilon_{\beta\gamma} = \delta^{\alpha}_{\gamma}$$

$$\chi^{\alpha} = \epsilon^{\alpha\beta}\chi_{\beta} \quad , \quad \bar{\chi}^{\dot{\alpha}} = \epsilon^{\dot{\alpha}\dot{\beta}}\bar{\chi}_{\dot{\beta}}$$

$$\chi\psi = \chi^{\alpha}\psi_{\alpha} \quad , \quad \bar{\chi}\bar{\psi} = \bar{\chi}_{\dot{\alpha}}\bar{\psi}^{\dot{\alpha}}$$

$$U^{A}W_{A} = U^{a}W_{a} + U^{\alpha}W_{\alpha} + U_{\dot{\alpha}}W^{\dot{\alpha}}$$

REFERENCES

1 For a review of supersymmetry see: B. Zumino, Proc. 17th Internat. Conf. on High-Energy Physics, 1974 (Ed. J.R. Smith) (Rutherford Lab., Chilton, Didcot, U.K., 1974), p. I-254 and A. Salam and J. Strathdee, Phys. Rev. D 11, 521 (1975);
S. Ferrara, Rivista Nuovo Cimento 6, 105 (1976);
P. Fayet and S. Ferrara, Supersymmetry, to appear in Phys.Reports.

2 D.Z. Freedman, P. van Nieuwenhuizen and S. Ferrara, Phys. Rev. D 13 (1976) 3214;
S. Deser and B. Zumino, Phys. Lett. 62B (1976) 335;
D. Z. Freedman and P. van Nieuwenhuizen, Phys. Rev. D 14 (1976) 912;
M.T. Grisaru, P. van Nieuwenhuizen and J.A.M. Vermaseren, Phys. Rev. Letters 37 (1976) 1662;
P. Breitenlohner, Phys. Letters 67B (1977) 49.

3 B. Zumino, Proc. Conf. on Gauge Theories and Modern Field Theory, Northeastern Univ., Boston, 1975 (Eds. R. Arnowitt and P. Nath), (MIT Press, Cambridge, Mass., 1976), p. 255;
J. Wess and B. Zumino, Phys. Letters 66B (1977) 361;
A formulation essentially equivalent to the above publications of Ref. 3 has been developed independently in V.P. Akulov, D.V. Volkov and V.A. Soroka, JETP Letters 22 (1975) 396.

4 H. Flanders, Differential Forms, Academic Press 1963.

Further Literature:

P. Nath and R. Arnowitt, Phys. Letters 56B (1975) 177;
R. Arnowitt, P. Nath and B. Zumino, Phys. Letters 56B (1975) 81;
R. Arnowitt and P. Nath, Phys. Rev. Letters 36 (1976) 1526, Nucl. Phys. B122 (1977) 301; Phys. Letters 65B (1976) 73;
S. Ferrara, F. Gliozzi, J. Scherk and P. van Nieuwenhuizen, Nucl. Phys. B117 (1976) 333;
S. Ferrara, D.Z. Freedman, P. van Nieuwenhuizen, P. Breitenlohner, F. Gliozzi and J. Scherk, Phys. Rev. D 15 (1977)1013;
A. Das, M. Fishler and M. Rocek, Preprints ITP-SP-77-15 and ITP-SB-77-38 (1977;
E. Cremmer and J. Scherk, Preprint DAMTP 77/7 (1977);
S. Ferrara, J. Scherk and P. van Nieuwenhuizen, Phys. Rev. Lett. 37 (1976) 1037;
D.Z. Freedman and J.H. Schwarz, Phys. Rev. D 15 (1977) 1007;
D.Z. Freedman, to be published in Phys. Rev. D;
F. Gliozzi, J. Scherk and D. Olive, Phys. Lett. 65B, (1976) 282; Nucl. Phys. B122 (1977) 253;
S. Ferrara and P. van Nieuwenhuizen, Phys. Rev. Lett. 66B (1976) 1669;
S. Ferrara, J. Scherk and B. Zumino, Phys. Letters 66B (1977) 35;
D.Z. Freedman, Phys. Rev. Lett. 38 (1976) 105;
A. Das, Preprint ITP-SB-77-4 (1977);
E. Cremmer and J. Scherk, Algebraic simplifications in supergravity theories, DAMTP preprint, Cambridge, England;

P. van Nieuwenhuizen and J.A.M. Vermaseren, Phys. Lett. 65B (1977) 263;
M.T. Grisaru, P. van Nieuwenhuizen and J.A.M. Vermaseren, Phys. Rev. Lett. 37 (1976) 1662;
S. Ferrara, J. Scherk and B. Zumino, Nucl. Phys. B121 (1977) 393;
E. Cremmer, J. Scherk and S. Ferrara, Phys. Lett. 68B (1977) 234;
D.Z. Freedman and A. Das, Nucl. Phys. B120 (1977) 221;
P.K. Townsend, to be published in Phys. Rev. D.
S.W. MacDowell and F. Mansouri, Phys. Rev. Lett. 38 (1977) 739;
P.K. Townsend and P. van Nieuwenhuizen, to be published in Phys. Lett. B;
S. Deser and B. Zumino, Phys. Rev. Lett. 38 (1977) 1433;
M.T. Grisaru, Phys. Lett. 66B (1977) 75;
S. Deser, J.H. Kay and K.S. Stelle, Phys. Rev. Lett. 38 (1977) 527;
E. Tomboulis, Princeton preprint (1976);
P. van Nieuwenhuizen and J.A.M. Vermaseren, to be published in Phys. Rev. D;
M. Gell-Mann, Lecture given at the 1977 Coral Gables Conference;
J.C. Romao, A. Ferber and P.G.O. Freund, Chicago preprint EFI 76/73 (1976);
A. Ferber and P.G.O. Freund, Chicago preprint EFI 77/36 (1977);
M. Kaku, P.K. Townsend and P. van Nieuwenhuizen, to be published in Phys. Lett. B;
S. Ferrara, M. Kaku, P. van Nieuwenhuizen and P.K. Townsend, Trieste preprint IC/77/55.

PERTURBATION SERIES AT LARGE ORDER AND VACUUM INSTABILITY

J. Zinn - Justin
Service de Physique Théorique CEN SACLAY
and
Institut de Physique Théorique
Université de Louvain

Table of Contents

PERTURBATION SERIES AT LARGE ORDER AND VACUUM INSTABILITY

J. Zinn-Justin*
Service de Physique Théorique CEN SACLAY
and
Institut de Physique Théorique
Université de Louvain

INTRODUCTION

In these lectures we want to show how a new method, first introduced by Lipatov, allows us to calculate the asymptotic behaviour of perturbation theory at large orders. This method relies on an estimate through steepest descent of the path integral, in euclidean form, giving the kth order of the perturbation series. For boson field theories (this includes Quantum Mechanics) in general one finds:

$$k!\, a^k k^b c\,(1 + O(1/k))$$

Lipatov's method is a very simple generalization of the method one uses to solve the similar problem for integrals with a finite number of variables. It shows that large orders of perturbation theory are dominated by the finite action solutions (in general complex) of the euclidean equations of motion or instantons. When real instantons exist, then one finds that the perturbation series has a pathological behaviour. At large order all terms in the series have the same sign. This shows that the series, which is divergent for all values of the coupling constant, does not define a unique function; it is not Borel summable. On the other hand, as it is well known, real instantons indicate a vacuum instability, and are responsible for the tunneling between various would-be vacuum states. The study of the large order behaviour shows that it is impossible to forget about this vacuum instability and to work only with the naive perturbation theory.

The set up of these lectures is as follows: In the first section we shall expose the method on the trivial example of an integral on one variable. In section II we shall show how this method generalizes to the anharmonic oscillator in Quantum Mechanics. One can reproduce in

---

* Permanent address: Service de Physique Théorique CEN Saclay, 91 Gif sur Yvette, France.

this way a result previously derived by Bender and Wu through the use of the W.K.B. method. In sections III and IV we shall work out the case of the $g\varphi^4$ field theory. In section V we shall indicate how the method generalizes to arbitrary scalar boson field theory. In section VI we shall examine the problem of vector field and fermions and list a set of results obtained.

I. A TRIVIAL EXAMPLE

Let us first consider a simple integral which has a structure similar to the functional integrals we want to study:

$$Z(g) = \frac{1}{\sqrt{2\pi}} \int_{-\infty}^{+\infty} dx \exp -\left(\frac{x^2}{2} + g\frac{x^4}{4}\right) \tag{1}$$

By rotating the integration contour in the x plane it is easy to verify that Z(g) is an analytic function in a cut plane. It can be calculated as a power series in g. Setting

$$Z(g) = 1 + \sum_{k=1}^{\infty} Z_k g^k \tag{2}$$

we find

$$Z_k = \frac{(-1)^k}{k!} \frac{1}{4^k} \frac{1}{\sqrt{2\pi}} \int_{-\infty}^{+\infty} e^{-\frac{x^2}{2}} x^{4k} dx \tag{3}$$

This integral can of course be calculated explicitly:

$$Z_k = \frac{(-1)^k}{k!} \frac{1}{\sqrt{\pi}} \Gamma(2k+\tfrac{1}{2}) \tag{4}$$

The series (2) is therefore divergent for all values of g. It is elementary to show that:

$$\left| Z(g) - \sum_0^k Z_k g^k \right| < \frac{1}{\sqrt{\pi}} \frac{\Gamma(2k+3/2)}{(k+1)!} \frac{|g|^{k+1}}{\left[\cos(\frac{1}{2}\operatorname{Arg} g)\right]^{2k+\frac{3}{2}}}$$

Therefore the series (2) is an asymptotic series in the whole cut plane.

In order to understand the divergence of the series, we have used our knowledge of the behaviour of the Γ function for large values of the argument which is given for example by the Stirling formula. We could instead estimate directly the large order behaviour of Z_k from the expression (3) through steepest descent.

$$Z_k = \frac{(-1)^k}{k!} \frac{1}{4^k} \frac{1}{\sqrt{2\pi}}\, 2\int_0^{\infty} dx \exp\left(-\frac{x^2}{2} + 4k \ln x\right) \tag{5}$$

The saddle point x_c is given by:

$$-x_c + \frac{4k}{x_c} = 0$$

or

$$x_c = 2\sqrt{k} \tag{6}$$

It is the fact that the saddle point x_c grows with k which is responsible for the divergence of the series. Integrating the small fluctuations around the saddle point yields

$$Z_k \sim \frac{(-1)^k}{k!} \frac{\sqrt{2}}{4k} \exp -2k + 2k \ln 4k \tag{7}$$

which is equivalent to

$$Z_k \sim \frac{(-1)^k}{\pi\sqrt{2}} 4^k \frac{k!}{k} . \tag{8}$$

In order to estimate Z_k we have used the fact that we could obtain en explicit integral representation for it by expanding $\exp -\frac{g}{4} x^4$. As we shall see, this is not necessary. We can write instead:

$$Z_k = \frac{1}{\sqrt{2\pi}} \frac{1}{2i\pi} \int dx \int_C \frac{dg}{g^{k+1}} e^{-\left(\frac{x^2}{2} + g\frac{x^4}{4}\right)} \tag{9}$$

where the contour C surrounds the origin. Now for k large, we can estimate Z_k by looking for a saddle point both in the x and g plane. The saddle point equations are

$$\begin{cases} 1 + g_c x_c^2 = 0 \\ \dfrac{k}{g_c} + \dfrac{x_c^4}{4} = 0 \end{cases} \tag{10}$$

which yield

$$\begin{cases} x_c = \pm 2\sqrt{k} \\ g_c = -\dfrac{1}{4k} \end{cases} . \tag{11}$$

Notice that g_c is as small as $\frac{1}{k}$ and negative. This will be a general and essential feature of our analysis. To justify the saddle point approximation and to evaluate completely the leading term, we have still to calculate the matrix M of the second derivatives at the saddle point,

$$M = \begin{bmatrix} -2 & \pm 8k^{3/2} \\ \pm 8k^{3/2} & -16k^3 \end{bmatrix} \tag{12}$$

It is clear from this expression that the fluctuations around the saddle point are small.

$$\det M = -32k^3 \tag{13}$$

The determinant is negative, because the saddle point corresponds to a negative value of g and does not therefore correspond to a minimum of the integrand. On the other hand, as the integration over small fluctuations around the saddle point yields a factor $(\det M)^{-1/2}$, the factor i in front of expression (9) is canceled and the result is real as it should be,

$$Z_k \sim \frac{(-1)^k}{\sqrt{\pi}} \frac{1}{\sqrt{k}} \exp -k + k \ln 4k . \tag{14}$$

It is easy to verify that this expression is equivalent to the expression

(8). Following Bender and Wu's argument (used in the case of the anharmonic oscillator), it is possible to understand directly why the saddle point value of g is small and negative. The function Z(g) being analytic in a cut plane can be written as

$$Z(g) = \frac{1}{\pi}\int_{-\infty}^{0} \frac{\mathrm{Im}\, Z(g')}{g'-g}\, dg' \quad . \tag{15}$$

This yields an expression for Z_k:

$$Z_k = \frac{1}{\pi}\int_{-\infty}^{0} \frac{\mathrm{Im}\, Z(g)}{g^{k+1}}\, dg \quad . \tag{16}$$

For k large, Z_k is dominated by the values of ImZ(g) for g small and negative. Actually, we can calculate directly ImZ(g) for g small. Let us estimate Z_k from equation (9) through steepest descent by looking for the saddle point in x at g fixed

$$1 + g x_c^2 = 0 \quad . \tag{17}$$

We then get:

$$Z_k \sim \frac{1}{\pi\sqrt{2}} \int \frac{dg}{g^{k+1}}\, e^{\frac{1}{4g}} \tag{18}$$

This shows that:

$$\mathrm{Im}\, Z(g) \underset{g \to 0_-}{\sim} \frac{1}{\sqrt{2}}\, e^{\frac{1}{4g}} \quad . \tag{19}$$

Integrating then over g on the negative real axis:

$$Z_k \sim \frac{(-4)^k}{\pi\sqrt{2}}(k-1)! \tag{20}$$

We shall generalize this last method to the case of functional integrals. But before considering this question we want to outline that this large order estimate is not only interesting but also useful.

Borel transformation

We have seen that Z(g) is singular at the origin and that the series in powers of g is only asymptotic. Therefore the question arises: how to reconstruct this function from its asymptotic expansion?

Of course, in general this question has no unique answer. It is always possible to add to Z(g) a function which vanishes faster than a power when g goes to zero in the cut plane, for example like $\exp -\frac{1}{\sqrt{g}}$. In some cases, one special function can be defined from the Taylor series through the use of the Borel (or generalized Borel) transformation. Let us consider the function

$$B(g) = \sum_{0}^{\infty} g^k \frac{Z_k}{\Gamma(k+b+1)} \tag{21}$$

where b is some inessential real number introduced only for convenience. The function is analytic in a circle whose radius is just given by the large order estimate (20), which shows that the nearest singularity of B(g) is located at the point $g=-\frac{1}{4}$. Now formally Z(g) can be written as:

$$Z(g) = \int_0^\infty e^{-t} t^b B(gt)\, dt \tag{22}$$

in the sense of a power series expansion in g. It defines the function, for example for g positive, only if B(g) can be continued outside the circle of convergence of its Taylor series, in some neighbourhood of the positive real axis, and if B(g) does not grow too fast for g large.

If these conditions are satisfied and if Z(g) is indeed given by the equation (22) it is said to be Borel summable. Of course part of the ambiguity concerning the function Z(g) has been fixed by choosing for B(g) the only function having the Taylor series expansion (21) being analytic in a circle. From the large order estimate, one cannot prove that Z(g) is Borel summable. But if one finds that at large orders all terms of the series have the same phase, then the Borel transform has a singularity on the positive real axis, and the function is certainly not Borel summable. For our example it is easy to calculate B(g) and to verify all the needed properties. For b half integer, B(g) is particularly simple

$$\begin{aligned} b = -\tfrac{1}{2} \quad &\Rightarrow \quad B(g) = \frac{1}{1+u}\sqrt{1-u} \\ b = \tfrac{1}{2} \quad &\Rightarrow \quad B(g) = \sqrt{1-u} \end{aligned} \tag{23}$$

with:

$$u = \frac{\sqrt{1+4g}-1}{\sqrt{1+4g}+1} \tag{24}$$

B(g) is therefore also analytic in a cut plane. Knowing this property of the Borel transform, and knowing from the large order behaviour the nearest singularity of B(g), here $g=-\frac{1}{4}$, it is possible to derive from the divergent series (2), a convergent expansion for Z(g). One first maps the plane cut at $g=-\frac{1}{4}$ onto a circle centred at the origin, the origin being left invariant. This conformal mapping is just realized by the transformation (24). Then the Taylor series expansion for B(g) is transformed in an expansion in powers of u, convergent in the circle image of the whole cut plane. The integral (22) yields therefore a convergent expansion for Z(g) :

$$Z(g) = \sum_k U_k \int_0^\infty e^{-t} t^b u^k(gt)\, dt \tag{25}$$

For $b=\frac{1}{2}$, the coefficients U_k decrease as $k^{-3/2}$, and the integrals behave as

$$\int_0^{\infty} e^{-t}\sqrt{t}\, u^k(gt)\, dt \underset{k\to+\infty}{\sim} e^{-3\left(\frac{k}{2\sqrt{g}}\right)^{2/3}} \tag{26}$$

This shows that this new expansion converges essentially as $\exp -3\left(\frac{k}{2\sqrt{g}}\right)^{\frac{2}{3}}$

II. THE ANHARMONIC OSCILLATOR

We shall study now the perturbation series of the anharmonic oscillator. We shall take the example of the ground state energy. As we intend to use the functional integral in euclidean form, we shall calculate the ground state energy from

$$E(g) = \lim_{\beta\to\infty} -\frac{1}{\beta} \ln \operatorname{tr} e^{-\beta H(g)} \tag{27}$$

where H(g) is the Hamiltonian

$$H = -\frac{1}{2}\left(\frac{\partial}{\partial x}\right)^2 + \frac{1}{2}x^2 + \frac{1}{4}g x^4 \tag{28}$$

The quantity $\operatorname{tr} e^{-\beta H}$ can be calculated from the Feynman-Kac formula as:

$$\operatorname{tr} e^{-\beta H} = \int da \int_{q(0)=a}^{q(\beta)=a} [dq(t)] \exp - \mathcal{A}[q(t)] \tag{29}$$

where $\mathcal{A}(q)$ is the euclidean action:

$$\mathcal{A}[q(t)] = \int_0^{\beta} dt \left[\frac{1}{2}\dot{q}^2(t) + \frac{1}{2}q^2(t) + \frac{1}{4} g\, q^4(t)\right] \tag{30}$$

Now one can look for the region in the complex g plane where the path integral is meaningful, in the naïve sense, if one allows for rotation of the contour in the complex $q(t)$ plane. Such an analysis indicates again that $\operatorname{tr}\exp -\beta H$ and therefore E(g) is analytic in a cut plane, with a singularity at the origin. Actually a rigorous analysis confirms this expectation. In order to study the large order behaviour of the perturbation series, it is then natural, following Bender and Wu, to consider the imaginary part of the energy for g small and negative, as we have already argued in the previous section. For g negative, the would-be ground state, obtained by expanding around the ground state of the harmonic oscillator, is unstable, and the imaginary part of the energy is given by the tunneling through the barrier. It is well known that for g small, this problem can be solved through WKB method. But the WKB method can hardly be generalized to Field Theory. In one dimension an equivalent method exists, based on the path integral representation calculated through steepest descent. As we know that the tunneling is associated with a classical propagation in imaginary time, it is clear that we should consider the euclidean form of the path integral. Following Lipatov, this is what we shall do here. The kth order of the perturbative expansion of $\operatorname{tr} e^{-\beta H}$ is given by:

$$\{ \mathrm{tr}\, e^{-\beta H} \}_k = \frac{1}{2i\pi} \int_{q(\beta)=q(0)} [dq(t)] \frac{dg}{g^{k+1}} \exp -A[q] \tag{31}$$

The saddle point equations for q(t) and g are:

$$\begin{cases} \ddot{q}_c(t) = q_c(t) + g_c\, q_c^3(t) \\ \dfrac{k}{g_c} = -\dfrac{1}{4} \displaystyle\int q_c^4(t)\, dt \end{cases} \tag{32}$$

These equations have a solution with g_c negative and of order $1/k$ and $q_c(t)$ of order $k^{1/2}$ as in the case of the simple integral of the previous section. Setting

$$q_c(t) = \frac{1}{\sqrt{-g}}\, y(t) \tag{33}$$

we obtain in the large β limit

$$y(t) = \frac{\sqrt{2}}{ch(t-\tau)} \tag{34}$$

$$g_c = -\frac{4}{3k} \tag{35}$$

where τ is an arbitrary constant.

As we did for the simple integral, we shall integrate now over q(t) through steepest descent at g negative fixed and integrate on g after, instead of calculating through steepest descent on q(t) and g together. The classical action on the trajectory is then:

$$A_c\,[q_c(t)] = -\frac{4}{3g} \tag{36}$$

We have to integrate the small fluctuations around the classical solution. We have therefore to calculate the second functional derivative $M(t_1, t_2)$ of the action $A[q(t)]$ taken at the saddle point:

$$M(t_1,t_2) = \left.\frac{\delta^2 A}{\delta q(t_1)\,\delta q(t_2)}\right|_{q = q_c(t)} \tag{37}$$

$$M(t_1,t_2) = \left[-\left(\frac{\partial}{\partial t}\right)^2 + 1 - \frac{6}{ch^2(t_1-\tau)}\right] \delta(t_2-t_1)$$

The gaussian integration around the saddle point will then give essentially $(\det M)^{-1/2}$. But we have to be careful about the following fact. Due to translational invariance in time, $A[q_c(t)]$ is independent on the parameter τ. Therefore, the fluctuations corresponding to an increase of τ are not small. This is reflected by the fact that the operator M has a zero eigenvalue corresponding to the vector $\dot{q}_c(t-\tau)$ which just describes an infinitesimal increase of τ. We have therefore to make a change of variable on q(t) taking τ as one of the new variables. Let $f_n(t-\tau)$ be a complete set of orthogonal and normed wave functions, eigenvectors of M, and orthogonal to $\dot{q}_c(t-\tau)$. We can then make the following change of variable:

$$q(t) = q_c(t-\tau) + \sum_1^{\infty} c_n f_n(t-\tau) \tag{38}$$

taking τ and the c_n as new variables. The condition that the functions f_n are orthogonal to $\dot{q}_c(t)$ insures that the total number of degrees is conserved. As the f_n are normed, the jacobian J comes entirely from the variation of $q_c(t-\tau)$ with respect to τ.

$$J = \left[\int \dot{q}_c^2(t)\, dt \right]^{1/2} = \frac{2}{\sqrt{3}} \left(\frac{1}{-g} \right)^{1/2}. \tag{39}$$

Furthermore, as we now see the determinant of M is the determinant in the subspace orthogonal to $\dot{q}_c(t)$ in which M has no zero eigenvalue. A convenient way of calculating this determinant is to take the limit:

$$(\det M)_{\perp} = \lim_{z \to \infty} \frac{\det(z+M)}{z} \quad . \tag{40}$$

It happens that $-\frac{6}{ch^2 t}$ is a reflexion free Bargman potential for which this determinant can be analytically calculated. It has of course to be normalized by dividing it by the determinant of the free problem, which we have chosen implictly to normalize the path integral:

$$\det \frac{z+M}{z+M_0} = \frac{(\sqrt{1+z}-1)(\sqrt{1+z}-2)}{(\sqrt{1+z}+2)(\sqrt{1+z}+1)} \quad . \tag{41}$$

As the saddle point is not stable in all directions, one of the eigenvalues is negative, so that the factor i in front of the integral is canceled. For what concerns the integration with respect to τ, as nothing depends on τ, it just gives a factor β in front of the result.

A last factor comes from the fact that we have normalized the path integral with respect to its value at g=o. But in the free problem, to each mode corresponds a gaussian integral, which brings in not only the inverse of the square root of the corresponding eigenvalue but also a factor $(2\pi)^{1/2}$. On the other hand, in the interacting problem, one mode has been treated separately, the τ integration which was not gaussian. So we have to divide the result by $(2\pi)^{1/2}$. Bringing all factors together we get:

$$\left\{ e^{-\beta E(g)} \right\}_k \underset{k \to +\infty}{\sim} 2\beta e^{-\beta/2} \left(\frac{2}{\pi^3} \right)^{1/2} (-1)^{k+1} \int_{-\infty}^{0} \frac{dg}{(-g)^{k+3/2}} e^{4/3g} \quad . \tag{42}$$

Setting

$$E(g) = \frac{1}{2} + \sum_{k=1}^{\infty} E_k g^k \tag{43}$$

we remark that if E_k behaves like k! then

$$\left\{ \exp - \beta E(g) \right\}_K \underset{K \to +\infty}{\sim} e^{-\beta/2} E_K \tag{44}$$

As a result we get:

$$E_K \sim (-1)^{K+1} 2 \left(\frac{2}{\pi^3} \right)^{1/2} \int_{-\infty}^{0} \frac{dg}{(-g)^{K+3/2}} e^{4/3g} \tag{45}$$

We see therefore that the imaginary part of E(g) for g small and negative behaves like:

$$\operatorname{Im} E(g) \underset{g \to 0_-}{\sim} \left(\frac{-\beta}{\pi g} \right)^{1/2} e^{4/3g} \tag{46}$$

Integrating on g we finally obtain:

$$E_K \underset{K \to \infty}{\sim} (-1)^{K+1} \left(\frac{6}{\pi^3} \right)^{1/2} \left(\frac{3}{4} \right)^K \Gamma(K + \tfrac{1}{2}) \tag{47}$$

This result agrees with Bender and Wu. The correction terms obtained by expanding around the saddle point would generate systematic corrections of the form of powers of $1/K$ to equation (47). It is also elementary to generalize equation (47) to successive excited states. It is sufficient to calculate the leading corrections to $\mathcal{A}[q_c]$ for β large:

$$\mathcal{A}[q_c] = -\frac{4}{3g} - \frac{16}{g} e^{-\beta} + O\left(\frac{e^{-2\beta}}{g} \right) \tag{48}$$

If we expand $\exp - \mathcal{A}$ in powers of $e^{-\beta}$, then by comparing to the harmonic oscillator, we see that the coefficient of $e^{-(N+1/2)\beta}$ corresponds to the Nth excited state. For k large, the leading contribution will be given by the largest negative power of g. It always comes, clearly, from the first power of $e^{-\beta}$ in the expansion of $\mathcal{A}[q_c]$ for β large

$$E_{K,N} \sim (-1)^{K+N+1} (16)^N \left(\frac{8}{\pi^3} \right)^{1/2} \int \frac{dg}{(-g)^{K+N+3/2}} e^{4/3g} \tag{49}$$

from which we derive

$$E_{K,N} \sim (-1)^{K+N+1} (12)^N \left(\frac{6}{\pi^3} \right)^{1/2} \left[\frac{3}{4} \right]^K \Gamma(K+N+\tfrac{1}{2}) \tag{50}$$

This result had also been obtained by Bender and Wu. A last remark is in order. We were looking for periodic trajectories of period β. We took a solution of periodic β exactly, but there exist solutions of period $\beta/2$, $\beta/3$ etc..., which are also solutions of our problem. These solutions have a classical action 2, 3 ... times larger so that they give only exponential small corrections to the result.

III. THE ϕ^4 FIELD THEORY

We shall now study the ϕ^4 field theory. We shall restrict ourselves first to 2 and 3 dimensions where the theory is superrenormalizable.

In 4 dimensions where the theory is just renormalizable, a few additional complications appear which we shall consider later. We shall follow exactly what we have done for the anharmonic oscillator. We shall calculate the large order behaviour of the N-point function in euclidean space:

$$\left\{ Z^{(N)}(x_1 \cdots x_N) \right\}_K = \frac{1}{2i\pi} \int [d\phi(x)] \frac{dg}{g^{K+1}} \phi(x_1) \cdots \phi(x_N) \exp - A[\phi] \tag{51}$$

with:

$$A[\phi] = \int d^d x \left[\frac{1}{2} (\partial_\mu \phi)^2 + \frac{1}{2} m^2 \phi^2 + \frac{1}{4} g \phi^4 \right] \tag{52}$$

As before the functional integral will be normalized by dividing by the vacuum amplitude of the free theory. Again naively we expect the theory to be analytic in a cut plane in terms of the bare coupling constant. We have of course now to add a mass counter-term to the action, but as there are only a finite number of primitively divergent diagrams, this should not modify the analytic structure.

The saddle point equation for g shows that if a real solution exists for ϕ, g_c will be negative:

$$\frac{K}{g_c} + \frac{1}{4} \int d^d x \, \phi_c^4(x) = 0 \tag{53}$$

The equation for $\phi_c(x)$ is

$$[-\Delta + m^2] \phi_c(x) + g \phi_c^3(x) = 0 \tag{54}$$

It is clear that in any dimension, g_c is of order $\frac{1}{k}$ and $\phi_c(x)$ of order $\sqrt{k}$.

Let us introduce a new function $\Psi(x)$

$$\phi_c(x) = \frac{m}{\sqrt{-g}} \Psi(mx) \tag{55}$$

The function $\Psi(x)$ then satisfies

$$(-\Delta + 1) \Psi(x) - \Psi^3(x) = 0 \tag{56}$$

A mathematical theory based on Sobolev inequalities tells us that the solution of minimal action is a function only of a radial variable, and zerofree. Let us take an origin at an arbitrary point x_o and set:

$$r = |(x - x_0)_\mu| \tag{57}$$

Then the equation (56) becomes

$$\left[-\left(\frac{d}{dr}\right)^2 - \frac{(d-1)}{r} \frac{d}{dr} + 1 \right] \Psi(r) - \Psi^3(r) = 0 \tag{58}$$

This equation has a solution even in r, finite at the origin, zerofree, and decreasing as e^{-r} at large distances. Indeed mechanically this equation is the equation of motion of a particle under the influence of two forces, a force deriving from a potential $-\frac{\Psi^2}{2} + \frac{\Psi^4}{4}$ and a dissipative force proportional to the velocity. Starting at time r=o, from a point $\Psi(o)$, with zero velocity, the particle has to reach the origin at time $+\infty$. It is clear that one and only one starting point $\Psi(o)$ satisfies all conditions and certainly the potential at $\Psi(o)$ has to be positive to compensate for the loss of energy due to the dissipative force:

$$\Psi(0) > \sqrt{2}$$

It is easy to find numerical solutions to equation (58).

At the saddle point we then obtain for the euclidean action:

$$\mathcal{A}[\phi_c] = -\frac{m^{4-d}}{g} a(d) \tag{59}$$

Where a is pure positive number only function of the dimension, and gm^{d-4} is the dimensionless coupling constant.

Using scaling arguments we shall obtain two useful relations. Let us calculate the action corresponding to the function $\mu\phi_c(\lambda x)$ and express that $\frac{\partial \mathcal{A}(\lambda)}{\partial\lambda}$ and $\frac{\partial \mathcal{A}}{\partial\mu}$ vanish for the solution $\phi_c(x)$, i.e. for $\lambda=\mu=1$. We obtain then:

$$\frac{\partial \mathcal{A}}{\partial\mu} = 0 \Rightarrow \int d^dx\left[(\partial_\mu\phi_c)^2 + \phi_c^2 + g\phi_c^4\right] = 0 \tag{60}$$

$$\frac{\partial \mathcal{A}}{\partial\lambda} = 0 \rightarrow \int d^dx\left[(2-d)(\partial_\mu\phi_c)^2 - d\,\phi_c^2 - \frac{dg}{2}\phi_c^4\right] = 0\ . \tag{61}$$

Solving these equations and using the definition of $\Psi(x)$ yields:

$$a = \frac{1}{4}\int \Psi^4(x)\, d^dx > 0 \tag{62}$$

$$\int\left[\partial_\mu \phi_c(x)\right]^2 d^dx = -\frac{m^{4-d}}{g} d\, a(d)\ . \tag{63}$$

We have to integrate over the fluctuations around the saddle point. By choosing an origin, we have broken the translation symmetry of the problem. The action $\mathcal{A}[\phi_c]$ being independent of the origin $x_{o\mu}$ that we have chosen, the fluctuations which only change $x_{o\mu}$ are not bounded, and therefore, as in the case of Quantum Mechanics, we have to take $x_{o\mu}$ as one of the new integration variables. Also the operator M:

$$M(x,x') = \frac{\delta^2 \mathcal{A}}{\delta\phi_c(x)\,\delta\phi_c(x')} \tag{64}$$

has now d zero eigenvalues corresponding to the eigenvector $\partial_\mu\phi_c(x)$. Indeed:

$$M(x,x') = \left[(-\Delta + m^2) + 3g\,\phi_c^2(x)\right]\delta(x-x') \tag{65}$$

Now $\phi_c(x)$ which is a function of $|x-x_0|$ satisfies the equation of motion:

$$(-\Delta+m^2)\,\phi_c(x)+g\,\phi_c^3(x)=0$$

Taking the derivative of this equation with respect to $x_{0\mu}$, yields then:

$$(-\Delta+m^2)\,\partial_\mu\phi_c(x)+3g\,\phi_c^2(x)\,\partial_\mu\phi_c(x)=0 \tag{66}$$

Again we shall make the following change of variables

$$\phi(x)=\frac{m}{\sqrt{-g}}\,\Psi(m(x-x_0))+\sum_n \phi_n\, f_n(x-x_0) \tag{67}$$

where the f_n are all the normalized eigenvectors of M orthogonal to $\partial_\mu\phi(x)$. The jacobien J of the change of variable will come entirely from the fact that the variation of $\phi(x)$ with respect to $x_{0\mu}$, $\partial_\mu\phi_c(x)$, is not properly normalized:

$$J=\|\partial_\mu\phi_c(x)\|^d$$

which gives

$$J=\left[\frac{1}{d}\int d^dx\,[\partial_\mu\phi_c(x)]^2\right]^{d/2}. \tag{68}$$

The factor $\frac{1}{d}$ comes from having had to integrate $[\partial_\mu\phi_c(x)]^2$ at μ fixed. As it is μ independent, we have summed over μ and divided by d. Using now the relation (63) we get

$$J=\left[-\frac{m^{4-d}}{g}\,a\right]^{d/2}. \tag{69}$$

We have then to calculate the determinant of M, $(\det M)_\perp$, in the subspace orthogonal to M

$$(\det M)_\perp=\lim_{z\to 0}\frac{\det(M+z)}{z^d}. \tag{70}$$

We have also to remember that for each translational mode we have to divide by a factor $(2\pi)^{1/2}$. Combining all factors we get:

$$\left\{Z^{(N)}(x_1\cdots x_N)\right\}_k\sim\frac{(-1)^{k+1}}{2i\pi}\left(\frac{a}{2\pi}\right)^{d/2}\left\{(\det M)_\perp\, m^{2d}\right\}^{-1/2}\times \\ F_N(x_1\cdots x_N)\, m^{(4-d)\frac{d}{2}+N}\int\frac{dg}{(-g)^{k+\frac{d}{2}+\frac{N}{2}+1}}\, e^{a\frac{m^{4-d}}{g}} \tag{71}$$

where $F_N(x_1\cdots x_N)$ is defined as

$$F_N(x_1\cdots x_N)=m^d\int d^dx_0\prod_{i=1}^{N}\Psi[m(x_i-x_0)] \tag{72}$$

and $((\det M)_\perp m^{2d})$ is dimensionless.

We see here explicitly that the integration over $x_{0\mu}$ restores the translational invariance of the theory. Furthermore, the Fourier transform of F_N is just the product of the Fourier transform $\tilde{\Psi}(p_i/m)$ of the functions $\Psi(mx_i)$. We can now integrate over g and finally obtain:

$$\left\{ Z^{(N)}(x_1 \cdots x_N) \right\}_k \sim \frac{(-1)^k}{2i\pi} \left(\frac{1}{2\pi}\right)^{d/2} \left\{ m^{2d} (\det M)_\perp \right\}^{-1/2} \times$$
$$m^{\frac{N}{2}(d-2)-k(4-d)} \, F_N(x_1 \cdots x_N) \, a^{-k-\frac{N}{2}} \, \Gamma(k + \tfrac{d}{2} + N/2) \quad . \tag{73}$$

Notice that the factor i in front in the expression is again canceled by one of the eigenvalues of the operator M which is negative. This is also a consequence of the Sobolev inequalities which we shall discuss at the end of the section.

We come now to an important problem which we have consciously ignored: The expression (73), as it stands, is infinite because the determinant of M is ultraviolet divergent. Also we have not taken into account the mass counterterm that we have to add to the action $\mathcal{A}(\phi)$ to render the theory finite. We shall examine this question now.

Renormalization

To give a meaning to the theory, we shall first add a cut-off to the action and then a mass counterterm:

$$\mathcal{A}_\Lambda(\phi) = \frac{1}{2} \int d^dx \left\{ \phi \left[-\Delta + \frac{\Delta^2}{\Lambda^2} + m^2 \right] \phi + \frac{g}{4} \phi^4 \right\} + \frac{1}{2} \delta m_0^2(\Lambda) \int \phi^2(x) \, d^dx \tag{74}$$

Two remarks are now in order:
First the cut-off term $\frac{1}{\Lambda^2}\int (\Delta\phi)^2 d^dx$ will of course modify the equation of motion at short distances, but it is elementary to see that this gives a negligible contribution to $\mathcal{A}(\phi_c)$ when Λ is large.
Second and more important, the mass counterterm starts at order g, but as g_c is of order $\frac{1}{k}$ and ϕ_c at order $\sqrt{k}$, the counterterm gives a contribution of order 1, where the classical action gives a contribution of order k. Therefore, the position of the saddle point is not modified by the counterterm. Furthermore, only the counterterm of order g has to be taken into account at leading order, the counterterms of order g^2 giving already a contribution of order $\frac{1}{k}$.

We shall now show how this mass counterterm renders the determinant finite. We shall write det M in the following way:

$$\det M = \exp \operatorname{tr} \ln M$$
$$\operatorname{tr} \ln M = \operatorname{tr} \ln \left(-\Delta + \frac{\Delta^2}{\Lambda^2} + m^2\right) + \operatorname{tr} \left\{ \left(-\Delta + \frac{\Delta^2}{\Lambda^2} + m^2\right)^{-1} 3m^2 \Psi^2(mx) \right\}$$
$$+ \sum_{n=2}^{\infty} \frac{(-1)^{n-1}}{n} \operatorname{tr} \left\{ \left[\left(-\Delta + \frac{\Delta^2}{\Lambda^2} + m^2\right)^{-1} 3m^2 \Psi^2(mx) \right]^n \right\} \tag{75}$$

The first term is canceled in the division of the path integral by the free vacuum amplitude. The second term, which is also divergent, can be rewritten as

$$tr\left\{(-\Delta+\frac{\Delta^2}{\Lambda^2}+m^2)^{-1}3m^2\Psi^2(mx)\right\}=\left[-\frac{3m^2}{g}\int d^dx\,\Psi^2(mx)\right]g\int\frac{d^dp}{p^2+\frac{p^4}{\Lambda^2}+m^2}\ . \quad (76)$$

It is now obvious that it is canceled by the mass counterterm. In practice it is often convenient to use the following method. Consider the operator O:

$$O = (-\Delta+m^2)^{-1}(3m^2\Psi^2(mx)) \quad (77)$$

Let λ_n be the eigenvalues of O. Then formally the determinant D(z)

$$D(z) = \det[1+zO] \quad (78)$$

reads

$$D(z) = \prod_1^\infty(1+\lambda_n z) \quad . \quad (77)$$

Now the first term in the expansion of $\ln$ D(z) in powers of z is just opposite to the divergent one loop counterterm:

$$\sum_n\lambda_n = +\infty$$
$$\sum_n\lambda_n^2 < \infty \qquad \text{for } d<4 \quad (78)$$

The renormalized determinant reads:

$$D_{ren}(z) = \prod_{n=1}^\infty(1+\lambda_n z)e^{-\lambda_n z} \quad (79)$$

which is now a convergent infinite product. Actually the λ_n behave as:

$$\lambda_n \sim \frac{1}{n^{2/d}} \quad (80)$$

and the expression (79) is just the infinite product representation of an entire function of order $\frac{d}{2}$. Of course the operator O has d eigenvalues equal to -1 which we have to cancel by dividing $D_{ren}(z)$ by $(1-z)^d$ before taking the limit z=1. But we have to be careful that the coefficient of z is no longer 1 in the determinant but $3m^2\Psi^2(mx)$. It is straightforward to verify that we have therefore to divide the result by an extra factor D' which is the determinant of the operator $3m^2\Psi^2(mx)$ in the subspace spanned by the eigenvectors $\partial_\mu\Psi(mx)$:

$$D' = \det D'_{\mu\nu} \quad (81)$$

with

$$D'_{\mu\nu} = \frac{\int[\partial_\mu\Psi(mx)\,\partial_\nu\Psi(mx)\,3m^2\Psi^2(mx)]d^dx}{\|\partial_\mu\Psi(mx)\|\,\|\partial_\nu\Psi(mx)\|} \quad (82)$$

As Ψ(mx) depends only on the radial variable this reduces to:

$$D' = \left\{3m^2\frac{\int\Psi^2(mx)[\partial_\mu\Psi(mx)]^2d^dx}{\int[\partial_\mu\Psi(mx)]^2d^dx}\right\}^d \quad (83)$$

Sobolev Inequalities

A mathematical result known as a Sobolev inequality tells us that if we consider the following ratio $R(\phi)$

$$R(\phi) = \frac{\left\{\int d^dx\left[\frac{1}{2}(\partial_\mu\phi(x))^2 + \frac{1}{2}m^2\phi^2(x)\right]\right\}^2}{\frac{1}{4}\int\phi^4(x)\,d^dx} \tag{84}$$

Then $R(\phi)$ has an absolute, non-vanishing minimum R which is obtained furthermore by choosing for ϕ a solution ϕ_c (x) of the differential equation

$$\frac{\delta R(\phi)}{\delta\phi_c(x)} = 0 \quad . \tag{85}$$

In addition it can be shown that the relevant solution depends only on a radial variable and is zerofree. The equation (85) is, up to a rescaling of ϕ, identical to the euclidean field equation (54). Following Parisi, we shall use this inequality to give arguments which indicate that the perturbation series is certainly bounded by the large order estimate that we have obtained, justifying in this way our choice of a rotationally invariant zerofree solution to the field equation.

The kth order of perturbation theory is given by

$$Z_k \sim \frac{(-1)^k}{4^k k!}\int[d\phi(x)]\left[\int d^dx\,\phi^4(x)\right]^k \exp -A_0(\phi) \tag{86}$$

with

$$A_0(\phi) = \frac{1}{2}\int d^dx\left[(\partial_\mu\phi(x))^2 + m^2\phi^2(x)\right] \quad . \tag{87}$$

Using the definition (84) we can write (86) as

$$Z_k \sim \frac{(-1)^k}{k!}\int[d\phi(x)]\frac{[A_0(\phi)]^{2k}}{[R(\phi)]^k}\exp -A_0(\phi) \quad . \tag{88}$$

Let us now imagine that we decompose the space of integration in cells Ω_i containing each only a finite number of degrees of freedom, and that we replace $R(\phi)$ by its minimum in the cell:

$$|Z_k| < \frac{1}{k!}\sum_i\int_{\Omega_i}[d\phi(x)]\frac{[A_0(\phi)]^{2k}}{[R(\Omega_i)]^k}\exp -A_0(\phi) \tag{89}$$

Now we can calculate the integrals:

$$\begin{aligned}\int_{\Omega_i}[d\phi(x)][A_0(\phi)]^{2k}\exp -A_0(\phi) &= \\ &= \frac{\Gamma(2k+\frac{n_i}{2})}{\Gamma(\frac{n_i}{2})}\int_{\Omega_i}[d\phi(x)]\exp -A_0(\phi)\end{aligned} \tag{90}$$

where n_i is the number of degrees of freedom in Ω_i.

It is now intuitive that the dominant contribution will come from the cell for which $R(\phi)$ is minimum, which corresponds to $\phi_c(x)$. Comparing to the zero dimensional case (the simple integral of the first section), we have in addition in the cell giving the largest contribution, the d degrees of freedom corresponding to the translational invariance. Therefore we obtain a bound:

$$|Z_k| < \frac{1}{k!} \Gamma(2k + \frac{d}{2} + \frac{1}{2}) R^{-k} C$$

which agrees with the large order estimate (73) for N=o. Of course we would have a hard time justifying each step in this argument, but it suggests strongly that our large order estimate is indeed correct. The Sobolev inequality can also be used to prove rigorously that the operator M has one and only one negative eigenvalue. As $\phi_c(x)$ is an absolute minimum of $R(\phi)$ the operator $\frac{\delta^2 R}{\delta\phi_c(x)\,\delta\phi_c(x')}$ is a positive operator. It is easy to verify that it is a linear combination of M and a projector $\phi_c^2(x)\,\phi_c^2(x')$, with positive coefficients. From this remark it is straightforward to derive that M has at most one negative eigenvalue (Assume that it has at least two negative eigenvalues and prove the contradiction.) Second, it is easy, using the equation of motion, to show that det M is up to a trivial positive factor $-\det \frac{\delta^2 R}{\delta\phi_c(x)\,\delta\phi_c(x')}$.

This shows that it has one negative eigenvalue. (It comes mainly from the fact that $R''(\phi)$ has zero eigenvalues corresponding to $\partial_\mu \phi_c$).

Summary

As this section was full of technical arguments, we want to summarize what we have obtained. At large orders the perturbative contribution to the N point function behaves as:

$$\left\{ Z^{(N)}(x_1 \cdots x_N) \right\}_{k \to +\infty} \sim C(x_1 \cdots x_N)\, a^{-k}\, \Gamma(k + \tfrac{d}{2} + \tfrac{N}{2})$$

where a depends only on the action, the function $C(x_1 \cdots x_n)$ is independent of k, and its Fourier transform depends only on the square of the external momenta.

IV. THE ϕ^4 THEORY IN 4 DIMENSIONS

The massless ϕ^4 theory was studied by Lipatov who was the firstto show , on this example, how a steepest descent calculation using the functional integral in euclidean form, allows us to estimate the large behaviour of perturbation theory. The remarkable feature of this example is that due to an extra dilatation invariance of the classical theory, the equation of motion can be solved analytically, and the determinant can be written as an explicit infinite product. As the details of the calculation have already been published in two papers, one by

Lipatov himself, the other by Brezin, Le Guillou and myself, I will not present the explicit calculation, but just outline the difference between this case and the previous one. A first remark is in order. Let us go back to the Sobolev inequality:

$$R = \min_{\{\phi(x)\}} \frac{\left\{\frac{1}{2}\int d^4x\left[(\partial_\mu\phi)^2+m^2\phi^2\right]\right\}^2}{\frac{1}{4}\int\phi^4 d^4x} \quad . \tag{91}$$

Let us now replace $\phi(x)$ by $\phi(\lambda x)$ and minimize on λ first at ϕ fixed. After a change of variable $\lambda x \to y$, we obtain

$$R = \min_{\{\phi(x)\}} \min_{\lambda} \frac{\left\{\frac{1}{2}\int d^4x\left[(\partial_\mu\phi)^2+\frac{m^2}{\lambda^2}\phi^2(x)\right]\right\}^2}{\frac{1}{4}\int\phi^4 d^4x} \quad . \tag{92}$$

We now see that the minimum on λ corresponds to the massless theory. We discover both that the Borel transform of the perturbation series in the massless and massive case have the same radius of convergence, and that the field equations of the massive case have no solution. This is a situation which, although non generic, can occur. In such a case the equation of motion can have a quasi solution, in the sense that one can find a set of functions depending on one parameter, for which the action has a limit, and the functional derivative of the action goes to zero, when the parameter tends towards some limit, although the functions themselves have no limit. Here in the massive case the quasi solution is the solution of the massless case, infinitely dilatated, and cut at a distance of order $\frac{1}{m}$. The calculation of the saddle point contribution goes then through the following procedure. One takes the dilatation parameter λ as one of the degrees of freedom, and integrates at the end on the translations and λ for λ large.

We shall now concentrate on the massless case:

$$A[\phi] = \int d^4x\left[\frac{1}{2}(\partial_\mu\phi)^2+\frac{1}{4}g\phi^4\right] \quad . \tag{93}$$

The equation of motion

The saddle point $\phi_c(x)$ is given by:

$$\Delta\phi_c(x) - g\phi_c^3(x) = 0 \quad . \tag{94}$$

For g negative, it has a real solution, entirely determined by the conformal invariance of the problem:

$$\phi_c(x) = \frac{2\sqrt{2}}{\sqrt{-g}} \frac{\lambda}{1+\lambda^2(x-x_0)^2} \quad . \tag{95}$$

We have now a solution which depends on five parameters instead of

four, dilatation and translations. In the functional change of variable, in order to calculate the small fluctuations around the saddle point we shall have to take both λ and $x_{o\mu}$ as new variables. A new factor $\sqrt{k}$ will come from the Jacobian. Now in order to see that the determinant can be calculated analytically, one can use the conformal invariance. Another method which makes use of our calculation of the determinant of the anharmonic oscillator, is the following:
The operator M(x,x') is now:

$$M(x,x') = \left[-\Delta + \frac{24}{(1+x^2)^2}\right]\delta(x-x') \quad . \tag{96}$$

To find the eigenvalues of M we have to solve a 4 dimensional Schrödinger problem with a radial potential. Diagonalizing the angular momentum leaves a radial Hamiltonian H:

$$H = -\left(\frac{d}{dx}\right)^2 + \frac{3}{x}\frac{d}{dx} + \frac{24}{(1+x^2)^2} + \frac{\ell(\ell+2)}{x^2} \quad . \tag{97}$$

Making the change of variable

$$x = e^t \tag{98}$$

and writing the wave function $\psi(x)$ as

$$\psi(x) = e^{-t}\chi(t) \tag{99}$$

leads to the new Hamiltonian H'

$$H' = -\left(\frac{d}{dt}\right)^2 - \frac{6}{ch^2 t} + (\ell+1)^2 \tag{100}$$

which is just the Hamiltonian we studied in the case of the anharmonic oscillator up a trivial constant shift.

Renormalization

The theory is now simply renormalizable. One has to introduce a cut-off:

$$A_\Lambda(\phi) = \frac{1}{2}\int d^4x \left\{ \phi\left[-\Delta + \frac{\Delta^2}{\Lambda^2} - \frac{\Delta^3}{\Lambda^4}\right]\phi + \frac{g}{4}\phi^4 \tag{101}$$

Again as g is of order $\frac{1}{k}$, only the one loop counterterms have to be taken into account to render the determinant finite. But we have now also a one-loop coupling constant counterterm. We have therefore to expand the determinant up to second order in $\phi_c^2(x)$ in order to make the corresponding substractions.

Another new feature also appears. As the theory is just renormalizable, it has an infinite number of primitive divergencies. So if the counterterms corresponding to a finite number of loops larger than one give a negligible contribution, we still have the counterterms of order k which should be important. As we have added no such counter-term to the action, we should find a divergence in our calculation. This is exactly what happens. It comes from the final integration over dilatations. The classical action at the saddle point has the form

$$\mathcal{A}(\lambda) = \frac{1}{g}\left[c_1 \frac{\lambda^2}{\Lambda^2} + c_2 \frac{\lambda^4}{\Lambda^4} - \frac{3}{8\pi^2}\right] \tag{102}$$

We see therefore that the contribution coming $\int \phi \Delta^2 \phi$ and $\int \phi \Delta^3 \phi$ provide a cut-off to the λ integration. The integral over λ has then the form

$$\int_0^\infty \frac{d\lambda}{\lambda}\, \lambda^{N+4}\, \phi_c(\lambda x_1) \cdots \phi_c(\lambda x_N)\, \exp -\frac{1}{g}\left(c_1 \frac{\lambda^2}{\Lambda^2} + c_2 \frac{\lambda^4}{\Lambda^4}\right). \tag{103}$$

It is easy to verify that the two-point function has a quadratic, and the four-point function a logarithmic divergency. The latter leads to a large order estimate of the renormalization group function $\beta(g)$.

V. LARGE ORDER ESTIMATE FOR GENERAL POTENTIALS

We shall consider now the same problem for the ground-state energy of potentials expanded starting from an harmonic oscillator approximation. More precisely we consider the Hamiltonian:

$$H = -\frac{1}{2}\left(\frac{\partial}{\partial x}\right)^2 + \frac{1}{\lambda^2} V(\lambda x) \tag{104}$$

where V(x) is an analytic function which for x small, behaves like

$$V(x) = \frac{1}{2} x^2 + O(x^3) \quad . \tag{105}$$

The perturbation series is an expansion in powers of λ^2 and we shall set:

$$g = \lambda^2 \quad . \tag{106}$$

Again we shall proceed in the same way as for the anharmonic oscillator:

$$E(g) = \lim_{\beta \to +\infty} -\frac{1}{\beta} \operatorname{tr} e^{-\beta H}$$

and use the functional integral formulation

$$\left\{ \operatorname{tr} e^{-\beta H} \right\}_K = \frac{1}{2i\pi} \int [dq(t)] \frac{dg}{g^{K+1}} \exp -\mathcal{A}[q] \tag{107}$$

with

$$\mathcal{A}[q] = \int_0^\beta \left[\frac{1}{2} \dot{q}^2(t) + \frac{1}{\lambda} V(\lambda q(t))\right] dt \quad . \tag{108}$$

The saddle point equations are now:

$$\ddot{q}_c(t) = \frac{1}{\lambda} V''(q_c(t))$$

$$\frac{2k}{\lambda_c} = \frac{2}{\lambda_c^3} \int dt\, V(\lambda_c q_c(t)) - \frac{1}{\lambda_c^2} \int dt\, q_c V'(\lambda_c q_c) \quad . \tag{109}$$

Introducing the function $y(t)$:

$$y(t) = \lambda q_c(t) \tag{110}$$

we can rewrite the equations:

$$\ddot{y}(t) = V'(y) \tag{111}$$

$$k g_c = \int dt \left[\tfrac{1}{2} \dot{y}^2(t) + V(y) \right] \quad .$$

This shows as expected that g_c is again of order $\frac{1}{k}$. But now, as we shall see on examples, the trajectories solution of these equations are in general complex, as well therefore as g_c. The only situations in which one can find real solutions, corresponds to cases in which either we have chosen to expand around a relative minimum of the potential, or there exist degenerate classical minima. In both cases, the classical minimum around which we expand, is quantum mechanically unstable, and the classical solutions correspond to a tunneling effect.

Indeed, in the large β limit, we are looking for solutions in which a particle leaves the origin at time $-\infty$, is reflected by a zero of the potential, and comes back to the origin at time $+\infty$. Clearly if the potential has a real zero which is not at the origin, we are in the unstable situation described before. We are not going now to repeat the analysis of the calculation of the explicit contribution to the large behaviour. The calculation can always be done explicitly in this case, using a quite different method, called the shifting method which one can find in various references. This method is specific of completely integrable Hamiltonian systems, as are one dimensional potential problems.

The final result reads:

$$E_k \sim \sum_{\substack{\text{leading} \\ \text{saddle points}}} - \frac{1}{2\pi^{3/2}} \frac{\Gamma(k+\frac{1}{2})}{[a(y_c)]^{k+1/2}}\, y_+ \cdot \exp\left[\int_0^{y_+} dy \left(\frac{1}{(2V(y))^{1/2}} - \frac{1}{y} \right) \right] \quad . \tag{112}$$

Let us rather examine what this result means on an example which illustrates the different possible situations.

Let us consider the potential:

$$V(q) = \frac{1}{2} q^2 - \gamma q^3 + \frac{1}{2} q^4 \quad . \tag{113}$$

a) $|\gamma| > 1$

In this case we have expanded the perturbation series around a relative minimum of the potential. The classical trajectory is real. The particle is reflected at the point

$$y_{+} = \gamma - (\gamma^{2}-1)^{1/2} \tag{114}$$

and the classical action is

$$a(y_c) = -\frac{2}{3} + \gamma^2 - \frac{1}{2}\gamma(\gamma^2-1)\ln\frac{\gamma+1}{\gamma-1} > 0 \tag{115}$$

The coefficients E_k then behave as

$$E_K \sim -\frac{1}{\pi^{3/2}} \frac{\Gamma(k+1/2)}{a^{k+1/2}} (\gamma^2-1)^{1/2} \quad . \tag{116}$$

As a is positive, we see that at large order all terms of the series are negative. As a consequence the Borel transform of the function E(g) has a singularity on the positive real axis. Therefore, either E(g) is not uniquely defined by its asymptotic series because it is not Borel summable, or more likely here, it has really a cut on the positive real axis, because the asymptotic expansion generated in this way is an expansion for the energy of an unstable state.

b) $|\gamma| < 1$

We have here expanded correctly the perturbation theory around the absolute minimum of the potential, and the classical trajectory is now complex. Actually there are two complex conjugate solutions:

$$E_K \sim \frac{2}{\pi^{3/2}} (1-\gamma^2)^{-1/2} \operatorname{Im}\left[a(\gamma)^{-k-1/2}\right] \Gamma(k+1/2) \tag{117}$$

The Borel transform of E(g) has two complex conjugate singularities located at $a(\gamma)$ and $a^*(\gamma)$. This result indicates that E(g) is Borel summable. In the limit γ=o, we find again the large order estimate of the anharmonic oscillator.

c) $\gamma = \pm 1$

The classical action $a(\gamma)$ has a limit which is $\frac{1}{3}$ and therefore positive. As we know that now the result for E(g) should be real for g positive, we see that the perturbation series is certainly not Borel summable. This is always the case when the classical minimum of the potential is degenerate, and at the same time there exists a quantum mechanical tunneling between the classical minima. Then one can find real instanton solutions, which dominate the large order of the perturbation series, and also give direct exponential small contributions to the perturbation theory. It seems that these direct contributions to the perturbation theory should be exactly the terms needed to cancel the singularity of the Borel transform, but this has still to be shown.

Let us illustrate this point on a mathematical example. Consider the function F(g):

$$F(g) = \int_0^\infty e^{-t/2}\left(e^{-t/2} - \frac{P(tg)}{P(1)}\, e^{-\frac{1}{2g}}\right)\frac{dt}{1-gt} \qquad (118)$$

where P(x) is some arbitrary polynomial. For g small and positive, F(g) has an asymptotic expansion:

$$F(g) = \sum_0^\infty k!\, g^k \qquad (119)$$

We see on the integral representation (118) that F(g) has an isolated singularity at the origin. If we try to resum the asymptotic series (119) we find instead a function cut on the positive real axis. Therefore the exponentially small contribution coming from:

$$\int_0^\infty \frac{P(tg)}{P(1)}\, e^{-1/2g}\, \frac{dt}{1-gt}$$

is needed to cancel the singularity. As P(x) is an arbitrary polynomial, not enough information is contained in the sole perturbative expansion to reconstruct the function.

Remark

When γ goes to ± 1, the classical action $a(\gamma)$ has a limit, but not the whole expression (112) which diverges. This comes from the fact that when the classical minimum is degenerate, the classical solution starts from zero at time $-\infty$ and reaches $q\gamma=1$, the other minimum at time $+\infty$. It can never come back so that there is no classical solution satisfying the boundary conditions. But there are quasi solutions, corresponding now to an instanton, anti-instanton pair largely separated. Let us call θ the separation. As explained before, one has to calculate the contribution to the path integral, at θ fixed and large, of such a configuration, and then integrate over θ.

VI. OTHER MODELS

Other models have been analyzed. They included the generalization to Quantum Field Theory of the models studied in the previous section, an abelian gauge field coupled to massless scalar bosons, non abelian gauge theories, fermions coupled through a Yukawa interaction, and Reggeon Field Theory.

We shall briefly review some of the features of these models.

A) Scalar Field Theory

Let us just indicate in this case when there exist no real

instantons. Consider the action $\mathcal{A}(\phi)$:

$$\mathcal{A}(\phi) = \int d^d x \left[\frac{1}{2} \partial_\mu \phi_\alpha M_{\alpha\beta}(\phi) \partial_\mu \phi_\beta + \frac{1}{g^2} V(g\phi) \right] \tag{120}$$

There will always exist real instantons if we expand around a relative minimum of the potential. So let us assume that $M_{\alpha\beta}$ is a positive matrix and $V(\phi) \geqslant 0$. Let $\phi_\alpha^c(x)$ be a real solution to the euclidean field equations, and let us calculate, following Derrick's argument, $\mathcal{A}[\phi^c(\lambda x)]$ as a function of λ:

$$\mathcal{A}[\phi^c(\lambda x)] = \lambda^{2-d} A + \lambda^{-d} B \tag{121}$$

where A comes from the term with derivatives, and B from $V(\phi)$. The quantity A is positive, and B positive, or can vanish in some cases.

Calculating $\frac{d\mathcal{A}(\lambda)}{d\lambda}$, and expressing that it vanishes at $\lambda = 1$ because at $\phi_c(x)$, $\mathcal{A}(\phi)$ is stationary, yields:

$$(2-d)A - dB = 0 \tag{122}$$

This equation shows that for $d > 2$, no real instantons can exist. At $d=2$, real solutions can only exist if B vanishes. In this case $\phi_c(x)$ will be a minimum of the potential for any x, which shows that $V(\phi)$ has an infinite set of minima continuously connected.

B) <u>Gauge Theories With Bosons Only</u>

a) Consider the euclidean action:

$$\mathcal{A}[\phi, A] = \int d^d x \left\{ \frac{1}{2} F_{\mu\nu}^\alpha F_{\mu\nu}^\alpha + \frac{1}{2} (D_\mu \phi_\alpha)^2 + V(\underset{\sim}{\phi}) \right\} \tag{123}$$

with standard notations for gauge theories.
Assume again that $V(\underset{\sim}{\phi})$ is non negative. Let $\phi_c(x)$ and $A_\mu^c(x)$ be real classical instantons and let us calculate $\mathcal{A}[\phi, A]$ for $\phi_c(\lambda x)$, $\lambda A_\mu^c(\lambda x)$. Expressing again that $\frac{d\mathcal{A}(\lambda)}{d\lambda}$ vanishes for $\lambda = 1$ yields:

$$(4-d) \int d^d x \, \frac{1}{4} (F_{\mu\nu}^\cdot)^2 + (2-d) \int d^d x \, (D_\mu \phi)^2 - d \int d^d x \, V(\phi) = 0 \tag{124}$$

We see first that there will exist no real instantons for $d > 4$.

In four dimensions we obtain two conditions:

$$D_\mu \phi_c = 0 \tag{125}$$

$$V(\phi_c) = 0 \tag{126}$$

If $\phi_c(x)$ does not vanish identically, then equation (125) implies that A_μ is a pure gauge. After a gauge transformation it then reduces to:

$$\partial_\mu \phi_c = 0 \tag{127}$$

which has no solution with finite action. So $\phi_c(x)$ must vanish identically. Then as we know there exist real instantons for any non abelian gauge group having SU(2) as a subgroup.

b) In one case also complex instantons have been found by Itzykson, Parisi and Zuber. Consider the action:

$$\mathcal{A}[A_\mu, \phi] = \int d^4x \left[\frac{1}{4} F_{\mu\nu}^2 + |\partial_\mu \phi + ie A_\mu \phi|^2 \right] .$$

Let us just give here the Ansatz which transforms the field equations, in scalar field equations which can then be solved numerically.

First one can choose $\phi(x)$ "real" by eliminating one of its components through a gauge transformation. Then the field equations read:

$$-\Delta \phi + e^2 (A_\mu)^2 \phi = 0 \tag{128}$$

$$[-\Delta \delta_{\mu\nu} + \partial_\mu \partial_\nu] A_\nu + 2e^2 A_\mu \phi^2 = 0 \quad . \tag{129}$$

Of course the idea is to look for a solution in which $\phi(x)$ is function of a radial variable only, and to find the corresponding Ansatz for $A_\mu(x)$. A function of the form $x_\mu A(|x|)$ is a pure gauge. One therefore tries

$$A_\mu(x) = M_{\mu\nu} x_\nu A(|x|) \tag{130}$$

where M is a numerical matrix satisfying

$$M^2 = \pm 1 \quad .$$

Writing then the two equations (128),(129) one checks that they are consistent with this Ansoltz if M is antisymmetrical:

$$\begin{cases} M_{\mu\nu} = - M_{\nu\mu} \\ M^2 = -1 \end{cases} \quad . \tag{131}$$

Writing then the corresponding two scalar equations, one finds that they have instanton solutions with $eA_\mu(x)$ and $e\phi(x)$ pure imaginary as in the case of the ϕ^4 theory.This solution yields a singularity of the Borel transform which lies therefore on the negative real axis.

c) <u>The Problem of Fermions</u>

Up to now we have only considered boson problems. Indeed, although it is possible to write Green's functions in a theory with fermions under the form of a functional integral over anticommuting variables which has algebraic properties very close to integrals over bosons,this integral has very different analytic properties. In particular it has no positivity. Let us again first consider the case of an integral over a finite number of variables:

$$Z(g) = \int \prod_a dx_a \prod_\alpha d\xi_\alpha d\bar{\xi}_\alpha \exp - \mathcal{A}[\underline{x}, \underline{\xi}, \underline{\bar{\xi}}] \tag{132}$$

with

$$\mathcal{A} = \frac{1}{2}\sum_a x_a^2 + \sum_\alpha \bar{\xi}_\alpha \xi_\alpha + g\sum_{\alpha\beta a} C_{\alpha\beta a}\, x_a \bar{\xi}_\alpha \xi_\beta \tag{133}$$

in which x_a are commuting, and $\{\xi_\alpha, \bar{\xi}_\alpha\}$ anticommuting variables. The action (133) imitates the action of a theory in which fermions $(\Psi, \bar{\Psi})$ are coupled through a Yukawa type interaction $\bar{\Psi}\Psi\varphi$, and the boson fields ϕ have no self interaction.

Now as a function of g, as long as there are only a finite number of anticommuting variables $(\xi_\alpha, \bar{\xi}_\alpha)$, Z(g) is just a polynomial. We already see that the structure of perturbation series with fermions will have different properties. We expect from this example that, as long as the bosons have no self interaction, the theory will be less divergent than in the case of a boson theory. We know also that the exactly soluble Schwinger and Thirring models have no essential singularity at the origin in the coupling constant. Following Parisi we shall examine more closely the Yukawa type theory with an action $\mathcal{A}(\bar{\Psi},\Psi,\varphi)$:

$$\mathcal{A}(\bar{\Psi},\Psi,\varphi) = \int d^d x \Big[\frac{1}{2}(\partial_\mu\phi)^2 + \frac{m^2}{2}\phi^2 + \bar{\Psi}(i\not{\partial} + m')\Psi + g\bar{\Psi}\Psi\varphi\Big] \quad . \tag{134}$$

In order to go around the problem of fermions, we shall integrate explicitly on Ψ and $\bar{\Psi}$:

$$Z = \int [d\phi(x)]\, \det[i\not{\partial} + m' + g\phi(x)]\, \exp -\frac{1}{2}\int d^d x[(\partial_\mu\phi)^2 + m^2\phi^2] \tag{135}$$

The key remark is now the following: The integration over fermons gives a determinant rather than the inverse of a determinant as in the case of bosons. On the other hand, for the class of fields $\phi(x)$ in which we are interested, the determinant is an entire function of g. Therefore, again, the singularities of the perturbation series come from the integration on large values of $\phi(x)$. We have therefore to evaluate the determinant for $\phi(x)$ large. It can be shown that:

$$\det(i\not{\partial} + m' + g\phi(x)) \underset{|\phi| \to +\infty}{\sim} \exp - C g^d \int \phi^d(x)\, d^d x \tag{136}$$

where C is some numerical constant depending only on the dimension d.

The large order behaviour of the perturbation series can therefore be calculated from the effective action $\mathcal{A}(\phi)$:

$$\mathcal{A}(\phi) = \int d^d x\Big[\frac{1}{2}(\partial_\mu\phi)^2 + \frac{1}{2}m^2\phi^2 + C g^d \phi^d(x)\Big] \tag{137}$$

From the form of $\mathcal{L}(\phi)$ we see that $g^{\frac{2d}{d-2}}$ plays the role of $\hbar$, i.e. of the parameter associated to the number of loops. As we have seen that in $\hbar$ the perturbation series behaves as K! at large order, we therefore obtain:

$$Z(g) = \sum Z_K g^{2K} \tag{138}$$

with

$$Z_K \sim (K!)^{\frac{d-2}{d}} \tag{139}$$

We therefore see from the large order behaviour that the perturbation series is asymptotic for d>2, but less divergent than in a boson field theory. For d=2, the constant C diverges and one obtains instead:

$$Z_K \sim (\ln K)^K \tag{140}$$

In this case a rigorous bound has been derived in constructive field theory which gives:

$$|Z_K| < (K!)^{\varepsilon} \qquad \forall \varepsilon > 0 \tag{141}$$

which is very much compatible with the estimate given by equation (140). It is believed that the result for QED with fermions will be similar.

REFERENCES

Sections II, III and IV

C. M. Bender and T. T. Wu, Phys. Rev. Lett. 27, 461 (1971). Phys. Rev. D7, 1620 (1972).

T. Banks, C. M. Bender and T. T. Wu, Phys. Rev. D8, 3346 (1973).

L. N. Lipatov, Zh. Eksp. Teor. Fiz. Red. 25, 116 (1974) and Leningrad Institute of Nuclear Physics Reports 253 and 255 (unpublished).

E. Brezin, J. C. Le Guillou and J. Zinn-Justin, Phys. Rev. D 15, 1544 (1977).

Section V

E. Brezin, J. C. Le Guillou and J. Zinn-Justin, Phys. Rev. D15, 1558 (1977).

E. Brezin, G. Parisi and J. Zinn-Justin, 'Perturbation Theory At Large Orders for Potential With Degenerate Minima', Phys. Rev. D16, 408 (1977).

Section VI

Same references as section V and

A. A. Belavin, A. M. Polyakov, A. S. Schwartz and Yu S. Tyupkin, Phys. Lett. 59B, 85 (1975).

G. 't Hooft, Phys. Rev. Lett. 37, 8 (1976).

C. Itkzykson, G. Parisi and J. B. Zuber, Phys. Rev. Lett. 38, 306 (1977) and preprint DPh-T/77/27 Phys.Rev D to be published.

For the problem of fermions, see:

G. Parisi, Phys. Lett. 66B, 167 (1977).

General Remarks

The method of using the Borel transformation and a conformal mapping to transform the perturbation series has first been proposed in the case of the anharmonic oscillator by:

J. J. Loeffel, Proceedings of the 1976 workshop on Padé approximants, D. Bessis, J. Gilewicz and P. Mery ed., CEA publication.

It has been extensively studied and applied successfully to the calculation of critical exponents starting from the ϕ^4 in 3 dimensions by:

J. C. Le Guillou and J. Zinn-Justin, 'Critical Exponents For the n-Vector Model from Field Theoretical Methods in 3 Dimensions', Phys. Rev. Lett. 39, 95 (1977).

The properties of analyticity and Borel summability of the anharmonic oscillator have been studied in:

J. J. Loeffel, A. Martin, B. Simon and A. S. Wightman, Phys. Lett. 30B, 1656 (1968).

S. Graffi, V. Grecchi and B. Simon, Phys. Lett. 32B, 631 (1970).

In the $[\phi^4]_3$ field theory by:

J. P. Eckmann, J. Magnen and R. Seneor, Comm. Math. Phys. 39, 251 (1975).

J. S. Feldman and K. Osterwalder, Ann. Phys. (N.Y.) 97, 80 (1976).

The method of collective coordinates has been exposed in:

J. S. Langer, Ann. Phys. (N.Y.) 41, 108 (1967).

J. L. Gervais and B. Sakita, Phys. Rev. D11, 2943 (1975).

V. E. Korepin, P. P. Kulish and L. D. Faddeev, Zh. Eksp. Teor. Fiz. Pisma Red. 21, 302 (1975), (J.E.T.P. Lett. 21, 138 (1975)).

See also:

N. H. Christ and T. D. Lee, Phys. Rev. D12, 1606 (1975).

and references in:

Phys. Rep. 23C, 237-374 (1976).

PHENOMENOLOGICAL ASPECTS OF THE MIT BAG

A.J.G. Hey,
Physics Department,
Southampton University

Table of contents | Page

PREFACE

This preface is intended primarily as an apology to the 'MIT Bag Company': much of the research described in these lectures has been performed by the originators of the MIT Bag theory and is contained in Refs. 1-3. Since there also exist several excellent reviews of the formulation of the Bag and its applications (Refs. 4-8), it is a somewhat daunting task to attempt yet another review. In view of this, these lectures are intended rather to supplement these other sources, and consequently concentrate on some less well-explored areas of Bag phenomenology. No attempt is made to review all applications with equal attention: in the main, the discussion will be restricted to the phenomenology of excited baryon states in the Bag theory. To introduce the Bag, a discussion of Bogolioubov's fixed radius model is included.

0. INTRODUCTION

To familiarize the reader with the physics of confined Dirac particles, we begin by considering the problem of confining a Dirac particle in a box. This will serve to introduce the Bag wavefunctions and the linear boundary condition that we shall require later. We first review some "well-known" results for non-relativistic and relativistic spin ½ particles in a central field: an appendix collects together these formulae and elaborates on their derivation.

(a) Non-Relativistic Central Field Spinors

In this situation the eigenfunctions may be characterised by the quantum numbers $\ell sj\mu$ (in the usual notation) and a quantum number κ (see below):

$$\psi_j^\mu = R_j(r)\ \chi_\kappa^\mu$$

The spin-angular functions χ_κ^μ are just the Clebsch-Gordan products of two component Pauli spinors χ^μ and the usual spherical harmonics Y_ℓ^m

$$\chi_\kappa^\mu = \sum_m C(j\mu;\ \ell\mu-m,\ \tfrac{1}{2}m)\ Y_\ell^{\mu-m}\ \chi^m$$

Notice that we may specify this product by the quantum number κ, instead of the more usual ℓ and j quantum numbers. This is defined by:

$$k\chi_\kappa^\mu = -\kappa\ \chi_\kappa^\mu$$

where k is the operator

$$k = \underline{\sigma}\cdot\underline{\ell} + 1$$

Thus:

$$\kappa = \begin{cases} \ell & \text{for } j = \ell-\frac{1}{2} \\ -\ell-1 & \text{for } j = \ell+\frac{1}{2} \end{cases}$$

The sign and magnitude of κ therefore determines both j and ℓ since:

$$j = |\kappa| - \tfrac{1}{2}$$

and

$$\ell = \begin{cases} \kappa & \text{for } \kappa>0 \\ -\kappa-1 & \text{for } \kappa<0 \end{cases}$$

Consequently, there is a one-to-one correspondence between the standard spectroscopic notation, L_J, and κ

$$S_{\frac{1}{2}} \leftrightarrow \kappa = -1$$

$$P_{\frac{1}{2}} \leftrightarrow \kappa = +1$$

$$P_{\frac{3}{2}} \leftrightarrow \kappa = -2$$

and so on.

Finally, we note an important property of these spin functions, namely:

$$\underline{\sigma}\cdot\underline{\hat{r}}\ \chi_{\kappa}^{\mu} = -\ \chi_{-\kappa}^{\mu}$$

The form of this relation is made evident with the observation that $\underline{\sigma}\cdot\underline{\hat{r}}$ is a pseudo-scalar operator which does not affect j or μ but does change the parity: this is equivalent to reversing the sign of κ.

(b) Dirac Particle in Central Field

For the free Dirac equation:

$$H\psi = (\underline{\alpha}.\underline{p} + \beta M)\ \psi = i\ \frac{\partial\psi}{\partial t}$$

it is easy to show that $\underline{\ell} = \underline{r} \times \underline{p}$ and $\underline{s} = \frac{1}{2}\underline{\sigma}$ are not separately constants of the motion. The angular momentum operator $\underline{j}$

$$\underline{j} = \underline{\ell} + \tfrac{1}{2}\sigma$$

and the relativistic analogue of k

$$K = \beta(\underline{\sigma}.\underline{\ell} + 1)$$

do commute with H and with each other. The eigenvalues, κ, of K may be deduced from the relation:

$$K^2 = \underline{j}^2 + \tfrac{1}{4}$$

from which we obtain

$$\kappa^2 = (j + \tfrac{1}{2})^2$$

leading to the result

$$j = |\kappa| - \tfrac{1}{2}$$

as for the non-relativistic case.

In the case of a central field W(r)

$$H\psi = (\underline{\alpha}.\underline{p} + \beta M + W(r))\psi = E\psi$$

the eigenfunctions may be specified by the eigenvalues of $\underline{j}^2$, j_z and κ:

$$\underline{j}^2\psi = j(j + 1)\psi$$

$$j_z\psi = \mu\psi$$

$$K\psi = -\kappa\psi$$

Writing ψ in terms of two two-component spinors

$$\psi = \begin{pmatrix} \phi_1 \\ \phi_2 \end{pmatrix}$$

it is evident from the eigenvalue equation for K that the general form of ψ is:

$$\psi \equiv \psi_\kappa^\mu = \begin{bmatrix} g(r)\ \chi_\kappa^\mu \\ if(r)\ \chi_{-\kappa}^\mu \end{bmatrix}$$

Substituting this form back into the Dirac equation written in polar co-ordinates, using:

$$\underline{\alpha} \cdot \underline{p} = -\, i\underline{\alpha}\cdot\underline{\hat{r}}\ \frac{\partial}{\partial r} + \frac{i}{r}\,\underline{\alpha} \cdot \underline{\hat{r}}\ \ (\beta K - 1)$$

we obtain coupled equations for g,f, $\frac{dg}{dr}$ and $\frac{df}{dr}$. In our applications we shall only need free particle solutions corresponding to W(r) = 0. Defining:

$$u_1 = rg$$

$$u_2 = rf$$

the coupled equations read:

$$\frac{d}{dr}\begin{bmatrix} u_1 \\ u_2 \end{bmatrix} = \begin{bmatrix} -\frac{\kappa}{r} & (E + M) \\ -(E - M) & +\frac{\kappa}{r} \end{bmatrix} \begin{bmatrix} u_1 \\ u_2 \end{bmatrix}$$

Eliminating u_2 and defining $p^2 = E^2 - M^2$, leads to the second-order equation:

$$\frac{d^2 u_1}{dr^2} + \left[p^2 - \frac{\kappa(\kappa + 1)}{r^2}\right] U_1 = 0$$

This is a standard form of Bessel's equation: the solution regular of the origin is:

$$u_1 = A\ r\ j_\ell\ (pr)$$

where j_ℓ (pr) are spherical Bessel functions. Defining:

$$S_\kappa \equiv \left|\frac{\kappa}{\kappa}\right| \text{ and } \bar{\ell} \equiv \ell_{-\kappa}$$

we arrive at the general form for free particle Dirac spinors in the spherical basis:

$$\psi_\kappa^\mu = A \begin{bmatrix} j_\ell\ (pr)\ \chi_\kappa^\mu \\ \frac{ip}{E+M}\, S_\kappa\ j_{\bar{\ell}}\ (pr)\ \chi_{-\kappa}^\mu \end{bmatrix}$$

For example, consider the solution for $\kappa = -1$. This corresponds to a $j = \frac{1}{2}$ state with $\ell = 0$ and $\bar{\ell} = 1$: we shall refer loosely to this solution as an "$S_{\frac{1}{2}}$" eigenfunction although this only reflects the ℓ-value of the upper components. For our Bag model calculations we shall need $\kappa = -1$, $\kappa = +1$ and $\kappa = -2$ eigenfunctions - $S_{\frac{1}{2}}$, $P_{\frac{1}{2}}$ and $P_{\frac{3}{2}}$ respectively. These are detailed explicitly in the appendix along with more details of the Dirac algebra needed to prove these results.

Bogolioubov's Quark Model

We are now in a position to return to physics and make some connection with the MIT Bag. We consider a quark model proposed by Bogolioubov [9] in 1968, which essentially reduces to confined Dirac particles in a 'scalar' box - i.e. a Lorentz scalar potential. The Dirac equation now reads:

$$E\psi = \left[\underline{\alpha} \cdot \underline{p} + \beta(M + V(r))\right] \psi$$

The form of the solution is as before

$$\psi_{\kappa}^{\mu} = \begin{bmatrix} g(r)\ \chi_{\kappa}^{\mu} \\ if(r)\ \chi_{-\kappa}^{\mu} \end{bmatrix}$$

where f and g now satisfy the coupled equations:

$$\frac{df}{dr} = \frac{(\kappa-1)}{r} f - (E-U(r))g$$

$$\frac{dg}{dr} = (E+U(r))f - \frac{(\kappa+1)}{r} g$$

with

$$U(r) = M + V(r)$$

Note that it is important that we choose a <u>scalar</u> potential so that U(r) acts like an effective mass term. This enables us to confine both positive and negative energy solutions without ambiguity and avoid the Klein paradox [10]. Bogolioubov envisaged that hadrons could be considered as comprising non-interacting quarks moving in an effective scalar potential generated by the quark-quark interactions, much in the same way as one considers the motion of nucleons in the nuclear shell model. The proton in this model is described by three quarks in their lowest energy eigenmode: this is the $j=\frac{1}{2}$, $\kappa=-1$, "$S_{\frac{1}{2}}$" mode.

Consider the following approximation to the potential U(r)

<u>Region I</u>: $r<R_o$: $V(r) = -M$
giving $U(r) = 0$

<u>Region II</u>: $r>R_o$: $V(r) = 0$
giving $U(r) = M$

Inside R_o, the 'potential' compensates the entire mass of the quark: at the end of the calculation we shall impose confinement by taking the limit $M \to \infty$.

What are the allowed energy levels for the quarks? To find these we must derive some eigenvalue condition. This we do in the usual way, by matching the wavefunction at the boundary of regions I and II, and looking for solutions with a finite energy $E<M$. For $\kappa=-1$ solutions we require

$$f = \frac{1}{E+U} \frac{dg}{dr}$$

and with $u_1 = rg$

$$\frac{d^2u_1}{dr^2} + (E^2 - U^2)\, u_1 = 0,\ r \neq R_o$$

Region I: $r<R_o$

We have:

$$\frac{d^2u_1}{dr^2} + E^2u_1 = 0$$

Demanding that g be regular at r = 0 requires

$$u_1 = rg = A \sin Er$$

Region II: $r>R_o$

We now have an equation with real exponential solutions:

$$\frac{d^2u_1}{dr^2} - (M^2 - E^2)\, u_1 = 0$$

Demanding $u_1 \to 0$ as $r \to \infty$ and matching at $r = R_o$, we obtain:

$$u_1 = rg = A \sin ER_o\, e^{-\sqrt{(M^2-E^2)}\,(r-R_o)}$$

The required eigenvalue condition for the allowed energies is now obtained by demanding that

$$f = \frac{1}{E+U}\,\frac{dg}{dr}$$

be continuous at $r = R_o$. This yields

$$\cos ER_o + \frac{\sqrt{1-(E/M)^2}}{1+E/M} \cdot \sin ER_o = \frac{\sin ER_o}{ER_o} \cdot \left[1 - \frac{E}{E+M}\right]$$

We may now take the limit $M \to \infty$, corresponding to complete quark confinement. We obtain

$$\frac{\sin ER_o}{ER_o} = \frac{1}{ER_o}\left[\frac{\sin ER_o}{ER_o} - \cos ER_o\right]$$

or

$$j_o(ER_o) = j_1(ER_o)$$

Notice that the $\kappa = -1$ solutions may be written (for massless quarks) as:

$$\psi^{\mu}_{S\frac{1}{2}} \equiv \psi^{\mu}_{-1} = N_{-1}\begin{bmatrix} j_o\, \chi^{\mu}_{-1} \\ -ij_1\, \chi^{\mu}_{1} \end{bmatrix}$$

and the boundary condition written as a condition on $\psi^{\mu}_{S\frac{1}{2}}$

$$- i \underline{\gamma} \cdot \hat{\underline{r}} \; \psi^{\mu}_{S\frac{1}{2}} = \psi^{\mu}_{S\frac{1}{2}}$$

This is just the linear boundary condition of the spherical cavity approximation to the MIT Bag. In covariant form it becomes:

$$i \; \gamma . n \; \psi^{\mu}_{-1} = \psi^{\mu}_{-1}$$

where

$$n^{\mu} = (0, \hat{\underline{r}})$$

Some comments are in order:

(i) The scalar potential, βV, avoids the Klein paradox and confines quarks and antiquarks (negative energy solutions) separately [10]. This is essentially because the scalar potential is equivalent to matching fermion fields corresponding to different masses. In the limit of complete confinement $\psi = 0$ outside R_o although $\psi \neq 0$ on the boundary.

(ii) We may use this fixed radius model of hadrons to predict the mass of the radially-excited proton state. If we define:

$$E_{n\kappa} = \frac{\omega_{n\kappa}}{R_o}$$

we may solve the boundary condition:

$$j_o \; (\omega_{n-1}) = j_1 \; (\omega_{n-1})$$

for the eigenfrequencies of the $\kappa = -1$ states. The first two positive roots are labelled n = 1, and n = 2 respectively

$$\omega_{1-1} = 2.04 \quad : \quad 1S_{\frac{1}{2}}$$

$$\omega_{2-1} = 5.40 \quad : \quad 2S_{\frac{1}{2}}$$

In our simple quark model, the hadron mass is just the sum of the quark kinetic energies. Thus the proton, with all three quarks in the $1S_{\frac{1}{2}}$ state, has a mass M_p given by

$$M_p = \frac{3\omega_{1-1}}{R_o}$$

Similarly, the first radial excitation, with one quark excited to the $2S_{\frac{1}{2}}$ state, has a mass $M_{p'}$ given by:

$$M_{p'} = \frac{2\omega_{1-1} + \omega_{2-1}}{R_o}$$

The ratio is independent of R_o and has the value:

$$\frac{M_{p'}}{M_p} = 1.55$$

The Roper resonance at 1470 MeV is a good experimental candidate for a radial excitation. Using this mass we obtain:

$$\left(\frac{M_{p'}}{M_p}\right)_{\text{Expt}} = 1.56$$

As Bogolioubov [9] comments, " une telle coincidence est même un peu surprenante"! However, there is more to be said concerning the Roper and the Bag as we shall see later in these lectures.

(iii) Consider the flux of 'quark' quantum numbers across the boundary. The quark number current is given by:

$$J_\mu (x) = \bar{\psi} (x) \gamma_\mu \psi (x)$$

Inside R_o, we find that this current is conserved by virtue of the free Dirac equation:

$$\partial^\mu J_\mu (x) = 0, \quad \text{inside } R_o$$

If, however, we want the charge Q to be conserved

$$Q = \int d^3x \;\; J_o(x)$$

we require

$$n^\mu J_\mu = 0, \text{ on the surface } r = R_o$$

The boundary condition:

$$i \gamma\cdot n \psi = \psi$$

with

$$i \bar{\psi} \gamma\cdot n = - \bar{\psi}$$

implies

$$i n^\mu J_\mu = \bar{\psi} i \gamma\cdot n \psi = \bar{\psi}\psi = - \bar{\psi}\psi$$

so that

$$\left.\begin{aligned} n^\mu J_\mu &= 0 \\ \text{and} \qquad & \\ \bar{\psi}\psi &= 0 \end{aligned}\right\} \quad \text{on the boundary } r = R_o$$

as required for confinement.

(iv) Consider now the flux of momentum across the boundary. The Dirac energy-momentum stress tensor may be written:

$$T_D^{\mu\nu}(x) = \frac{i}{2}\,\bar{\psi}(x)\,\gamma^\mu\partial^\nu\psi(x) - \frac{i}{2}\,(\partial^\nu\,\bar{\psi}(x))\,\gamma^\mu\psi(x)$$

Inside R_o, we have

$$\partial_\mu T_D^{\mu\nu} = 0$$

but the momentum flow through the surface is given by

$$n_\mu T_D^{\mu\nu}\big|_{r=R_o} = \tfrac{1}{2}\bar{\psi}(i\,\gamma\cdot n)\partial^\nu\psi - \tfrac{1}{2}(\partial^\nu\bar{\psi})\,i\,\gamma\cdot n\,\psi\,\Big|_{r=R_o}$$

$$= -\tfrac{1}{2}\,\partial^\nu(\bar{\psi}\psi)\big|_{r=R_o}$$

Since $\bar{\psi}\psi = 0$ on the boundary

$$\partial^\nu(\bar{\psi}\psi)\;\alpha\;n^\nu$$

and we have

$$n_\mu T_D^{\mu\nu} = n^\nu P_D \neq 0$$

where the Dirac pressure P_D is defined by

$$P_D = \tfrac{1}{2}(n\cdot\partial)(\bar{\psi}\psi)$$

Thus in this scalar potential calculation with R_o held fixed, momentum is clearly not conserved at the boundary: no attempt is made to determine R_o dynamically. These considerations lead us naturally on to the MIT Bag model which does attempt to determine the confining surface in a self-consistent manner.

I. EL SACO DEL MIT*

There is currently much optimism that hadron dynamics may be described in terms of quarks and massless vector gluons with interactions prescribed by a local SU(3) colour gauge field theory.[13,14]: this is the scheme of "Quantum Chromo-Dynamics" (QCD). So far all attempts to solve such a theory have been unsuccessful but there are grounds for hope that the solution (if any solution indeed exists) will exhibit two key features:

(i) At large separations the inter-quark forces become so strong that isolated free quarks cannot exist, and

(ii) At short distances the quark-quark interactions become very weak and perturbation theory is applicable.

These properties are usually referred to as "quark-confinement" and "asymptotic freedom", respectively: they may be summarized by an imaginative sketch (Fig. 1.1) of the variation of the running coupling constant of QCD, $\alpha_c = g^2/4\pi$ (g is the quark-gluon coupling), with Q^2, the momentum scale at which we are probing the interactions. This reflects the belief that at large distances (low Q^2)

FIG. 1.1: Hypothetical Variation of QCD Coupling Constant with Q^2

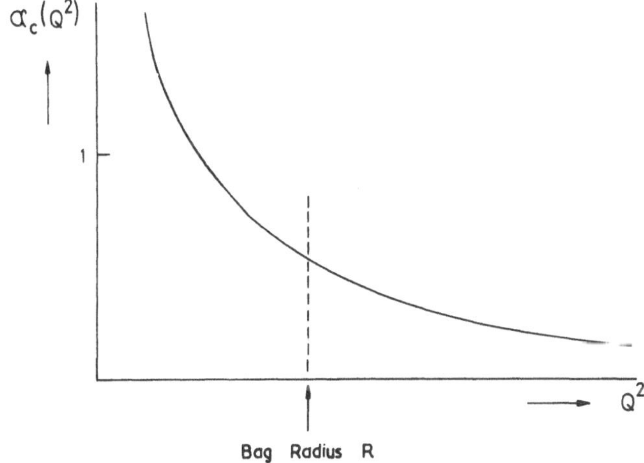

* The MIT Bag.

the interactions are very strong, $\alpha_c >> 1$, while at short distances (high Q^2), the coupling is weak, $\alpha_c << 1$.

The MIT Bag model[1-3] is an attempt to incorporate these features - quarks and glue confined ab initio, with quarks approximately massless and weakly interacting at short distances - in an explicit, relativistic fashion. The Bag, therefore, aims more at a formulation of deep inelastic 'parton' aspects of quarks and current algebra, than at the non-relativistic "massive" quark model of hadron spectroscopy.[15] The bag picture of quark bound states is one in which the quarks are a priori confined to an extended region of space-time, inside which they move almost freely. Outside this region the quarks have effectively infinite mass and zero penetration. A constant energy density is associated with the volume of the "bag" and the bag configuration is determined by the local balance of the field pressure against this "bag pressure". Quark confinement, in the sense that all hadrons are colour singlets, is achieved by demanding that no colour flux, carried by the quarks and gluons, leaves the bag. The MIT Bag theory formulates these ideas in a manifestly Lorentz covariant way. Before writing down the Bag equations, it is worthwhile to elaborate a little on a possible interpretation of the Bag pressure term B, which contributes a term BV to the total energy, where V is the volume of the bag. In Weisskopf's view of the bag model[5] the introduction of some confining region, characterized by a radius R, is the simplest possible way to incorporate the low q^2, long range confining forces. Inside this volume, the interactions between quarks and gluons may be treated perturbatively: this is the picture indicated schematically in fig. 1.1. The precise path from a large quark coupling constant $\alpha_c(Q^2)$ to a Bag pressure term, BV, in the energy is nevertheless somewhat obscure. The viewpoint of these lectures, (though not necessarily shared by the original authors), is that the Bag provides a plausible and tractable approximation to QCD which yields a very interesting picture of the hadron spectrum.

The Bag Equations

In Bogolioubov's fixed boundary model, momentum was not conserved at the boundary of the 'box'. To balance the Dirac field pressure P_D, a universal constant 'Bag pressure' B is introduced. The resulting stress tensor of the hadron has the form:

$$T_H^{\mu\nu} = \theta_B(x)\,(T_D^{\mu\nu} + g^{\mu\nu} B)$$

where:

$$\theta_B(x) = \begin{cases} 1 & \text{'inside' hadron} \\ 0 & \text{'outside' hadron} \end{cases}$$

The θ-function defines the region over which the quark (and later, gluon) fields are non-zero, i.e. the Bag. We have:

$$\frac{\partial}{\partial x_\mu} \theta_B (x) = n^\mu \delta_s (x)$$

Where n^μ is the surface normal and $\delta_s(x)$, a surface δ-function. We find:

$$\partial_\mu T_H^{\mu\nu} = \delta_s (x) n_\mu (T_D^{\mu\nu} + g^{\mu\nu} B)$$

and conservation of momentum at the boundary requires:

$$B = P_D$$

i.e. $B = \frac{1}{2} n.\partial (\bar{\psi}\psi)$ on surface

This equation in principle determines the shape of the surface - the surface variables are functions of the fermion field ψ. This is the essential difference between the problem of a fermion in a rigid 'box' and a soft 'bag'.

We summarize the Bag equations for a free massless Dirac field:

$$T_H^{\mu\nu} = \theta_B (T_D^{\mu\nu} + g^{\mu\nu} B)$$

Inside Bag:		$i\partial\!\!\!/\psi = 0$
On Surface:	(1)	$i\gamma\cdot n\psi = \psi$
	(2)	$\frac{1}{2}n\cdot\partial (\bar{\psi}\psi) = B$

A realistic model of hadrons must of course involve interacting quark fields - this is clearly necessary to achieve confinement of quarks only in colour singlet hadrons. Gluon fields A_a^μ (where a = 1, ..., 8, is the SU(3) colour index) may be introduced and the boundary condition:

$$n_\mu F_a^{\mu\nu} = 0 \qquad \text{on surface}$$

imposed on the gluon field tensor $F_a^{\mu\nu}$

$$F_a^{\mu\nu} = \partial^\mu A_a^\nu - \partial^\nu A_a^\mu + g f_{abc} A_b^\mu A_c^\nu$$

(the f_{abc} are the usual SU(3) structure constants and g is the quark gluon coupling constant). This condition leads directly to the result that only colour singlet hadrons states can exist[4]. In fact, since we shall be mainly interested in the gross features of the spectrum of excited states, we shall continue to discuss the free quark bag equations: the only effect of the gluon interactions that remains in this approximation is the restriction to colour singlet hadron states.

Ground State Baryons in the Round Hadron Approximation[16]

We assume that low-lying hadron states are approximately described by the bag equations appropriate to quarks confined to a rigid spherical cavity of radius R_o. The quadratic boundary condition (2) will be implemented merely as a restriction on the allowed modes in this cavity: the solutions will thus be semi-classical in the sense that we are ignoring the fact that R_o is an operator which may have fluctuations. The radius of the cavity is determined by balancing field pressure and bag pressure B via the quadratic boundary condition. This glueless, rigid spherical cavity model for hadrons will be designated the 'Round Hadron Approximation'.

The linear boundary condition on the quark wavefunction $q_\alpha(\underline{x},t)$ in this approximation reduces to the condition we derived before:

$$- i \underline{\hat{r}} \cdot \underline{\gamma}\, q_\alpha (x) = q_\alpha (x) \text{ at } r = R_o$$

now valid for each quark flavour and colour index α. The general solution is a superposition with coefficients a_α of the central field free Dirac spinors $\psi_{n\kappa jm}(\underline{x},t)$

$$q_\alpha (\underline{x},t) = \sum_{n\kappa jm} a_\alpha (n\kappa jm)\, \psi_{n\kappa jm} (\underline{x},t)$$

The linear boundary condition determines the allowed eigenfrequencies $\omega_{n\kappa}$. In our applications we need only the following solutions:

(i) $\underline{\kappa = -1}$: These are the $j = \frac{1}{2}$ '$S_{\frac{1}{2}}$' eigenmodes, with eigenfunction:

$$\psi_{n-1\frac{1}{2}m} (\underline{x},t) = \frac{N_{n-1}}{\sqrt{4\pi}} \begin{bmatrix} i\, j_o \left(\frac{\omega_{n-1}\, r}{R_o} \right) u_m \\ - j_1 \left(\frac{\omega_{n-1}\, r}{R_o} \right) \underline{\sigma}\cdot\underline{\hat{r}}\, u_m \end{bmatrix} e^{-iE_{n-1}t}$$

where $E_{n\kappa} \equiv \omega_{n\kappa}/R_o$. The linear boundary condition reduces to:

$$j_o(\omega_{n-1}) = + j_1(\omega_{n-1})$$

and the first two eigenfrequencies are

$$\omega_{1-1} = 2.04$$

$$\omega_{2-1} = 5.40$$

(ii) $\underline{\kappa = +1}$: There are $j = \frac{1}{2}$ '$P_{\frac{1}{2}}$' eigenmodes satisfying the linear boundary condition:

$$j_o(\omega_{n+1}) = - j_1 (\omega_{n+1})$$

The lowest eigenfrequency is:

$$\omega_{1+1} = 3.81$$

(iii) $\underline{\kappa = -2}$: These are $j = 3/2$ '$P_{\frac{3}{2}}$' solutions with:

$$j_1(\omega_{n-2}) = + j_2(\omega_{n-2})$$

and lowest frequency:

$$\omega_{1-2} = 3.20$$

By convention, positive n values label the positive energy solutions and negative n, the negative energy solutions, obeying the relation

$$\omega_{n\kappa} = -\omega_{-n-\kappa}$$

The normalization of the eigenfunctions is fixed by requiring

$$\int_{\text{Bag}} d^3x\, \psi^{\dagger}\psi = 1$$

For example, we find for $\kappa = -1$ states

$$N_{n-1} = \left[\frac{\omega_{n-1}^{3}}{2R_o^{3}\,(\omega_{n-1} - 1)\,\sin^2 \omega_{n-1}} \right]^{\frac{1}{2}}$$

We must now implement the quadratic boundary condition for the static spherical cavity:

$$-\tfrac{1}{2} \sum_{\alpha} \frac{\partial}{\partial r} (\bar{q}_\alpha\, q_\alpha) = B \text{ at } r = R_o$$

where B is the bag constant. This constraint results in the following severe restrictions on the allowed modes:

(i) The angular independence of the boundary condition admits only $j = \frac{1}{2}$ solutions with $\kappa = \pm 1$.

(ii) The time independence of the surface requires:

$$\sum_{\alpha} a^*_\alpha (n\kappa j = \tfrac{1}{2}m)\, a_\alpha (n'\kappa' j = \tfrac{1}{2}m') = 0$$

unless:

$$n = n' \qquad \kappa = \kappa'$$

or:

$$n = -n' \qquad \kappa = -\kappa'$$

This constraint is implemented by requiring that in physical baryon states only one mode $(n\kappa)$ is occupied for each internal degree of freedom α. For mesons, the possibility of the $(-n, -\kappa)$ modes in addition is allowed: this is necessary to allow meson states containing quarks and anti-quarks with the same flavour and colour. With these restrictions the quadratic boundary condition reads:

$$\sum_{\alpha n \kappa m} \omega_{n\kappa}\, a^*_\alpha (n\kappa j = \tfrac{1}{2}m)\, a_\alpha (n\kappa j = \tfrac{1}{2}m) = 4\pi B\, R_o^{4}$$

where, for baryons, it is understood that only one a_α $(n\kappa j = \frac{1}{2}m)$ is different from zero for each α. In order to avoid problems with negative energy states - the left hand side of this equation is not positive definite - one must use the Dirac sea and hole hypothesis to associate the negative energy solutions with anti-particles. Invocation of the filled Dirac sea of negative energy states converts the single particle theory to a many particle theory and is essentially equivalent to second quantization. In keeping with this interpretation it is convenient to introduce an operator formalism for the expansion coefficients a_α which incorporates Fermi-Dirac statistics for multi-quark states. We define:

$$a_\alpha (n\kappa jm) \equiv b_\alpha (n\kappa jm) \qquad n>0$$

$$a_\alpha (n\kappa jm) \equiv d_\alpha^\dagger (-n,-\kappa,j,m) \qquad n<0$$

with:

$$\{b_\alpha (n\kappa jm), b_\alpha^\dagger (n\kappa jm)\} = 1$$

$$\{d_\alpha (n\kappa jm), d_\alpha^\dagger (n\kappa jm)\} = 1$$

and all other anticommutators zero. The quark number operator N_α now becomes

$$N_\alpha = \int d^3 x : q_\alpha^\dagger (x)\, q_\alpha (x):$$

$$= \sum_{n\kappa jm} \Big[b_\alpha^\dagger (n\kappa jm)\, b_\alpha (n\kappa jm) - d_\alpha^\dagger (n\kappa jm)\, d_\alpha (n\kappa jm) \Big]$$

Baryon states are now defined by quark creation operators acting on the bag vacuum. For the ground state baryons all three quarks are in the n=1, κ= -1 mode and the mass is obtained by computing the expectation value of the operator

$$P^o = \int d^3x\, (T_D^{oo} + B)$$

The result we obtain is

$$M(R) = \frac{3\omega_{1-1}}{R} \quad + \quad \frac{4\pi}{3} R^3 B$$

Quark kinetic energy — Bag volume energy

where the equilibrium radius R_o must be determined by the quadratic boundary condition. However, it is easy to verify that this condition is equivalent to minimizing M with respect to R:

i.e. $$\frac{\partial M}{\partial R} = 0 \qquad \text{at } R = R_o$$

requires:

$$3\omega_{1-1} = 4\pi B\, R_o^4$$

This is just the constraint imposed by the quadratic boundary condition for this state: it requires that the volume energy BV is one third of the total quark kinetic energy. Notice the difference between this picture of the mass of a hadron and the non-relativistic massive quark model : in the Bag with massless quarks most of the hadron mass is quark kinetic energy. The quadratic boundary condition leads to a direct relation between the mass of the state and the size of the bag:

$$M = 4\ BV = \frac{16\pi}{3}\ B\ R_o^{\ 3}$$

Without the pressure term, it is clear that the system can have no minimum of the energy for a finite bag radius.

Phenomenological Implications

Let us now discuss some of the predictions of this picture. In the present approximation - the 'basic bag' - the nucleon and delta states are degenerate since they both comprise three non-strange '$S_{\frac{1}{2}}$' quark states. If we take an average mass for the N-Δ system (weighted by the spin-isospin degeneracy of the states) to be:

$$\bar{M} \simeq 1200\ \text{MeV}$$

we obtain:

$$B^{1/4} \simeq 120\ \text{MeV}$$

and:

$$R_o^{\ -1} \simeq 140\ \text{MeV} \sim m_\pi$$

These parameters and the bag wavefunctions may now be used to calculate the static parameters of the nucleon - the magnetic moment, charge-radius-squared, and axial-vector coupling constant, and also Bjorken scaling functions.[17,18] For the ratio of magnetic moments of neutron and proton one obtains the famous SU(6) result:

$$\frac{\mu^n}{\mu^p} = -\frac{2}{3}$$

but the Bag is also able to predict the magnitude of the magnetic moment. The "basic bag" described above predicts the proton gyromagnetic ratio to be 2.6 which compares well with the experimental value of ∿ 2.8. For the weak axial vector coupling constant the model predicts:

$$|g_A/g_V| = \frac{5}{3}\ \eta$$

where 5/3 is the usual SU(6) result: in the Bag, η is less than one and calculable. The "basic bag" predicts $|g_A/g_V| \sim 1.1$ which again compares well with the experimental value ∿ 1.2.

Given the level of our approximations - nucleon and delta degenerate - the agreement is remarkable. Instead of discussing all these calculations in detail, we refer the reader to the original papers, and here describe only the magnetic moment application, as a prelude to our later study of photon transitions to excited states.

Magnetic Moment Calculation in the Bag

To calculate the nucleon magnetic moment we must first construct the appropriate ground state wavefunctions. The wavefunction of a three quark state has the following structure:

$$|\psi\rangle = |SU(3)\rangle_{\text{flavour}}\ |SU(2)\rangle_{\text{"spin"}}\ |SU(3)\rangle_{\text{colour}}$$

The overall wavefunction is required to be antisymmetric under exchange of quarks. Since physical baryons are required to be colour singlet states - i.e. antisymmetric in the colour indices - they must be symmetric in "SU(6)" indices, made up of SU(3) flavour and SU(2) "spin" indices. The "spin" degree of freedom in the Bag is actually the total angular momentum of the relativistic spin ½ quarks: with this important proviso the symmetric 'SU(6)' wavefunctions in the Bag, are identical in form to the usual non-relativistic SU(6) wavefunctions. These are constructed in the standard manner which we sketch here: some more details are contained in the Appendix.

For the spin degrees of freedom we have the well-known decomposition:

$$SU(2):\quad \tfrac{1}{2}\otimes\tfrac{1}{2}\otimes\tfrac{1}{2} = \tfrac{3}{2}_s \oplus \tfrac{1}{2}_\alpha \oplus \tfrac{1}{2}_\beta$$

Here the spin 3/2 state is totally symmetric (s) under interchange of quark labels 1, 2 and 3. The two spin ½ states have mixed symmetry (α,β) under interchange of the quarks, roughly corresponding to quarks 2 and 3 coupled to spin 1 (α) and spin 0 (β). Explicitly we have:

$$|\tfrac{1}{2}\tfrac{1}{2}\rangle_\alpha \equiv |\uparrow\uparrow\downarrow\rangle_\alpha = \frac{1}{\sqrt{6}}\{|\uparrow\uparrow\downarrow\rangle + |\uparrow\downarrow\uparrow\rangle - 2|\downarrow\uparrow\uparrow\rangle\}$$

$$|\tfrac{1}{2}\tfrac{1}{2}\rangle_\beta \equiv |\uparrow\uparrow\downarrow\rangle_\beta = \frac{1}{\sqrt{2}}\{|\uparrow\uparrow\downarrow\rangle - |\uparrow\downarrow\uparrow\rangle\}$$

Similarly, we can reduce the products of three quarks into irreducible representations of SU(3) and SU(6):

$$SU(3):\quad \underline{3}\otimes\underline{3}\otimes\underline{3} = \underline{10}_s \oplus \underline{8}_\alpha \oplus \underline{8}_\beta \oplus \underline{1}_A$$

$$SU(6):\quad \underline{6}\otimes\underline{6}\otimes\underline{6} = \underline{56}_s \oplus \underline{70}_\alpha \oplus \underline{70}_\beta \oplus \underline{20}_A$$

Again the symmetry properties of these representations have been emphasized by subscripts: we note the possibility of totally antisymmetric representations for SU(3) and SU(6). (These properties and the decomposition of products of representations are most easily verified by the Young tableaux techniques outlined in the appendix). We now need to combine the spin and SU(3) wavefunctions to produce the wavefunctions for the symmetric $\underline{56}$ representation. One obtains directly the decomposition:

$$\underline{56}_s = {}^4\underline{10} + {}^2\underline{8}$$

where the superscript labels the quark spin multiplicity (2S + 1). Thus the $^4 10$ wavefunctions of the delta decuplet are the product:

$$|\underline{56}_s : {}^4\underline{10}\rangle = |\underline{10}\rangle_s \, |\tfrac{3}{2}\rangle_s$$

while the $^2\underline{8}$ wavefunctions are given by the symmetric combination of α and β states:

$$|\underline{56}_s : {}^2\underline{8}\rangle = \frac{1}{\sqrt{2}} \left[|\tfrac{1}{2}\rangle_\alpha \, |8\rangle_\alpha + |\tfrac{1}{2}\rangle_\beta \, |8\rangle_\beta \right]$$

In this way we obtain explicit SU(6) symmetric wavefunctions. We give two examples:

(i) <u>Nucleon Wavefunctions</u> (Proton : $S_z = \frac{1}{2}$)

$$|\underline{56} \; ; P\uparrow\rangle = \frac{1}{\sqrt{18}} \{2 \left[|u\uparrow u\uparrow d\downarrow\rangle + |u\uparrow d\downarrow u\uparrow\rangle + |d\downarrow u\uparrow u\uparrow\rangle \right] - \left[|u\uparrow d\uparrow u\downarrow\rangle + |u\uparrow u\downarrow d\uparrow\rangle + |d\uparrow u\uparrow u\downarrow\rangle + |d\uparrow u\downarrow u\uparrow\rangle + |u\downarrow d\uparrow u\uparrow\rangle + |u\downarrow u\uparrow d\uparrow\rangle \right] \}$$

In an obvious shorthand we write this as:

$$|\underline{56} \; ; P\uparrow\rangle \equiv \frac{1}{\sqrt{18}} \{2 \left[|u\uparrow u\uparrow d\downarrow\rangle + \text{A.P.} \right] - \left[|u\uparrow d\uparrow u\downarrow\rangle + \text{A.P.} \right] \}$$

where A.P. represents all permutations of quarks inside the relevant bracket.

(ii) <u>Delta Wavefunction (Δ^+; $S_z = \frac{1}{2}$)</u>

$$|\underline{56} : \Delta^+\uparrow\rangle = \frac{1}{\sqrt{9}} \{ \left[|u\uparrow u\uparrow d\downarrow\rangle + \text{A.P.} \right] + \left[|u\uparrow d\uparrow u\downarrow\rangle + \text{A.P.} \right] \}$$

In non-relativistic quark model calculations the ratio of proton to neutron magnetic moments comes directly from assuming that the quark magnetic moment operator $\underline{\mu}$ transforms like $Q\underline{\sigma}$. Here Q is the quark charge matrix:

$$Q = \begin{bmatrix} \frac{2}{3} & 0 & 0 \\ 0 & -\frac{1}{3} & 0 \\ 0 & 0 & -\frac{1}{3} \end{bmatrix}$$

and $\underline{\sigma}$ are the usual Pauli matrices. Using the explicit SU(6) wavefunctions it is easy to verify that:

$$\langle p\uparrow | \sum_i (\sigma_z Q)_i | p\uparrow\rangle = +1$$

$$\langle n\uparrow | \sum_i (\sigma_z Q)_i | n\uparrow\rangle = -\frac{2}{3}$$

where the sum is over all three quark contributions. This leads directly to the famous magnetic moment ratio. One similarly obtains the result $g_A/g_V = 5/3$ by assuming that the weak axial current $\underline{A}^{(3)}$ transforms like $\tau_3\underline{\sigma}$ at the quark level.

In the Bag we may directly calculate these static properties in a consistent way. To calculate the magnetic moment we work in a static, no-recoil approximation and perform a multipole expansion of the Bag electromagnetic current. The magnetic moment operator is the coefficient of the term linear in the photon three momentum : this may be written:

$$\underline{\mu} = \int_{Bag} d^3x \; \tfrac{1}{2} \; \underline{r} \times \{q^\dagger \underline{\alpha} Q q\}$$

where Q is the quark charge matrix as before. For the ground state baryons, we evaluate matrix elements of this operator between states containing three quarks, all in the $1S_{\frac{1}{2}}$ state:

$$<(1S_{\frac{1}{2}})^3 \mid \underline{\mu} \mid (1S_{\frac{1}{2}})^3>$$

Only the diagonal terms corresponding to the $n = 1$ $\kappa = -1$ $1S_{\frac{1}{2}}$ mode survive in the expansion of the quark fields q and $q^\dagger$. We obtain:

$$\underline{\mu} = \frac{R_o}{12} \sum_{\alpha m' m} f(\omega_{1-1}) \; u^\dagger_{m'} \; \underline{\sigma} \; u_m \; b^\dagger_\alpha \; (1\text{-}1m') \; Q \; b_\alpha \; (1\text{-}1m)$$

where:

$$f(\omega_{1-1}) = \frac{4\omega_{1-1} - 3}{\omega_{1-1}(\omega_{1-1} - 1)} = 2.43$$

which results from integrating the wavefunctions over the volume of the bag. It is evident that the factor

$$\sum_{\alpha m' m} u^+_{m'} \; \underline{\sigma} \; u_m \; b^+_\alpha \; (m') \; Q \; b_\alpha(m)$$

sandwiched between nucleon states yields the SU(6) Clebsch-Gordan coefficients of the non-relativistic model. Numerically one obtains[2]:

$$\mu_p = \frac{1}{2M_p} \; 2.6$$

and:

$$\mu_n = - \frac{2}{3} \mu_p$$

in good agreement with experiment.

The difference between the non-relativistic quark spin matrix elements and those of the relativistic quarks of the bag is best illustrated by the expectation value of the quark spin operator. In the bag the one quark matrix element is easily evaluated:

$$\langle 1S_{\frac{1}{2}} : \uparrow | \int_{Bag} d^3x\, q^\dagger \sigma_z q | 1S_{\frac{1}{2}} : \uparrow \rangle$$

$$= 1 - \frac{1}{3} \frac{(2\omega_{1-1} - 3)}{(\omega_{1-1} - 1)} = 0.65$$

The analogous matrix element in the non-relativistic model is of course unity : the difference is caused by the presence of the lower components in the bag wavefunction which are not negligible.

Refinements to the Basic Bag

The Bag model as developed so far has scored some notable successes but is clearly in need of improvement. At present, the nucleon octet and delta decuplet are degenerate, as are the π and ρ nonets in the meson sector. Quark mass and lowest order gluon corrections have been incorporated by the MIT company[3] and other authors[19-21] and we briefly outline these modifications here. No attempt will be made to justify these in detail.

(i) Strange Quark Mass

SU(3) breaking may be introduced by allowing the strange quark to have a mass, m_s, different from zero. The previous analysis of the boundary conditions may be repeated for non-zero mass quarks inside the Bag and the resulting eigenfrequencies for strange quarks are no longer equal to those of the non-strange quarks:

$$\omega_{n\kappa} \text{ (s-quark)} \neq \omega_{n\kappa} \text{ (u,d-quarks)}$$

This will be the origin of some important kinematic mixing effects in excited Y* states when we consider the $[\underline{70}, 1^-]$ resonances in the next section.

(ii) Gluon Corrections

(a) Zero-Point Energy

The zero-point energy of quark and gluon fields confined to a finite cavity is a function of the bag radius. Calculations[3] of the zero-point energy for quarks and gluons confined to a slab-shaped cavity suggested that in addition to an (infinite) renormalization of the Bag constant B, the Hamiltonian should be modified to include a term of the form:

$$\Delta H_o = -\frac{Z}{R}$$

where Z was presumed to be finite and calculable in principle for Bag geometry. However, recent detailed calculations of this effect for a spherical cavity indicate that things are not so simple.[22] At present the above form is assumed and the parameter Z determined by fitting the mass spectrum : the physics of this parameter is now somewhat obscure.

(b) Gluon Exchange Corrections

Gluon interactions are included at lowest order in α_c : the relevant gluon exchange diagram is shown in Fig. 1.2. From their analysis the MIT group conclude[3] that the dominant effect of these corrections is a static magnetic colour-spin

FIG. 1.2: Gluon Exchange Between Quarks i and j

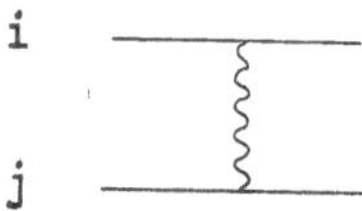

interaction which contributes a term to the energy of the form:

$$\Delta H_g = \frac{8\alpha_c}{3R} \sum_{i>j} \mu_{ij}\, \underline{\sigma}_i \cdot \underline{\sigma}_j$$

where μ_{ij} is a calculable function of the quark modes i and j. This term in the Hamiltonian is responsible for splitting the N and Δ, and the ρ and π. Such a contribution also arises in the potential model of De Rújula, Georgi and Glashow[23] again from single gluon exchange but derived under rather different physical assumptions. The phenomenological validity of such a term and the mechanism for producing the Σ-Λ mass difference has been discussed in detail by Professor Dalitz at this meeting.

The refined version of the Bag model predicts that the masses of hadrons derive from four types of contributions:

$$M = H_{KE} + BV + \Delta H_o + \Delta H_g$$

Quark mass effects are now included in H_{KE}, the total quark kinetic energy. There are four parameters to the model - B,Z,α_c and m_s - which may be determined (for example) by fitting the N,Δ,Ω^- and ω meson masses. One obtains:

$B^{\frac{1}{4}}$ = 145 MeV $\qquad$ α_c = 0.55

Z = 1.84 $\qquad$ m_s = 280 MeV

and the Bag radius varies from R_o^{-1} = 200 MeV for the proton to R_o^{-1} = 300 MeV for the pion. Since this pion radius is not too different from the pion Compton wavelength one should probably treat the pion results with some caution. The resulting mass spectrum for the ground state baryons and mesons is illustrated in Fig. 1.3. Apart from the π (and η and η' mesons not shown on the figure) the

FIG. 1.3: Mass Spectrum for the Ground State Baryons and Mesons in the Refined Bag

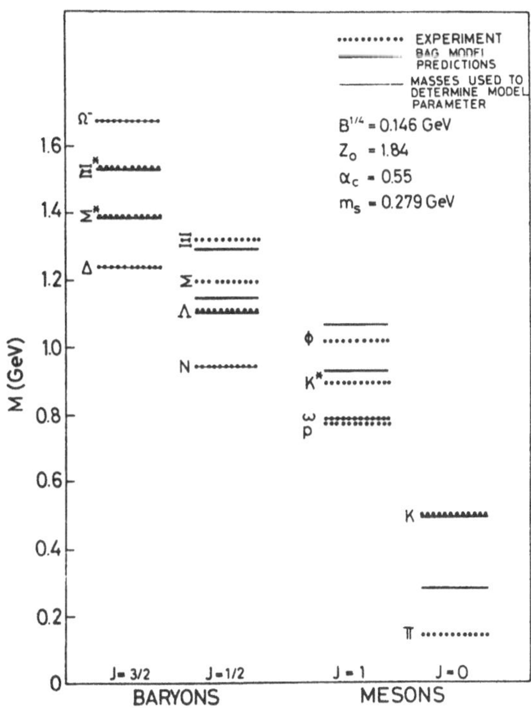

qualitative features of this fit are quite acceptable. The pion has always been a problem for quark models both because of its extreme lightness and because of its dual role as an "almost Goldstone boson" for chiral symmetry. The η and η' mesons are also problematical from the point of view of chiral symmetry and also because they may receive important contributions to their masses from quark-antiquark annihilation processes.[3,23]

The predictions for nucleon magnetic moments in the improved model are disappointing. Inclusion of the lowest order gluon corrections considerably worsens the agreement found for the basic bag model[3,20,21,24] : the situation merits further study.

For detailed discussion of these results and other interesting applications of the Bag, including deep inelastic scattering,[17,18] current algebra sum rules[25,26] and so on, the reader is urged to consult the literature.

2. EXCITED STATES I

In this section we attempt to use the round hadron approximation to examine the first excited states predicted by the Bag. Since the basic parameters have been determined by the fit to the ground state hadrons there is very little freedom for the Bag predictions. In order to set the scene, and for comparison, we first discuss the spectrum of non-relativistic oscillator models and the experimental situation.

Harmonic Oscillator Quark Model Spectrum

For the ground state baryons in the non-relativistic SU(6) scheme, the wavefunctions were easy to construct. All the quarks were in their lowest $\ell = 0$ state and the corresponding spatial wavefunction was symmetric. For colour singlet states this requires that the SU(6) part of the wavefunction is also symmetric under permutation of the quark labels : this leads to the usual $[\underline{56}, L^P = 0^+]$ representation of SU(6) x O(3). It is helpful to indicate the symmetries of each piece of the wavefunction by the corresponding Young Tableaux:

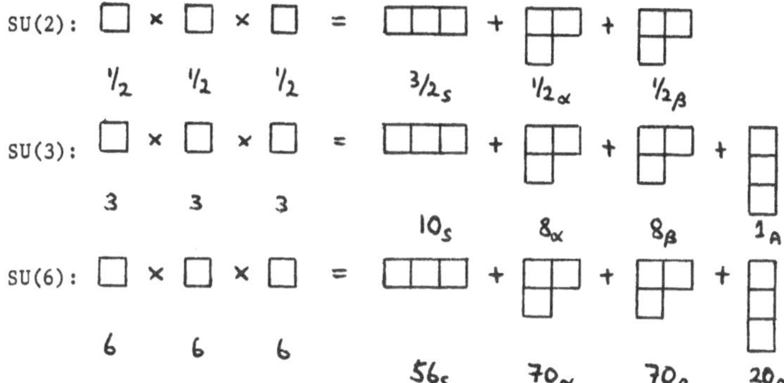

A brief introduction to these techniques is included in the Appendix. (This contains a statement of the Littlewood-Richardson rule for enumerating the allowed tableaux of a given product, and of the "Hook" rule for SU(n) for evaluating the dimensionality of any given tableau.).

In making a parallel between the non-relativistic quark model states and those of the Bag, it is helpful to indicate the permutation symmetry of the O(3) wavefunctions also by Young tableaux. Each quark i is assumed to be in an orbital angular momentum state ℓ, z-component m denoted by $|\ell m\rangle_i$. We consider only the ground state and the first excited state.

(i) L=0 State

All three quarks are in ℓ=0 states:

$$|00\rangle_1 \; |00\rangle_2 \; |00\rangle_3$$

The permutation symmetry of three objects in one possible state is trivially represented by a symmetric (S) tableau:

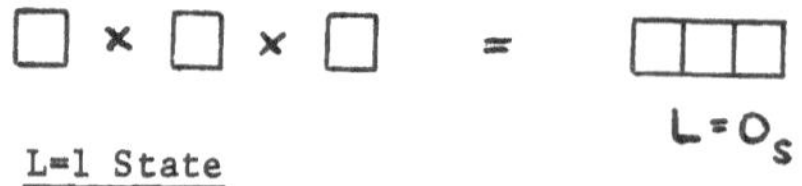

$\square \times \square \times \square = \square\square\square$

$L = 0_S$

(ii) L=1 State

Any one of the three quarks may be excited to an $\ell=1$ state

e.g. $|1m\rangle_1 \; |00\rangle_2 \; |00\rangle_3$

The possible symmetry combinations for three quarks in one of two possible states (i.e. $\ell=0$ or $\ell=1$) are given by:

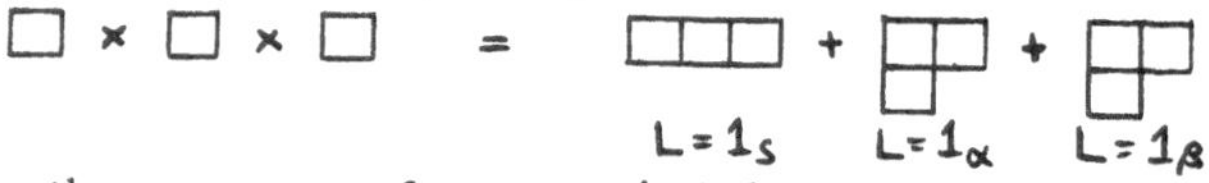

$\square \times \square \times \square = \square\square\square + \begin{smallmatrix}\square\square\\\square\end{smallmatrix} + \begin{smallmatrix}\square\square\\\square\end{smallmatrix}$

$L = 1_S \quad L = 1_\alpha \quad L = 1_\beta$

Notice the occurrence of a symmetric L=1 state. In the harmonic oscillator quark model this corresponds explicitly to excitation of the centre of mass. To see this, consider a simple example of two identical one-dimensional harmonic oscillators. The oscillator co-ordinates x_i, and momenta p_i, are transformed to creation and annihilation operators a_i^+, a_i in the usual way

$$x_1, p_1 \rightarrow a_1^+, a_1$$

$$x_2, p_2 \rightarrow a_2^+, a_2$$

The ground state has both oscillators unexcited:

$$|0\rangle = |0\rangle_1 \; |0\rangle_2$$

The first excited state is doubly degenerate and we may take symmetric and anti-symmetric combinations:

$$|1\rangle_S = \frac{1}{\sqrt{2}} \left\{ |1\rangle_1 \; |0\rangle_2 + |0\rangle_1 \; |1\rangle_2 \right\}$$

$$|1\rangle_A = \frac{1}{\sqrt{2}} \left\{ |1\rangle_1 \; |0\rangle_2 - |0\rangle_1 \; |1\rangle_2 \right\}$$

Now perform a change of variable to centre of mass (CM) and relative (Rel) co-ordinates and corresponding operators

$$\text{CM:} \quad X = \frac{x_1 + x_2}{2} \rightarrow A^+, A$$

$$\text{Rel:} \quad x = x_1 - x_2 \rightarrow a^+, a$$

We find explicitly:

$$|1\rangle_S = |0\rangle_{Rel} \; |1\rangle_{CM}$$

$$|1\rangle_A = |1\rangle_{Rel} \; |0\rangle_{CM}$$

Since we are not interested in CM motion, it is customary to factor this out from the start and always use relative co-ordinates. This is also the case with the three quark, three dimensional harmonic oscillator system : one is left with two independent harmonic oscillators in relative co-ordinates, corresponding to mixed permutation symmetry α and β.[15] The L=1 symmetric state in this model therefore corresponds to excitation of the CM and not to an excitation of the relative motions of the three quarks.

The allowed Baryon states are found by combining the O(3) wavefunctions with those of SU(6) to produce overall symmetric SU(6) x O(3) product wavefunctions. In terms of Young Tableaux we have:-

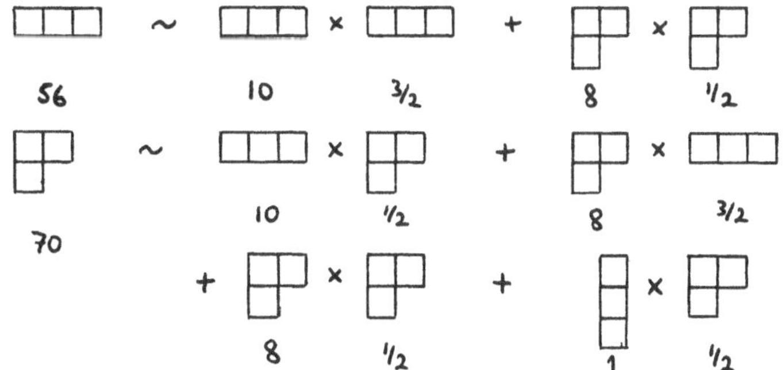

Adding the O(3) spatial part we obtain the following SU(6) x O(3) symmetric multiplets:

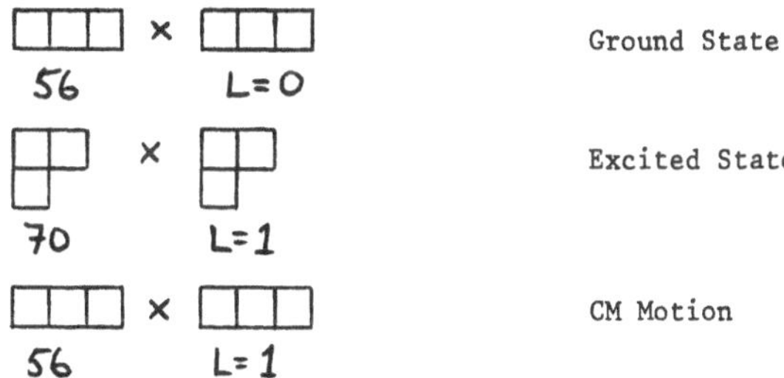

In the harmonic oscillator model, the $[\underline{56},1^-]$ is explicitly an L=1 CM excitation of the ground state $[\underline{56},0^+]$: things are not so simple in the Bag.

Comparison with Experimental Spectrum

It is a remarkable fact that all well-established negative parity resonances below about 1.9 GeV in mass may be accomodated in a $[\underline{70},1^-]$ SU(6) x O(3) classification scheme.[27] Table 2.1 shows the assignment of states deduced from $SU(6)_W$ decay analyses (mixing effects are ignored for the present). All save one of the S=0 states are three or four star rated by the Particle Data Group.[28] For the Y*'s there are some gaps, but one must remember that the Y* phase shift analyses have not yet reached the same level of sophistication as the N* analyses. It is significant that there are no unwelcome 'extra' states that cannot be assigned to this SU(6) multiplet.

TABLE 2.1: Experimental Candidates for $\underline{70},1^-$ States

$^4 8$	$\frac{5}{2}^-$	N* (1670)	Σ* (1770)	Λ* (1800)
	$\frac{3}{2}^-$	N* (?)	Σ* (?)	Λ* (?)
	$\frac{1}{2}^-$	N* (1690)	Σ* (?)	Λ* (1670)
$^2 8$	$\frac{3}{2}^-$	N* (1520)	Σ* (1675)	Λ*(1685)
	$\frac{1}{2}^-$	N* (1515)	Σ* (?)	Λ* (?)
$^2 10$	$\frac{3}{2}^-$	Δ* (1675)	Σ* (?)	-
	$\frac{1}{2}^-$	Δ* (1615)	Σ*(?)	-
$^2 1$	$\frac{3}{2}^-$	-	-	Λ* (1520)
	$\frac{1}{2}^-$	-	-	Λ* (1402)

(?) indicates no three or four star candidate.

What of other SU(6) multiplets? In Table 2.2 we list the predicted multiplet structure of the non-relativistic oscillator quark model up to the N=3 oscillator level.[29] In the most naive version of this model, all multiplets of a given N are

TABLE 2.2: Predicted Multiplet Structure of the Non-Relativistic Oscillator Quark Model.

EXCITATION NUMBER	SU(6) x O(3) MULTIPLETS
N=0	$[\underline{56},0^+]$
N=1	$[\underline{70},1^-]$
N=2	$[\underline{56},2^+]$, $[\underline{70},2^+]$ $[\underline{20},1^+]$ $[\underline{56},0^+]$, $[\underline{70},0^+]$

TABLE 2.2 (Cont'd)

EXCITATION NUMBER	SU(6) x O(3) MULTIPLETS
N=3	$[56,3^-]$, $[\underline{70},3^-]$, $[\underline{20},3^-]$ $[\underline{70},2^-]$ $[\underline{56},1^-]$, $[\underline{70},1^-]$, $[\underline{70},1^-]$, $[\underline{20},1^-]$

degenerate: this degeneracy is lifted in more sophisticated versions of the model which incorporate explicit symmetry breaking terms in the mass operator. One such fit[30] to the experimental mass spectrum is shown in Fig. 2.1.

Fig.2.1. SU(6) Mass Fit to the Hadron Spectrum.

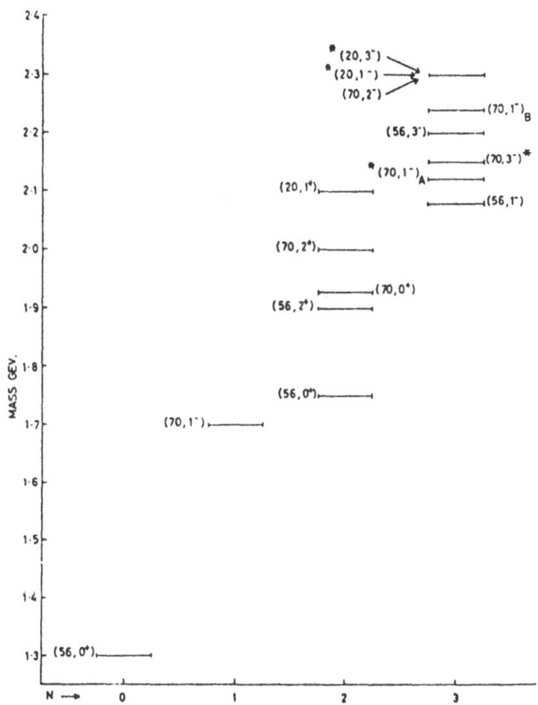

The experimental situation on other SU(6) multiplets may be summarized[31] as follows:

(i) $[\underline{56},0^+]_{N=2}$

This is a radial excitation of the ground state $\underline{56}$ containing $\underline{8}_{\frac{1}{2}}$ and $\underline{10}_{3/2}$ resonances. The (four star rated) Roper resonance NP11 (1470) is a good candidate for this multiplet and there is also a possible Δ partner in the P33 (1690).

(ii) $[\underline{56},2^+]$

Around 2 GeV and below there are many well-established states which may be assigned to this multiplet. Moreover, these assignments are consistent both with $SU(6)_W$ decay analyses[32] and with SU(6) mass fits.[33]

(iii) $[\underline{70},0^+]$ and $[\underline{70},2^+]$

At present there is little evidence for the existence of these multiplets at low masses. There are some possible candidates for these multiplets but, if we accept these as evidence, there remain very many resonances (some with high spin) at comparatively low masses which have apparently remained undetected up to now.[31,33] The question is not closed.

(iv) $[\underline{70},3^-]$ and $[\underline{56},4^+]$

At higher masses, roughly 2.2 to 2.4 GeV, there exist some well-established high spin states which may plausibly be assigned to these multiplets.

(v) $[\underline{56},1^-]$

There is some evidence for a D35 state at about 1900 MeV, though not all recent analyses agree on this.[34] Cutkosky[35] has suggested that this may be evidence for a low mass $[\underline{56},1^-]$ multiplet. (Such a state could also belong to a $[\underline{70},3^-]$ multiplet but has rather low mass for such an assignment). Cutkosky interprets this $[\underline{56},1^-]$ as an excitation of quarks relative to glue: a "Polish Bag" with glue inside and quarks outside! (This is actually not far from the picture we shall arrive at in the MIT Bag). Notice that harmonic oscillator SU(6) mass fits (Fig. 2.1) predict their lowest $[\underline{56},1^-]$ multiplet - a genuine N=3 three quark excitation - to have a mass around 2.1 GeV.

Low Mass, Negative Parity States in the Bag

The MIT Bag model was originally formulated with little reference to baryon spectroscopy. It is clear that the detailed experimental spectrum will provide a severe test of the Bag dynamics or at least of the usefulness of the 'round hadron' approximation to Bag theory.

If we construct negative parity states by exciting a single quark to a 'P-state', the lowest available eigenmodes are, $\kappa = -2$, with $\omega_{1-2} = 3.20$, and $\kappa = +1$, with $\omega_{1+1} = 3.81$. In our crude version of the model the mass of these states is rather low for the physical negative parity states

$(1S_{\frac{1}{2}})^2\ 1P_{3/2} \rightarrow M \sim 1350$ MeV

$(1S_{\frac{1}{2}})^2\ 1P_{1/2} \rightarrow M \sim 1400$ MeV

but gluon and zero-point energy effects have all been ignored in our approximation. We are immediately faced with a problem: the quadratic boundary condition allows only $j=\frac{1}{2}$ modes. The lower lying $j=3/2$ mode cannot satisfy this condition locally over the surface. Minimization of the energy, $\partial M/\partial R = 0$, is merely a statement of global pressure balance. As a first, clearly dangerous, approximation we try closing our eyes and ignoring these states. We must therefore construct symmetric $(S_{\frac{1}{2}})^2\ P_{\frac{1}{2}}$ quark wavefunctions to determine the allowed physical states. Since our quarks are relativistic, we have a (jj) coupling problem rather than the usual (ℓs) scheme. The spatial wavefunction may be decomposed into two parts, "spin" - really the total $j=\frac{1}{2}$ angular momentum of each quark - and "pseudo-spin"[36] - really the permutation freedom of arranging three quarks in two possible states, $S_{\frac{1}{2}}$ or $P_{\frac{1}{2}}$.

"Spin"

For three quarks, each with $j=\frac{1}{2}$, we form the usual product:

$$\square \times \square \times \square = \square\square\square + \begin{smallmatrix}\square\square\\ \square\ \end{smallmatrix} + \begin{smallmatrix}\square\square\\ \square\ \end{smallmatrix}$$

$$3/2_S \qquad 1/2_\alpha \qquad 1/2_\beta$$

"Pseudospin"

For three quarks, each in one of two possible states, S or P, the symmetry combinations are given by:

$$\square \times \square \times \square = \square\square\square + \begin{smallmatrix}\square\square\\ \square\ \end{smallmatrix} + \begin{smallmatrix}\square\square\\ \square\ \end{smallmatrix}$$

$$M_S \qquad M_\alpha \qquad M_\beta$$

We now combine these to obtain spatial wavefunctions with definite symmetry:

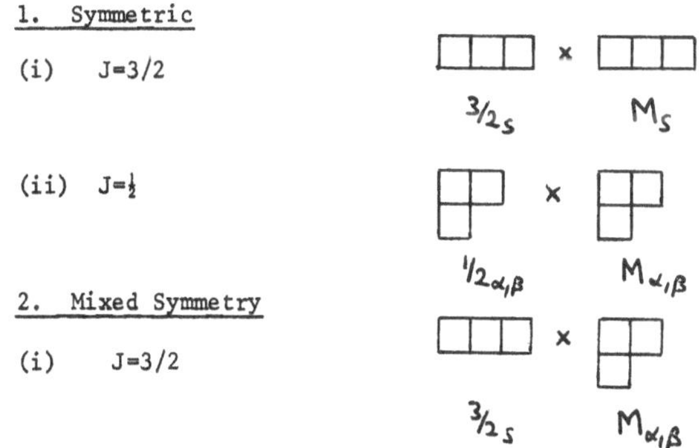

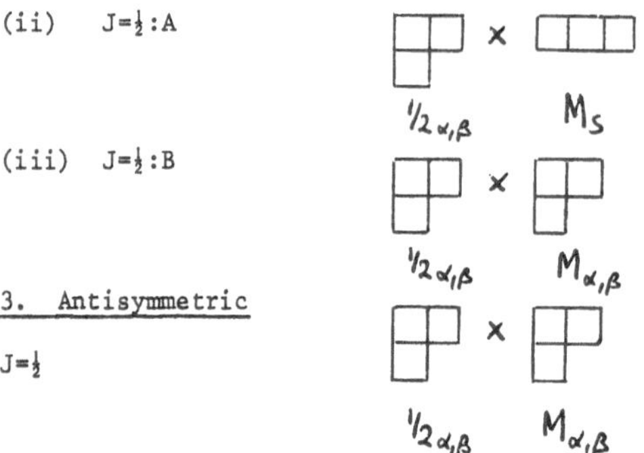

These are now combined with the usual SU(3) wavefunctions to yield overall symmetric (jj) coupled wavefunctions:

e.g. $\underline{10}_{\frac{1}{2}}$ $\qquad 10_S \times \left\{ \frac{1}{2}_{\alpha,\beta} \times M_{\alpha,\beta} \right\}_S$

In this way we find the allowed $(S_{\frac{1}{2}})^2\ P_{\frac{1}{2}}$ states:

$$(S_{\frac{1}{2}})^2\ P_{\frac{1}{2}} : \begin{cases} 10_{\frac{1}{2}}\ ,\ 10_{3/2} \\ 8_{\frac{1}{2}}\ ,\ 8_{\frac{1}{2}}\ ,\ 8_{3/2} \\ 1_{\frac{1}{2}} \end{cases}$$

This is to be compared with the observed states of the $[70,1^-]$:

$$[\underline{70},1^-] \begin{cases} 10_{\frac{1}{2}},\ 10_{3/2} \\ 8_{\frac{1}{2}},\ 8_{\frac{1}{2}},\ 8_{3/2},\ 8_{3/2},\ 8_{5/2} \\ 1_{\frac{1}{2}},\ 1_{3/2} \end{cases}$$

It is clear that the approximation of ignoring the $P_{3/2}$ states altogether is not viable - this was the origin of the bad disagreement observed in first applications of the Bag to excited states.[25,36]

As a second attempt, we try including the $(S_{\frac{1}{2}})^2\ P_{3/2}$ states even though these states only satisfy the quadratic boundary condition in an average sense. The counting problem for these $(S_{\frac{1}{2}})^2\ P_{3/2}$ states is more complicated since there are now two j=½ quarks and one j=3/2 quark. The two 'spin' ½ quarks can be combined symmetrically to spin 1, and antisymmetrically to spin 0: the j=3/2 quark must then be added and appropriate overall symmetry combinations formed. We quote the result:[36-38]

$$(S_{\frac{1}{2}})^2\ P_{3/2} : \begin{cases} 10_{\frac{1}{2}},\ 10_{3/2},\ 10_{5/2} \\ 8_{\frac{1}{2}},\ 8_{3/2},\ 8_{3/2},\ 8_{5/2} \\ 1_{3/2} \end{cases}$$

If we combine these with the $(S_{\frac{1}{2}})^2\ P_{\frac{1}{2}}$ states we now have too many states for the $[\underline{70},1^-]$. In fact, together, they contain the same states as a $[\underline{70},1^-]$ and a $[56,1^-]$ multiplet.

TABLE 2.3: Negative Parity Excited States in (jj) and (LS) Schemes

(a) (jj) Coupling

$(S_{\frac{1}{2}})^2\ P_{\frac{1}{2}}$: $\underline{10}_{\frac{1}{2}}$; $\underline{10}_{3/2}$
$\underline{8}_{\frac{1}{2}}$; $\underline{8}_{\frac{1}{2}}$; $\underline{8}_{3/2}$
$\underline{1}_{\frac{1}{2}}$

$(S_{\frac{1}{2}})^2\ P_{3/2}$: $\underline{10}_{\frac{1}{2}}$; $\underline{10}_{3/2}$; $\underline{10}_{5/2}$
$\underline{8}_{\frac{1}{2}}$; $\underline{8}_{3/2}$; $\underline{8}_{3/2}$; $\underline{8}_{5/2}$
$\underline{1}_{3/2}$

(b) LS Coupling

$[\underline{70},1^-]$: $\underline{10}_{\frac{1}{2}}$; $\underline{10}_{3/2}$
$\underline{8}_{\frac{1}{2}}$; $\underline{8}_{\frac{1}{2}}$; $\underline{8}_{3/2}$; $\underline{8}_{3/2}$; $\underline{8}_{5/2}$
$\underline{1}_{\frac{1}{2}}$; $\underline{1}_{3/2}$

$[\underline{56},1^-]$: $\underline{10}_{\frac{1}{2}}$; $\underline{10}_{3/2}$; $\underline{10}_{5/2}$
$\underline{8}_{\frac{1}{2}}$; $\underline{8}_{3/2}$

It is clear that the Bag has three genuine quark degrees of freedom since all three quarks may move independently relative to the fixed cavity. Unlike the harmonic oscillator model, it is not easy to identify a CM excitation of the quarks since we are considering a fixed cavity: to attack this problem it is necessary to consider motion of the surface.

Rebbi's Calculations

Rebbi[38] has considered the problem of small surface oscillations about the S-wave spherical cavity configuration. The problem is mathematically rather involved and he is forced to make several technical approximations in order to obtain a tractable Hamiltonian. In particular, one of his assumptions is that we are considering small fluctuations about a basically spherical equilibrium situation: the case when one quark is excited and the remaining (n-1) quarks are in the S-wave ground state, with n>>1. For the physical situation, n=3. Nevertheless,

Rebbi finds results which are physically appealing: namely, there is a zero-frequency mode of the small oscillation Hamiltonian. This corresponds to uniform translation of quarks and surface with no deformation and no restoring force. In the small oscillation approximation, one expects (and finds) that this translation mode contains just the states required for a $[\underline{56},1^-]$ multiplet, and furthermore, it is made up mainly from the lowest $(1P_{\frac{1}{2}})\,(1S_{\frac{1}{2}})^2$ and $(1P_{3/2})\,(1S_{\frac{1}{2}})^2$ configurations with only a small admixture of more highly excited modes. The first physical $[\underline{56},1^-]$ multiplet is predicted to exist at rather low masses, probably less than 2 GeV. The resulting picture is indicated in Fig. 2.2. The masses of the states

Fig.2.2. Static and Small Deformation Spectra of the Bag (after Jaffe ref.8)

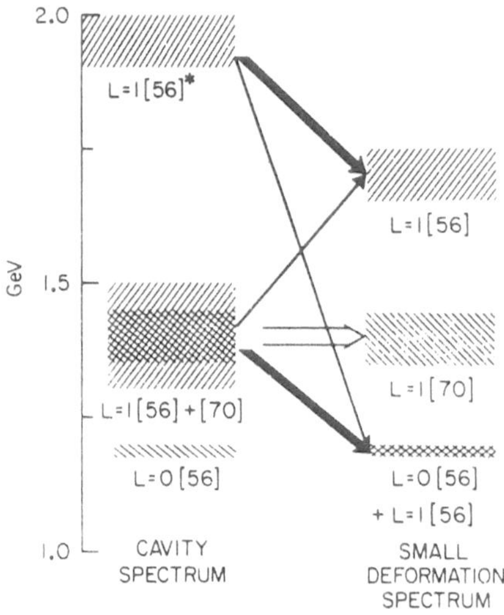

corresponding to the $[\underline{70},1^-]$ multiplet are largely unaffected by the small oscillations of the surface.

The $[\underline{70},1^-]$ Multiplet in the Bag

Rebbi's calculations suggest an obvious approximation for the $[\underline{70},1^-]$ states in the Bag. DeGrand[37] explicitly projects out the spurious $[\underline{56},1^-]$ component of the $(S_{\frac{1}{2}})^2\,P_{\frac{1}{2}}$ and $(S_{\frac{1}{2}})^2\,P_{3/2}$ modes: the remaining states then correspond to a good approximation to the physical low-lying negative parity $[\underline{70},1^-]$states. Obtaining the explicit $[\underline{70},1^-]$ wavefunctions is a tedious exercise in transforming from a jj-coupling scheme to an LS one: the results are displayed in Table 2.4.

TABLE 2.4: Transformation Matrices Between (jj) and (LS) Bases

$$|\underline{10},\ 5/2^-;\ P_{3/2}\rangle = |\underline{56};\ {}^4\underline{10}_{5/2}\rangle$$

$$|\ \underline{8},\ 5/2^-;\ P_{3/2}\rangle = |\underline{70};\ {}^4\underline{8}_{5/2}\ \rangle$$

$$\begin{bmatrix} |\underline{10},\ 3/2^-;\ P_{3/2}\rangle \\ |\underline{10},\ 3/2^-;\ P_{\frac12}\ \rangle \end{bmatrix} = \begin{bmatrix} -2/3 & +\frac{\sqrt5}{3} \\ +\frac{\sqrt5}{3} & 2/3 \end{bmatrix} \begin{bmatrix} |\underline{56};\ {}^4\underline{10}_{3/2}\rangle \\ |\underline{70};\ {}^2\underline{10}_{3/2}\rangle \end{bmatrix}$$

$$\begin{bmatrix} |\underline{10},\ \tfrac12^-;\ P_{3/2}\ \rangle \\ |\underline{10},\ \tfrac12^-;\ P_{\frac12}\ \rangle \end{bmatrix} = \begin{bmatrix} \frac13 & -\frac{\sqrt8}{3} \\ -\frac{\sqrt8}{3} & \frac13 \end{bmatrix} \begin{bmatrix} |\underline{56};\ {}^4\underline{10}_{\frac12}\rangle \\ |\underline{70};\ {}^2\underline{10}_{\frac12}\rangle \end{bmatrix}$$

$$\begin{matrix} |\underline{1},\ 3/2^-;\ P_{3/2}\ \rangle \\ |\underline{1},\ \tfrac12^-;\ P_{\frac12}\ \rangle \end{matrix} = \begin{matrix} |\underline{70};\ {}^2\underline{1}_{3/2}\rangle \\ |\underline{70};\ {}^2\underline{1}_{\frac12}\ \rangle \end{matrix}$$

$$\begin{bmatrix} |\underline{8},\ 3/2^-\ (S=1)\ P_{3/2}\rangle \\ |\underline{8},\ 3/2^-;(S=0)\ P_{3/2}\rangle \\ |\underline{8},\ 3/2^-;\ P_{\frac12}\ \rangle \end{bmatrix} = \begin{bmatrix} \sqrt{\frac{5}{18}} & -2/3 & \sqrt{\frac{5}{18}} \\ -\sqrt{\frac12} & 0 & \sqrt{\frac12} \\ \frac{\sqrt2}{3} & \frac{\sqrt5}{3} & \frac{\sqrt2}{3} \end{bmatrix} \begin{bmatrix} |\underline{56};\ {}^2\underline{8}_{3/2}\rangle \\ |\underline{70};\ {}^4\underline{8}_{3/2}\rangle \\ |\underline{70};\ {}^2\underline{8}_{3/2}\rangle \end{bmatrix}$$

$$\begin{bmatrix} |\underline{8},\ \tfrac12^-;\ (S=1)\ P_{3/2}\ \rangle \\ |\underline{8},\ \tfrac12^-;\ A\ P_{\frac12}\ \rangle \\ |\underline{8},\ \tfrac12^-;\ B\ P_{\frac12}\ \rangle \end{bmatrix} = \begin{bmatrix} -2/3 & +1/3 & -2/3 \\ +1/3 & -2/3 & -2/3 \\ -2/3 & -2/3 & +1/3 \end{bmatrix} \begin{bmatrix} |\underline{56};\ {}^2\underline{8}_{\frac12}\rangle \\ |\underline{70};\ {}^4\underline{8}_{\frac12}\rangle \\ |\underline{70};\ {}^2\underline{8}_{\frac12}\rangle \end{bmatrix}$$

We should stress at this point that although we have used, and shall continue to use, non-relativistic SU(6) terminology to describe these multiplets, the SU(6) of the Bag is not the usual static SU(6). The relation, if any, between the SU(6) structure of the Bag and other theoretical SU(6) schemes - such as the Melosh null-plane $SU(6)_W$ [39] - is an interesting problem.

DeGrand[37] has looked in some detail at the phenomenological consequences for the $[\underline{70},1^-]$ states in the Bag. He uses the appropriate combinations of $(S_{\frac12})^2\ P_{\frac12}$ and $(S_{\frac12})^2\ P_{3/2}$ wavefunctions and also incorporates gluon exchange corrections. In this way he is able to predict both the masses and mixing (between states of the same spin-parity) of all the states. In order to raise the average mass predicted for the $\underline{70}$ to a more reasonable value, he considers the possiblity that the zero-point energy paramater Z may be different from its value for ground state hadrons. Fitting Z_{70} to the mass of the ND15 (1670) resonance he obtains the spectrum shown in Fig. 2.3. Quantitatively, the model does not do too well but it must be remembered that this is not intended as a best fit for the parameter Z_{70}. Moreover, if one disregards the view that B is a universal parameter fixed in some fundamental Lagrangian, it may in fact be more reasonable to allow the Bag presssure B to be different for these states.[40] The observed masses, and the mixing of pure 'SU(6)' states derive from three sources:

(i) the $[\underline{70},1^-]$ states are mixtures of the $P_{\frac{1}{2}}$ and $P_{3/2}$ eigenmodes.

(ii) the $P_{\frac{1}{2}}$ and $P_{3/2}$ wavefunctions themselves may undergo 'kinematic mixing' (see below).

(iii) there is SU(6) breaking due to gluon exchange corrections.

Fig.2.3: The $[\underline{70},1^-]$ Spectrum of the Bag (ref.37).

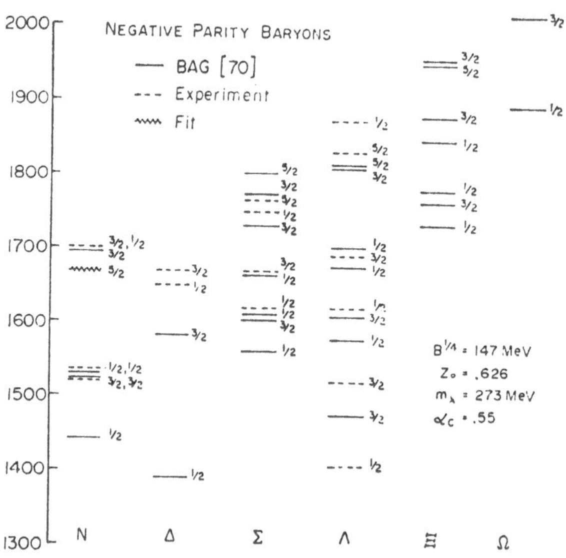

The actual numerical agreement between the predicted Y* mixing matrices and those obtained from $SU(6)_W$ decay analyses[32] or from SU(6) fits to the mass spectrum,[33] is again not spectacularly good. However, in view of the numerous approximations that have been made, it seems to us remarkable that there is even some qualitative agreement between observed and predicted masses and mixing matrices. It is a considerable achievement to have extracted even qualitatively reasonable wavefunctions for the $[\underline{70},1^-]$ from a model with relativistic quarks. Rather than discuss the predictions in, what seems to us, inappropriate detail, we end this section by highlighting a rather nice feature of the model in providing a simple mechanism for generating mixing between Y* resonances.

Kinematic Mixing for Y*'s[36]

In the Bag model, the dominant mechanism for inducing mixing between Y* resonances of the same spin-parity is purely kinematic. This may be illustrated by considering an $(S_{\frac{1}{2}})^2 P_{\frac{1}{2}}$ Σ-state wavefunction. We consider a $\Sigma^+(\underline{10}, J=3/2, J_z=3/2)$ state, built from two u quarks and one strange λ quark[+]:-

$$\Sigma^+(\underline{10}_{3/2}) = \frac{1}{3}\left\{\left[u_{S\uparrow}\, u_{S\uparrow}\, \lambda_{P\uparrow} + \text{A.P.}\right] + \left[u_{S\uparrow}\, \lambda_{S\uparrow}\, u_{P\uparrow} + \text{A.P.}\right]\right\}$$

We define:

$$\Sigma_1 = \frac{1}{\sqrt{3}}\,(u_{S\uparrow}\, u_{S\uparrow}\, \lambda_{P\uparrow} + \text{A.P.})$$

$$\Sigma_2 = \frac{1}{\sqrt{6}}\,(u_{S\uparrow}\, \lambda_{S\uparrow}\, u_{P\uparrow} + \text{A.P.})$$

and rewrite the Σ wavefunction as

$$\Sigma^+(\underline{10}_{3/2}) = \sqrt{\frac{1}{3}}\,\Sigma_1 + \sqrt{\frac{2}{3}}\,\Sigma_2$$

The orthogonal combination corresponds to the octet Σ^+ state:

$$\Sigma^+(\underline{8}_{3/2}) = \sqrt{\frac{2}{3}}\,\Sigma_1 - \sqrt{\frac{1}{3}}\,\Sigma_2$$

Since the strange quark has non-zero mass, these $\Sigma(\underline{10})$ and $\Sigma(\underline{8})$ states are <u>not</u> diagonal in quark kinetic energy. In fact:

$$H_{KE}\,\Sigma_1 = \frac{1}{R_o}\left[2\omega_S(0) + \omega_P(m_\lambda)\right]\Sigma_1$$

and

$$H_{KE}\,\Sigma_2 = \frac{1}{R_o}\left[\omega_S(0) + \omega_S(m_\lambda) + \omega_P(0)\right]\Sigma_2$$

These diagonal states, Σ_1 and Σ_2, are clearly mixtures of $\Sigma(\underline{10})$ and $\Sigma(\underline{8})$ states. As Jaffe and DeGrand remark, this mechanism is independent of the bag language.

[+] We apologise for the hybrid notation (u,d,λ) for quarks but wish to reserve S and P to label the different quark eigenmodes.

3. PHOTOPRODUCTION OF THE $[\underline{70},1^-]$

One of the favourite testing grounds for conventional quark models has been (and remains) the photoproduction of baryon resonances, in particular the negative parity resonances of the $[\underline{70},1^-]$

$$\begin{array}{ccccc} \gamma & + & N & \rightarrow & N^* \\ & & [\underline{56},0^+] & & [\underline{70},1^-] \end{array}$$

The usual oscillator quark models[41-43] assume this process proceeds via single-quark excitation of a quark in the ground state nucleon. Since the quark motion is usually assumed to be non-relativistic in these models, a non-relativistic reduction of the quark electromagnetic current is made. One obtains an electromagnetic current operator containing three terms:

$$J^{em}_{\pm} \sim AL_{\pm} + BS_{\pm} + CS_zL_{\pm}$$

where the subscripts $\pm$ refer to the photon helicity. The first and second terms correspond to orbital excitation and spin flip of the quark, respectively. The third operator corresponds to a spin-orbit type term and is usually neglected because of the non-relativistic assumptions of the models. Such models expect:

$$C << A,B$$

and in fact in most analyses C is set to zero. The parameters A and B are directly related to the parameters of the potential. Recently, stimulated by the work of Melosh[39] and others[44] it has proved useful to discuss these transitions in terms of their general $SU(6)_W$ structure rather than in terms of any specific model. In this approach, one merely assumes the most general algebraic structure for a single-quark transition operator. For excitation of the $[\underline{70},1^-]$, this amounts to allowing all three terms A,B,C, to be present, but with no prejudice as to the relative importance of the various terms. It may be, of course, that experiment reflects the non-relativistic model assumptions and one finds phenomenologically that C is negligible. In fact this is not the case: this term is of greater or comparable importance to the spin-flip B term.[32] This may be seen by examining the predictions for the best known resonance, the D13 (1520). We assume (as is indicated by $SU(6)_W$ π Decay analyses and SU(6) mass fits) that this resonance is almost a pure $[\underline{70},1^-]$ $^28_{3/2}$ state of SU(6) x O(3).

Consider the helicity 3/2 excitation amplitude, $A_{3/2}$, corresponding to the matrix element:

$$A_{3/2} \sim \langle N^* \lambda' = 3/2 | J^{em}_+ | N \lambda = \tfrac{1}{2} \rangle$$

The $SU(6)_W$ predictions for this amplitude are simple:

$$A^p_{3/2} = A + C \qquad \text{from proton targets (p)}$$

$$A^n_{3/2} = -A - \frac{1}{3} C \qquad \text{from neutron targets (n)}$$

in a suitable normalization. Simple quark models with C=0, therefore predict these amplitudes to be equal in magnitude

$$|A^p_{3/2}| = |A^n_{3/2}|$$

The PDG 'world average'[28] for these amplitudes is:

$$A^p_{3/2} = 171 \pm 15$$

$$A^n_{3/2} = -129 \pm 10$$

Despite the fact this it is not clear precisely what reliance to place on the 'errors' of the different analyses, it is reasonable to conclude that C is non-zero: these amplitudes suggest that A:C $\sim$ 9:5. In fact a best fit to all the available data with the predicted algebraic structure of Table 3.1 leads to the result:

$$A : B : C \sim 8 : 2 : 5$$

TABLE 3.1:

$SU(6)_W$ Structure for $[\underline{70}, 1^-]$ Photoproduction in the Melosh Parameterization

SU(6) State	Target	$A_{3/2}$	$A_{1/2}$
$^2 8_{1/2}$	p	-	$+\frac{1}{6} A + \frac{1}{6} B - \frac{1}{6} C$
	n	-	$-\frac{1}{6} A - \frac{1}{18} B + \frac{1}{18} C$
$^4 8_{1/2}$	p	-	0
	n	-	$-\frac{1}{18} B + \frac{1}{18} C$
$^2 8_{3/2}$	p	$+\frac{1}{2}\sqrt{\frac{1}{6}} A + \frac{1}{2}\sqrt{\frac{1}{6}} C$	$+\frac{1}{6}\sqrt{\frac{1}{2}} A - \frac{1}{3}\sqrt{\frac{1}{2}} B - \frac{1}{6}\sqrt{\frac{1}{2}} C$
	n	$-\frac{1}{2}\sqrt{\frac{1}{6}} A - \frac{1}{6}\sqrt{\frac{1}{6}} C$	$-\frac{1}{6}\sqrt{\frac{1}{2}} A + \frac{1}{9}\sqrt{\frac{1}{2}} B + \frac{1}{18}\sqrt{\frac{1}{2}} C$
$^4 8_{3/2}$	p	0	0
	n	$-\frac{1}{6}\sqrt{\frac{3}{5}} B + \frac{1}{3}\sqrt{\frac{1}{15}} C$	$-\frac{1}{18}\sqrt{\frac{1}{5}} B + \frac{2}{9}\sqrt{\frac{1}{5}} C$
$^4 8_{5/2}$	p	0	0
	n	$+\frac{1}{6}\sqrt{\frac{2}{5}} B + \frac{1}{6}\sqrt{\frac{2}{5}} C$	$+\frac{1}{6}\sqrt{\frac{1}{5}} B + \frac{1}{6}\sqrt{\frac{1}{5}} C$
$^2 10_{1/2}$	p/n	-	$+\frac{1}{6} A - \frac{1}{18} B + \frac{1}{18} C$
$^2 10_{3/2}$	p/n	$+\frac{1}{2}\sqrt{\frac{1}{6}} A - \frac{1}{6}\sqrt{\frac{1}{6}} C$	$+\frac{1}{6}\sqrt{\frac{1}{2}} A + \frac{1}{9}\sqrt{\frac{1}{2}} B + \frac{1}{18}\sqrt{\frac{1}{2}} C$

From this we must conclude that the C term is <u>not</u> a small relativistic correction to a basically non-relativistic situation. Given this, it is of interest to examine these photon transitions in a model with light, relativistic confined quarks, and the Bag is the only tractable model of this type.

<u>Photoproduction of the Δ in the Bag</u>

Before we consider the $[\underline{70},1^-]$ case, it is useful to practice on the $\gamma N \to \Delta$ ground state transition. In general, resonance photoproduction may be parameterized by two helicity amplitudes:

$$A_{\frac{1}{2}} = \frac{1}{\sqrt{2q}} \langle N^*; \tfrac{1}{2} | \underline{\varepsilon}^{(+1)} \cdot \underline{J}^{em} | N; -\tfrac{1}{2} \rangle$$

describing the absorption of a helicity +1 photon to produce a $\lambda=\frac{1}{2}$ final state, and, for resonances with $J \geq 3/2$

$$A_{3/2} = \frac{1}{\sqrt{2q}} \langle N^*; 3/2 | \underline{\varepsilon}^{(+1)} \cdot \underline{J}^{em} | N; +\tfrac{1}{2} \rangle$$

for a helicity 3/2 final state. Here $\underline{J}^{em}$ is the electromagnetic current, q the photon three momentum and $\underline{\varepsilon}^{(+1)}$ the polarization vector for a helicity +1 photon:

$$\underline{\varepsilon}^{(+1)} = -\frac{1}{\sqrt{2}} (1, +i, 0)$$

In the Bag model we are unable (as yet) to treat the problem of moving bag states in a covariant manner and all transition amplitudes must be evaluated in a static approximation. The electromagnetic current operator in the bag may be written:

$$\underline{\varepsilon}^{(+1)} \cdot \underline{J}^{em} = \int_{Bag} d^3x \, \underline{\varepsilon}^{(+1)} \cdot \psi^\dagger \underline{\alpha} Q \psi e^{i\underline{q}\cdot\underline{r}}$$

where the quark fields have the usual expansion in terms of bag eigenstates as in our calculation of the nucleon magnetic moment. In the static approximation we sandwich this operator between the relevant Bag wavefunctions and develop the current operator in a multipole expansion:

$$e^{i\underline{q}\cdot\underline{r}} = j_o(qr) + 3i\, j_1(qr)\, P_1(\cos\theta) - 5 j_2(qr)\, P_2(\cos\theta) + \ldots$$

For transitions within the $\left[\underline{56},0^+\right]$ multiplet only the diagonal terms in the quark fields corresponding to $(1S_{\frac{1}{2}})$ states are retained. Since the Bag current is a single-quark operator for the N-Δ transition we are concerned with the single quark matrix element

$$\langle S_{\frac{1}{2}\frac{1}{2}} | \underline{\varepsilon}^{(+1)} \cdot \underline{J}^{em} | S_{\frac{1}{2}-\frac{1}{2}} \rangle$$

This selects only the magnetic dipole (M1) piece of the current operator

$$M1 = \int_{Bag} d^3x \, (3i\, j_1(qr)) \left[\psi^\dagger \underline{\varepsilon}^{(+1)} \cdot \underline{\alpha} Q \psi P_1(\cos\theta)\right]_{J=1}$$

Following this through, one finds that the Bag reproduces one of the standard quark model successes in predicting that there should be no electric quadrupole (E2) excitation of the Δ:

$$A^{\gamma N\Delta}_{3/2} = \sqrt{3}\, A^{\gamma N\Delta}_{\frac{1}{2}} \qquad \text{No E2 condition.}$$

The results obtained are structurally identical to that of the Melosh algebraic approach in which only the spin-flip B term can contribute between L=0 states. The connection with this formalism and the original calculation of the nucleon magnetic moment[2] is made evident by expanding the M1 operator in powers of q and re-arranging terms. We find:

$$[\psi^\dagger \underline{\varepsilon}^{(+1)} \cdot \underline{\alpha} Q\psi \; 3j_1(qr)\; P_1(\cos\theta)]_{J=1} = \tfrac{1}{2}\psi(\underline{q}\mathrm{x}\underline{\varepsilon})\cdot(\underline{r}\mathrm{x}\underline{\alpha})Q\psi + O(q^3)$$

For the γN-Δ transition, the 'No E2 condition' is in good agreement with experiment. The Bag also predicts the magnitudes of the individual amplitudes[25] (Table 3.2) and

TABLE 3.2: Photoproduction of the Δ(1236)

	Bag[25]	Experiment[28]
$A_{\frac{1}{2}}$	-102	-141 ± 3
$A_{3/2}$	-176	-259 ± 5

we see that there is reasonable qualitative agreement. Can we do as well for the $[\underline{70},1^-]$?

The $[\underline{70},1^-]$ Photo-Transitions in the Bag[45]

For transitions to these excited states we must keep the off-diagonal terms connecting $S_{\frac{1}{2}}$ states to $P_{\frac{1}{2}}$ and $P_{3/2}$ states. Since the current acts at the quark level we see immediately that there are three basic amplitudes:

$S_{\frac{1}{2}} \rightarrow P_{\frac{1}{2}}$: E1 only allowed

$S_{\frac{1}{2}} \rightarrow P_{3/2}$: E1 and M2 allowed

In general, some of the physical transitions would permit an E3 amplitude: this is predicted to be zero in this model, in agreement with the single-quark transition operator $SU(6)_W$ analysis.[44,46]

We must evaluate the matrix elements of the E1 and M2 pieces of $\underline{\varepsilon}^{(+1)} \cdot \underline{J}^{em}$ between the explicit $S_{\frac{1}{2}}$, $P_{\frac{1}{2}}$, $P_{3/2}$ wavefunctions listed in the Appendix.

The E1 operator is defined by:

$$E1^{+} = \int_{Bag} d^3x \, \{ j_o(qr) \, \psi^{\dagger} \, \underline{\varepsilon}^{(+1)} \cdot \underline{\alpha} Q \psi - 5 j_2(qr) \, [\psi^{\dagger} \underline{\varepsilon}^{(+1)} \cdot \underline{\alpha} Q \psi \, P_2(\cos\theta)]_{J=1} \}$$

and the M2 operator by projecting out the J=2 piece:

$$M2^{+} = \int_{Bag} d^3x \, \{ (-5 \, j_2(qr)) \, [\psi^{\dagger} \underline{\varepsilon}^{(+1)} \cdot \underline{\alpha} Q \psi \, P_2 \, (\cos\theta)]_{J=2} \}$$

The helicity structure of the transitions may be parameterized in terms of three reduced matrix elements, namely:

$$\alpha \equiv \langle P_{\frac{1}{2}\frac{1}{2}} | E1^{+} | \, S_{\frac{1}{2}\,-\frac{1}{2}} \rangle$$

$$\beta \equiv \langle P_{3/2 \;\, 3/2} | E1^{+} | \, S_{\frac{1}{2}\frac{1}{2}} \rangle$$

$$\gamma \equiv \langle P_{3/2 \;\, 3/2} | M2^{+} | \, S_{\frac{1}{2}\frac{1}{2}} \rangle$$

This structure is shown in Table 3.3: there is an exact correspondence between the reduced matrix elements A,B and C and the Bag parameters α,β and γ.

TABLE 3.3: Algebraic Structure for $[\underline{70},1^{-}]$ Photoproduction in Terms of Bag Matrix Elements.

SU(6) State	Target	$A_{3/2}$	$A_{1/2}$
$^{2}8_{1/2}$	p	-	$-\sqrt{\frac{1}{3}}\,\alpha$
	n	-	$+\frac{4}{27}\sqrt{2}\,\beta + \frac{5}{9}\sqrt{\frac{1}{3}}\,\alpha$
$^{4}8_{1/2}$	p	-	0
	n	-	$-\frac{4}{27}\sqrt{\frac{1}{2}}\,\beta + \frac{2}{9}\sqrt{\frac{1}{3}}\,\alpha$
$^{2}8_{3/2}$	p	$-\sqrt{\frac{1}{3}}\,\beta - \sqrt{\frac{1}{3}}\,\gamma$	$-\frac{1}{3}\,\beta + \gamma$
	n	$+\frac{7}{9}\sqrt{\frac{1}{3}}\,\beta + \frac{1}{3}\sqrt{\frac{1}{3}}\,\gamma + \frac{1}{9}\sqrt{2}\,\alpha$	$+\frac{7}{27}\,\beta - \frac{1}{3}\,\gamma + \frac{1}{9}\sqrt{\frac{2}{3}}\,\alpha$
$^{4}8_{3/2}$	p	0	0
	n	$-\frac{5}{9}\sqrt{\frac{2}{15}}\,\beta + \frac{1}{3}\sqrt{\frac{2}{15}}\,\gamma + \frac{1}{9}\sqrt{5}\,\alpha$	$-\frac{5}{27}\sqrt{\frac{2}{5}}\,\beta - \frac{1}{3}\sqrt{\frac{2}{5}}\,\gamma + \frac{1}{9}\sqrt{\frac{5}{3}}\,\alpha$
$^{4}8_{5/2}$	p	0	0
	n	$-\frac{4}{3}\sqrt{\frac{1}{5}}\,\gamma$	$-\frac{4}{3}\sqrt{\frac{1}{10}}\,\gamma$
$^{2}10_{1/2}$	p/n	-	$-\frac{8}{27}\sqrt{2}\,\beta - \frac{1}{9}\sqrt{\frac{1}{3}}\,\alpha$
$^{2}10_{3/2}$	p/n	$-\frac{5}{9}\sqrt{\frac{1}{3}}\,\beta + \frac{1}{3}\sqrt{\frac{1}{3}}\,\gamma - \frac{2\sqrt{2}}{9}\,\alpha$	$-\frac{5}{27}\,\beta - \frac{1}{3}\,\gamma - \frac{2}{9}\sqrt{\frac{2}{3}}\,\alpha$

Explicitly we have the relation:

$$\begin{bmatrix} A \\ B \\ C \end{bmatrix} = \begin{bmatrix} -\frac{2}{\sqrt{3}} & -4\frac{\sqrt{2}}{3} & 0 \\ -\frac{2}{\sqrt{3}} & +2\frac{\sqrt{2}}{3} & -2\sqrt{2} \\ +\frac{2}{\sqrt{3}} & -2\frac{\sqrt{2}}{3} & -2\sqrt{2} \end{bmatrix} \begin{bmatrix} \alpha \\ \beta \\ \gamma \end{bmatrix}$$

It is worthwhile remembering that although we are performing a non-relativistic expansion in the photon momentum, the quark motion is relativistic. Thus, if we ignore all recoil corrections and set q=0, the parameter γ vanishes and the Bag predicts:

$$B = -C + O(q^2)$$

It is clear that the Bag model does produce a non-zero 'spin-orbit' type C term of the same order of magnitude as the other two terms. What of the phenomenology? If we try to ignore all recoil corrections as above, we find for the D13 the relation:

$$A^p_{3/2} = \sqrt{3}\, A^p_{\frac{1}{2}}$$

Since the 3/2 helicity amplitude is large and the helicity ½ amplitude almost zero[28] it is evident that we must keep $O(q^2)$ recoil "corrections" if we wish for a viable phenomenology. Numerically, we find the results shown in Table 3.4, where the results of a fit to experiment are also given for comparison.

TABLE 3.4: Bag Matrix Elements: Numerical Results

Photon 3-Momentum q (in GeV/C)	$\tilde{\alpha} \equiv \xi \frac{\sqrt{2q}}{ie} \alpha$	$\tilde{\beta} \equiv \xi' \frac{\sqrt{2q}}{ie} \beta$	$\tilde{\gamma} \equiv \xi' \frac{\sqrt{2q}}{ie} \gamma$
0.00	- 0.43	- 0.47	0.00
0.35	- 0.45	- 0.15	-0.14
0.55	- 0.52	+ 0.06	-0.20
Fit to Expt.	- 0.30	- 0.50	-0.20

How should these results be regarded? We must remember that we have used a very crude version of the Bag model for our predictions. For example we have not actually used mass eigenstates since the $(S_{\frac{1}{2}})^2P_{\frac{1}{2}}$ and $(S_{\frac{1}{2}})^2P_{3/2}$ states are only approximately degenerate (within 50 MeV), nor have we included gluon corrections to bring the masses up to a more realistic value. Most importantly, we evaluate these transitions in a no-recoil approximation and it is clear that the detailed predictions are very sensitive to q^2 "corrections". We therefore believe that these results represent a qualitative success for the Bag - certainly a great improvement over earlier Bag calculations which included only the $(S_{\frac{1}{2}})^2 P_{\frac{1}{2}}$ modes.[25] We regard the Bag model, with its relativistic internal motion for the quarks, which predicts:

$$|C| \sim |B| \sim |A|$$

in terms of order of magnitude, as significantly superior to the non-relativistic models for which

$$|C| << |B| \sim |A|$$

4. EXCITED STATES II

In our discussion of P-wave states in the Bag, we have had to take some liberties with the quadratic boundary condition in using the round hadron approximation. The treatment by Rebbi of small surface oscillations showed, however, that the spurious translation mode of the ground state is largely made up of the first negative parity P-states. We may therefore hope to avoid problems involving spurious states, orbital motion and distorted bags by considering spherical 'radial excitations' in the Bag.[47] The first such state is the $(2S_{\frac{1}{2}})(1S_{\frac{1}{2}})^2$ configuration with a predicted mass of around 1600 MeV. This compares well with the masses of the Roper P11 (∿ 1500 MeV) and the P33 (∿ 1700 MeV), which are experimental candidates for radial excitations of the nucleon and delta. Despite this encouraging sign, it may be that our spherical cavity approximation, even after the inclusion of small surface oscillations a la Rebbi, is inadequate to describe states at this high mass. Nevertheless, since the physical $[\underline{70},1^-]$ states are at about the same mass and the Bag model is qualitatively successful in describing these states, it is certainly interesting to examine the predictions for radial excitations in the same approximation.

Again we must construct all the physical state wavefunctions. Clearly we have states corresponding to a $[\underline{56},0^+]_R$ radial excitation

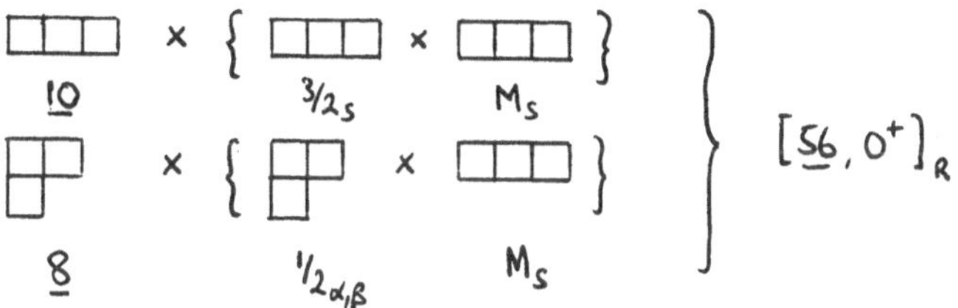

where M_S is the symmetric combination of 1S and 2S states (see before for the P-states). However, we also have the mixed symmetry combinations, $M_{\alpha,\beta}$, which give rise to states corresponding to a $[\underline{70},0^+]$ multiplet:

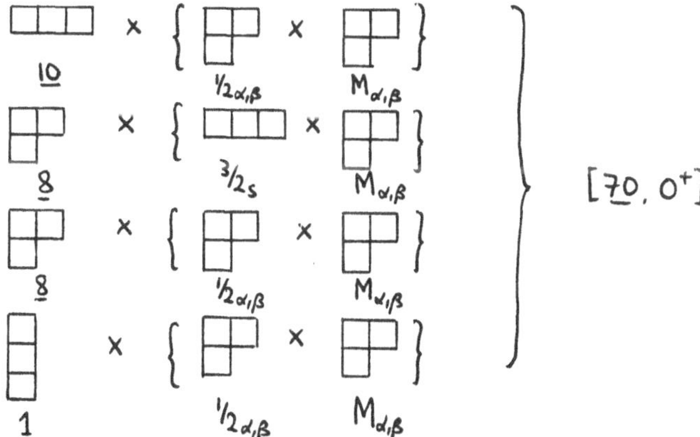

In the S=0 sector, this $\underline{70}$ multiplet contains P11, P13 and P31 states. The masses of all these states are given by the Bag Hamiltonian:

$$M = \frac{(2\omega_{1-1} + \omega_{2-1})}{R} + B.\frac{4\pi R^3}{3} - \frac{Z}{R} + \Delta H_g$$

where $\omega_{1-1} = 2.04$ and $\omega_{2-1} = 5.40$. Ignoring the gluon corrections ΔH_g, we see that the Bag predicts two P11's degenerate in mass.

Before we discuss the experimental implications of these results, we must examine the possibility that one or both of these states are weakly produced. In the absence of a reliable method to predict pion couplings (by fissioning of Bags or by PCAC?) we consider the photproduction of these states. The calculation proceeds in a similar fashion to that for Δ photoproduction and we obtain[47] the results of Table 4.1. The

TABLE 4.1: Photoproduction Matrix Elements for the $[\underline{56},0^+]$ and $[\underline{70},0^+]$ Bag States

State	Target	$A'_{3/2}$	$A'_{1/2}$
$(\underline{56},0^+)$			
$^2 8_{1/2}$	p	-	$+\sqrt{\frac{1}{3}}\,I$
	n	—	$-\frac{2}{3}\sqrt{\frac{1}{3}}\,I$
$^4 10_{3/2}$	p/n	$-\frac{1}{3}\sqrt{2}\,I$	$-\frac{1}{3}\sqrt{\frac{2}{3}}\,I$
$(\underline{70},0^+)$			
$^2 8_{1/2}$	p	-	$-\sqrt{\frac{1}{3}}\,I$
	n	-	$+\frac{1}{3}\sqrt{\frac{1}{3}}\,I$
$^4 8_{3/2}$	p	0	0
	n	$+\frac{1}{3}\,I$	$+\frac{1}{3}\sqrt{\frac{1}{3}}\,I$
$^2 10_{1/2}$	p/n	-	$+\frac{1}{3}\sqrt{\frac{1}{3}}\,I$

Notation: $A'_\lambda = \sqrt{2q}\,A_\lambda$ (see text)

P11 states from the $\underline{56}$ and $\underline{70}$ both couple to photons with comparable strengths, dictated by the same integral I over Bag wavefunctions.

Before we can draw any conclusions we must consider the effect of gluon interactions which may mix and split the states (for the N-Δ system, this splitting is

Fig.4.1: Gluon Corrections, a) Direct Interaction, b) Exchange Interaction

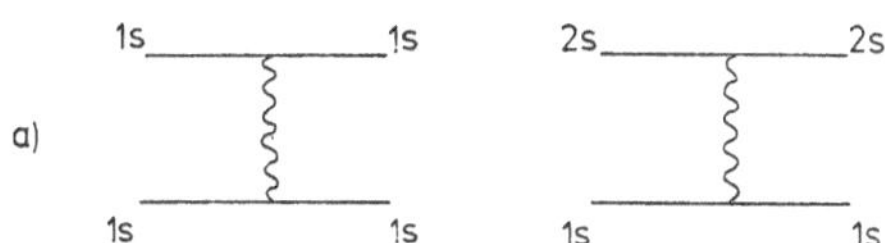

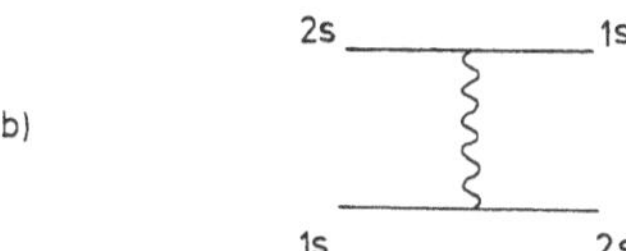

$\sim$ 300 MeV). The effect of including a direct magnetic gluon interaction (Fig. 4.1(a)) of the form:

$$\Delta H_g = \frac{8\alpha_c}{R} \sum_{i>j} \mu_{ij} \underline{\sigma}_i \cdot \underline{\sigma}_j$$

is to produce physical P11 states which are the following mixtures of the (previously degenerate) 56 and 70 P11 states:

$$|P11\ (\text{lower})\rangle = \frac{1}{\sqrt{2}} \{|P11;\ \underline{56}\rangle - |P11;\ 70\rangle\}$$

$$|P11\ (\text{upper})\rangle = \frac{1}{\sqrt{2}} \{|P11;\ 56\rangle + |P11;\ 70\rangle\}$$

The mass splitting of these states is estimated to be of the order of 100 MeV if one includes only the direct gluon interactions of Fig. 4.1 (a), and ignores the exchange graphs of Fig. 4.1 (b) which one expects to have little effect. The P31 state of the 70 is predicted to be not far away in mass and the P33 and P13 states to be somewhat higher and still degenerate.

The mixing is interesting in that it causes the upper P11 state to decouple completely from photoproduction off proton targets and to be only comparatively weakly excited off neutrons. Of course the observability of all these states in both $\gamma N \rightarrow \pi N$ and in purely hadronic πN reactions depends ultimately on their pion couplings which cannot be calculated with the same degree of certainty.

What of the experimental situation? Previously we have remarked on the problems of the non-relativistic oscillator quark model with the $[\underline{70},0^+]$ and the $[\underline{70},2^+]$, but it is worth examining the situation more closely. Remarkably enough, one fairly recent phase shift analysis[48] does indeed claim that the Roper is split into two states - one at 1413 MeV and the other at 1532 MeV! So far though this finding has not been substantiated by the next generation of phase shift analyses.[34] It is nevertheless an interesting question as to how best to detect the presence of two states which are split by less than their widths, and which differ in quantum number only in their SU(6) classification. It seems easier to establish the presence of the $\underline{70}$ by looking for states unique to this multiplet. Unfortunately, the most recent πN analyses disagree on the question of the existence of a low mass P31 state,[34] and there is insufficient data in this region for a definitive Y* analysis. In principle, such Y* searches could establish the $\underline{70}$ rather easily by the observation of an SU(3) singlet ΛP01 state.

At present, one must conclude that, rather surprisingly, the question of a low mass $[\underline{70},0^+]$ is not completely closed experimentally. On the theoretical side, the establishment of these states is clearly of crucial importance to the Bag model - or at least to the usefulness of the static cavity approximation.

5. OPEN PROBLEMS

This concluding section merely lists some of my favourite problems that remain to be elucidated in bag theory.

First of all, chiral symmetry, PCAC and predictions for pionic decays of resonances seem worthy of further investigation. Even in the case of massless quarks, the Bag axial vector current, although conserved within the Bag, is non-conserved by virtue of a non-vanishing surface contribution to the divergence. Chodos and Thorn[51] and Broadhurst[26] have studied some of the implications of this picture of chiral symmetry breaking. It remains to be seen whether such an approach can yield reasonable results for pion decays of excited states, or whether some more ambitious approach is required.

The Bag predictions for the meson spectrum are also very interesting. Here new problems arise for the states with one quark excited to a P-wave mode. Apart from the worry that such states do not strictly satisfy the quadratic boundary condition,[36] the problem with the spurious translation mode is more acute since Rebbi's approximation scheme is no longer valid. Nevertheless, if a picture of a meson as a spherical cavity plus small oscillations of the surface is at all relevant to these states, one expects that the symmetric combination of the first P-wave states should contribute a substantial component to the translation mode.[39] It would be interesting to see if any reasonable phenomenology could be built up by merely excluding these states. The situation is further complicated, however, by the prediction of low mass $(qq\bar{q}\bar{q})$ Bag states in the meson spectrum[52,53] which may obscure identification of the genuine P-states. For example, several features of the observed scalar mesons are elegantly accounted for by assignment to a $(qq\bar{q}\bar{q})$ configuration[52], but the outstanding difficulty is then the apparent absence of candidates for a second scalar nonet corresponding to a $(q\bar{q})$ excited level. Exotic states, such as an I=2 $\pi\pi$ resonance at about 1150 MeV, are also predicted. Such effects have not been observed and the whole question of the correct quark configuration for excited meson states is confused.[54]

The incorporation of charm into the Bag also presents some problems for the round hadron approximation. First attempts to include the J/ψ and ψ' as $c\bar{c}$ states were not successful. Jaffe and Kiskis[55] identified the ψ' as a $(1P_{\frac{1}{2}})^2$ configuration because of problems with the quadratic boundary condition: Donoghue and Golowich[56] identified it as a $(1S_{\frac{1}{2}})(2S_{\frac{1}{2}})$ state and ignored this problem. Both attempts failed to reproduce the observed $J/\psi - \psi'$ mass splitting. Again questions such as the correct quark configuration or the possibility of distorted bags would seem to be important: Kuti and co-workers have recently made some encouraging progress in the treatment of deformed bags containing heavy quarks.[57]

There also exist other interesting quark Bags which are worthy of mention. These are the "Budapest Bag", which includes a surface tension term in the energy as well as a volume term,[58] and the "CERN Bag" of Preparata and collaborators[59], which has an attractive phenomenology. It would be interesting to know whether any of these Bags can be derived in some realistic approximation to QCD.

In conclusion, I hope that these lectures have demonstrated the utility of the MIT-Bag model as an example of an explicit, tractable framework for investigating the consequences of relativistic, confined quarks. There are certainly some problems but there are also some encouraging successes.

ACKNOWLEDGEMENTS

Thanks are first due to Bob Jaffe for convincing me of the utility of the MIT Bag model for hadron spectroscopy, both by conversations on the ski slopes and in bars, and by his stimulating reviews of meson and baryon spectroscopy predictions in the Bag. I also wish to thank John Bell, Ken Bowler and Patrick Walters for several illuminating discussions on the Bag and for a critical reading of some portions of these lectures. Finally my apologies again to the MIT Bag company for any mis-representations I may have made concerning their theory.

APPENDIX A

1. Pauli Spinors in a Central Field

In the absence of spin-orbit coupling the Hamiltonian

$$H_o = \frac{\underline{p}^2}{2m} + V(r) \; ; \; \underline{p} = -i\underline{\nabla} \; ; \; \hbar=1$$

commutes with ℓ_z and s_z.

The wavefunctions which simultaneously diagonalize H_o, $\underline{\ell}^2$, ℓ_z, $\underline{s}^2$, and s_z are of the form:

$$\psi_o = R_j(r)\; Y_\ell^{\,m}\,(\underline{\hat{r}})\; \chi^{m'}$$

where $\chi^{m'}$ are the usual 2-component Pauli spinors, and $Y_\ell^{\,m}(\underline{\hat{r}})$ the standard spherical harmonics.

Introduction of a spin-orbit coupling term of the form $\underline{\ell}.\underline{s}$ into the Hamiltonian requires a change of basis which diagonalizes H, $\underline{j}^2$, $\underline{\ell}^2$, $\underline{s}^2$, j_z. The results are well known: they are the central field Pauli spinors.

$$\psi^\mu_{\ell+\frac{1}{2}} = \frac{R_{j=\ell+\frac{1}{2}}(r)}{\sqrt{2\ell+1}} \begin{pmatrix} \sqrt{\ell+\mu+\frac{1}{2}} & Y_\ell^{\,\mu-\frac{1}{2}} \\ \sqrt{\ell-\mu+\frac{1}{2}} & Y_\ell^{\,\mu+\frac{1}{2}} \end{pmatrix}$$

and

$$\psi^\mu_{\ell-\frac{1}{2}} = \frac{R_{j=\ell-\frac{1}{2}}(r)}{\sqrt{2\ell+1}} \begin{pmatrix} -\sqrt{\ell-\mu+\frac{1}{2}} & Y_\ell^{\,\mu-\frac{1}{2}} \\ +\sqrt{\ell+\mu+\frac{1}{2}} & Y_\ell^{\,\mu+\frac{1}{2}} \end{pmatrix}$$

This is just the usual Clebsch-Gordan coupling of $\ell \otimes \frac{1}{2}$ to $j=\ell+\frac{1}{2}$ and $j=\ell-\frac{1}{2}$

$$\psi^\mu_j = R_j(r) \sum_m C(j\mu;\; \ell\mu-m,\; \tfrac{1}{2}m)\; Y_\ell^{\,\mu-m}\; \chi^m$$

The ψ^μ_j are eigenfunctions of $\underline{j}^2$, $\underline{\ell}^2$ and $\underline{s}^2$ and consequently also of the operator

$$k \equiv 2\underline{s}.\underline{\ell}+1 = \underline{\sigma}.\underline{\ell} + 1 = \underline{j}^2-\underline{\ell}^2-\underline{s}^2+1$$

with eigenvalues

$$k\;\psi^\mu_j = -\kappa\;\psi^\mu_j$$

where

$$j(j+1) - \ell(\ell+1)+\tfrac{1}{4} = (j+\tfrac{1}{2})^2 - \ell(\ell+1) = -\kappa$$

Thus

$$\kappa=\begin{cases} \ell & \text{for } j=-\frac{1}{2} \\ -\ell-1 & \text{for } j=+\frac{1}{2} \end{cases}$$

i.e. κ has all integer values except zero, and

$$j = |\kappa| - \tfrac{1}{2}$$

Thus the specification of the quantum number κ results in an economy of notation, since its magnitude and sign determines both j and ℓ, and also the parity $(-1)^{\ell}$.

$$\ell = \begin{cases} \kappa & \text{for } \kappa>0 \\ -\kappa-1 & \text{for } \kappa<0 \end{cases}$$

We therefore rewrite the eigenfunctions ψ_j^{μ} as

$$\psi_j^{\mu} = R_j(r)\ \chi_\kappa^{\mu}$$

where the spin-angular functions χ_κ^{μ} are defined by

$$\chi_\kappa^{\mu} = \sum_m C(j\mu;\ \ell\mu-m,\ \tfrac{1}{2}m)\ Y_\ell^{\ \mu-m}\ \chi^m$$

These spinors have the following properties:

(i) $k\chi_\kappa^{\mu} = (\underline{\sigma}.\underline{\ell} + 1)\ \chi_\kappa^{\mu} = -\ \kappa\chi_\kappa^{\mu}$

(ii) $\underline{\sigma}.\underline{\hat{r}}\ \chi_\kappa^{\mu} = -\ \chi_{-\kappa}^{\mu}$

The last result may be verified explicitly but the form of the result is explained by observing that $\underline{\sigma}.\underline{\hat{r}}$ is a pseudoscalar operator which (a) cannot change j or μ, and (b) changes the parity. Since $|\kappa|$ is fixed, this corresponds to changing the sign of κ.

There is a one-to-one correspondence between the spectroscopic notation L_J for a state and the eigenvalue of κ:

$S_{\frac{1}{2}} \quad \leftrightarrow \quad \kappa = -1$

$P_{\frac{1}{2}} \quad \leftrightarrow \quad \kappa = +1$

$P_{3/2} \quad \leftrightarrow \quad \kappa = -2$

and so on.

2. Dirac Equation and Constants of the Motion

Let us consider the Dirac free particle Hamiltonian and the corresponding constants of the motion.

$$H\psi = (\underline{\alpha}.\underline{p} + \beta m)\psi = i\,\frac{\partial\psi}{\partial t}$$

We are interested in the angular momentum of the electron and we first consider the orbital angular momentum operator

$$\underline{\ell} = \underline{r} \times \underline{p} = -i\, \underline{r} \times \underline{\nabla}$$

We find explicitly:

$$[\beta, \underline{\ell}] = 0$$

but

$$[\underline{\ell}, \underline{\alpha}.\underline{p}] = i(\underline{\alpha} \times \underline{p}) \neq 0$$

and similarly:

$$[\underline{\ell}^2, \underline{\alpha}.\underline{p}] \neq 0$$

Thus the orbital angular momentum operator does not commute with the kinetic term in the free Hamiltonian and is not a constant of the motion. However, if we consider the operator:

$$\underline{j} = \underline{\ell} + \tfrac{1}{2}\underline{\sigma}$$

where $\underline{\sigma}$ is understood (by its context) to be the 4 x 4 matrix

$$\underline{\sigma} = \begin{pmatrix} \underline{\sigma} & 0 \\ 0 & \underline{\sigma} \end{pmatrix}$$

we may show that this <u>does</u> commute with H. Explicitly with

$$\gamma_5 = \begin{pmatrix} 0 & 1 \\ 1 & 0 \end{pmatrix} \qquad \underline{\alpha} = \begin{pmatrix} 0 & \underline{\sigma} \\ \underline{\sigma} & 0 \end{pmatrix}$$

and

$$\underline{\sigma} = \gamma_5\, \underline{\alpha} = \underline{\alpha}\, \gamma_5$$

we find

$$\tfrac{1}{2}\, [\underline{\sigma}, \underline{\alpha}.\underline{p}] = -i(\underline{\alpha} \times \underline{p})$$

so that

$$[\underline{j}, \underline{\alpha}.\underline{p}] = 0$$

Thus $\underline{j}$ is conserved and we may identify $\underline{s} = \tfrac{1}{2}\underline{\sigma}$ as the spin operator in the relativistic theory. Now $\underline{j}^2$ and $\underline{s}^2$ commute with H but $\underline{\ell}^2$ does not, thus the spin-orbit coupling operator $\underline{s}.\underline{\ell}$ cannot commute with H. However, there exists an operator which is the relativistic analogue of the non-relativistic operator:

$$k = (\underline{\sigma}.\underline{\ell} + 1)$$

which <u>does</u> commute with both H and $\underline{j}$. This is the operator

$$K = \beta(\underline{\sigma}.\underline{\ell} + 1)$$

We can calculate its commutator with H

$$[K,H] = [\beta\underline{\sigma}.\underline{\ell}, \underline{\alpha}.\underline{p}] + [\beta, \underline{\alpha p}] = \beta\{\underline{\sigma}.\underline{\ell}, \underline{\alpha p}\} + 2\beta\underline{\alpha}.\underline{p}$$

Using identities for 4 x 4 matrices $\underline{\sigma}$ and $\underline{\alpha}$

$$\underline{\sigma}.\underline{A}\ \underline{\sigma}.\underline{B} = \underline{A}.\underline{B} + i\underline{\sigma}.(\underline{A} \times \underline{B}) \ ; \qquad \underline{\alpha}.\underline{A}\ \underline{\sigma}.\underline{B} = \underline{\sigma}.\underline{A}\ \underline{\alpha}.\underline{B} = \gamma_5 \underline{A}.\underline{B} + i\underline{\sigma}.(\underline{A} \times \underline{B})$$

$$\underline{\alpha}.\underline{A}\ \underline{\alpha}.\underline{B} = \underline{\sigma}.\underline{A}\ \underline{\sigma}.\underline{B}$$

it is straightforward to prove

$$\{\underline{\sigma}.\underline{\ell},\underline{\alpha}.\underline{p}\} = -2\gamma_5\ \underline{\sigma}.\underline{p}$$

and hence that

$$[K,H] = 0$$

Similarly, one can show

$$[K,\underline{j}] = 0$$

and there is a connection between κ^2 and $\underline{j}^2$

$$K^2 = (\underline{\sigma}.\underline{\ell} + 1)^2 = \underline{\ell}^2 + i\underline{\sigma}.\underline{\ell} \times \underline{\ell} + 2\underline{\sigma}.\underline{\ell} + 1 = \underline{\ell}^2 + \underline{\sigma}.\underline{\ell} + 1 = \underline{j}^2 + \tfrac{1}{4}$$

Thus the eigenvalue of K is κ , where

$$\kappa^2 = j(j+1) + \tfrac{1}{4} = (j+\tfrac{1}{2})^2$$

and

$$\kappa = \pm(j+\tfrac{1}{2})$$

As for the non-relativistic case we have:

$$j = |\kappa| - \tfrac{1}{2}$$

Finally we note that the parity operator

$$P = \beta I_s$$

Where I_s is the operation of spatial inversion, commutes with H,$\underline{j}$ and K. As before, κ gives the eigenvalue of $\underline{j}^2$ and P.

3. Dirac Particle in Central Field

We now consider the problem of finding solutions of the Dirac equation for a particle in a spherically symmetric potential W(r). We first write down the Dirac equation for a stationary state of energy E with an additional potential energy term W(r)

$$H\psi = (\underline{\alpha}.\underline{p} + \beta m + W(r))\psi = E\psi$$

To obtain the equation in polar co-ordinates we use the identity

$$\underline{\nabla} = \underline{\hat{r}}\ (\underline{\hat{r}}.\underline{\nabla}) - \underline{\hat{r}} \times (\underline{\hat{r}} \times \underline{\nabla}) = \underline{\hat{r}}\frac{\partial}{\partial r} - i\frac{\underline{\hat{r}}}{r} \times \underline{\ell}$$

Thus the Kinetic energy term becomes:

$$\underline{\alpha}.\underline{p} = -i\underline{\alpha}.\underline{\hat{r}}\ \frac{\partial}{\partial r} - \frac{1}{r}\ \underline{\alpha}.\underline{\hat{r}} \times \underline{\ell} = -i\underline{\alpha}.\underline{\hat{r}}\ \frac{\partial}{\partial r} + \frac{i}{r}\ \underline{\alpha}.\underline{\hat{r}}\ \underline{\sigma}.\underline{\ell} = -i\underline{\alpha}.\underline{\hat{r}}\ \frac{\partial}{\partial r} + \frac{i}{r}\ \underline{\alpha}.\underline{r}\ (\beta K-1)$$

and we have

$$H\psi = \left[-i\gamma_5\, \underline{\sigma}\cdot\underline{\hat{r}}\, \left(\frac{\partial}{\partial r} + \frac{1}{r} - \frac{\beta K}{r}\right) + W + \beta m\right]\psi = E\psi$$

The operators $\underline{j}^2$, j_z and K all commute with W(r) and hence with H; thus we look for solutions characterized by their eigenvalues

$$\underline{j}^2\psi = j(j+1)\psi$$
$$j_z\psi = \mu\psi$$
$$K\psi = -\kappa\psi$$

If we write:

$$\psi = \begin{pmatrix}\phi_1\\ \phi_2\end{pmatrix}$$

we have

$$K = \beta(\underline{\sigma}\cdot\underline{\ell} + 1)$$

and consequently

$$(\underline{\sigma}\cdot\underline{\ell} + 1)\ \phi_1 = -\kappa\ \phi_1$$
$$(\underline{\sigma}\cdot\underline{\ell} + 1)\ \phi_2 = +\kappa\ \phi_2$$

showing that the upper and lower components correspond to different orbital parities. Moreover, since they are 2-component spinors they must be proportional to the usual central-field Pauli spinors χ^μ_κ and $\chi^\mu_{-\kappa}$. Thus we may write:

$$\psi \equiv \psi^\mu_\kappa = \begin{bmatrix} g(r)\ \chi^\mu_\kappa \\ if(r)\ \chi^\mu_{-\kappa}\end{bmatrix}$$

where g and f are radial functions which will in general depend on κ. The phase i is introduced to make the radial equations, obtained by inserting this form into the Dirac equation and equating components, explicitly real:

$$(E - W - M)\ g\ \chi^\mu_\kappa = \left[-\left(\frac{df}{dr} + \frac{f}{r}\right) + \frac{\kappa f}{r}\right]\chi^\mu_\kappa$$

$$(E - W + M)\ f\ \chi^\mu_\kappa = \left[\frac{dg}{dr} + \frac{g}{r} + \frac{\kappa g}{r}\right]\chi^\mu_{-\kappa}$$

We have used the result:

$$\underline{\sigma}\cdot\underline{\hat{r}}\ \chi^\mu_\kappa = -\chi^\mu_{-\kappa}$$

Thus the final radial equations for f and g are:

$$\frac{df}{dr} = \frac{(\kappa-1)}{r}\ f - (E - M - W)\ g$$

$$\frac{dg}{dr} = (E - W + M)\ f - \frac{(\kappa+1)}{r}\ g$$

It is often convenient to define:

$$u_1 = rg$$

$$u_2 = rf$$

and these equations become:

$$\frac{d}{dr}\begin{bmatrix} u_1 \\ u_2 \end{bmatrix} = \begin{bmatrix} -\frac{\kappa}{r} & E + M - W \\ -(E - M - W) & +\frac{\kappa}{r} \end{bmatrix}\begin{bmatrix} u_1 \\ u_2 \end{bmatrix}$$

4. Free Particle Solutions of the Dirac Equation in the Angular Momentum Representation.

To obtain the angular momentum representation for free Dirac particles we set $W = 0$ in the radial equations for (g,f) or (u_1,u_2). We obtain a second order differential equation by eliminating u_2

$$\frac{d^2u_1}{dr^2} + \frac{dW/dr}{E-W+M}\cdot\frac{du_1}{dr} + \left[(E-W)^2 - M^2 - \frac{\kappa(\kappa+1)}{r^2} + \frac{\kappa}{r}\,\frac{dW/dr}{E-W+M}\right]u_1 = 0$$

which simplifies for $W=0$ to

$$\frac{d^2u_1}{dr^2} + \left[p^2 - \frac{\kappa(\kappa+1)}{r^2}\right]u_1 = 0$$

with $p^2=E^2-M^2$. This is one of the standard forms of Bessel's equation: the solution regular at $r=0$ is:

$$u_1 = Ar\, j_\ell(pr)$$

where $\kappa=\ell$ for $j=\ell-\frac{1}{2}$; $\kappa=-\ell-1$ for $j=\ell+\frac{1}{2}$. The $j_\ell(pr)$ are spherical Bessel functions.

$$j_\ell(x) = \sqrt{\frac{\pi}{2x}}\; J_{\ell+\frac{1}{2}}(x)$$

with

$$j_o(x) = \frac{\sin x}{x}$$

and

$$j_n(x) = (-1)^n x^n \left(\frac{1}{x}\frac{d}{dx}\right)^n \left(\frac{\sin x}{x}\right)$$

It can be shown that

$$j_n(x) \sim \frac{x^n}{(2n+1)!!}$$

as $x\to 0$ for n integer. They also obey the relations:

$$j'_\ell(x) = \frac{\ell}{x}\, j_\ell - j_{\ell+1} = -\frac{(\ell+1)}{x}\, j_\ell + j_{\ell-1}$$

Now

$$u_1 = Ar\, j_\ell\,(pr)$$

and

$$u_2 = \frac{A}{E+M}\left\{\frac{d}{dr} + \frac{\kappa}{r}\right\} u_1$$

If we introduce the sign of κ, S_κ

$$S_\kappa = \frac{\kappa}{|\kappa|}$$

and define

$$\bar{\ell} = \ell_{-\kappa}$$

we have

$$\kappa>0 : \ell = \kappa \quad ; \bar{\ell} = \kappa-1$$
$$\kappa<0 : \ell = -\kappa-1 \ ; \bar{\ell} = -\kappa$$

Using all these results, the solution for u_2 for both signs of κ may be written:

$$u_2 = AS_\kappa \frac{pr}{E+M}\, j_{\bar{\ell}}\,(pr)$$

Thus the free particle angular momentum Dirac eigenfunctions are:

$$\psi_\kappa^\mu = A_\kappa \begin{bmatrix} j_\ell\, \chi_\kappa^\mu \\ \dfrac{ipS_\kappa}{E+M}\, j_{\bar{\ell}}\, \chi_{-\kappa}^\mu \end{bmatrix}$$

Let us write down some examples:

(1) $\underline{\kappa = -1}$

$\kappa<0$ so that $\ell=0$ and $\bar{\ell}=1$ with $j=\ell-\frac{1}{2}, S_\kappa = \frac{1}{2}$.

Thus the solutions for $\kappa = -1$ represents a Dirac particle with $j=\frac{1}{2}$ which we may designate loosely as "$S_{\frac{1}{2}}$", remembering that "S" applies only to the upper components: the bottom components have $\bar{\ell}=1$.

$$\psi_{-1}^\mu = A_{-1} \begin{bmatrix} j_o\ \chi_{-1}^\mu \\ \dfrac{-ip}{E+M}\, j_1\, \chi_1^\mu \end{bmatrix}$$

We can simplify this further using the relation:

$$\underline{\sigma}\cdot\underline{\hat{r}}\ \chi_\kappa^\mu = -\chi_{-\kappa}^\mu$$

and rewriting ψ_{-1}^μ in terms of χ_{-1}^μ which are just χ^μ, i.e. two component Pauli spinors:

$$\psi_{-1}^{\mu} = A_{-1}\begin{bmatrix} j_o\ \chi^{\mu} \\ \\ \frac{ip}{E+M}\ j_1\ \underline{\sigma}\cdot\underline{\hat{r}}\ \chi^{\mu}\end{bmatrix}$$

2. $\underline{\kappa = +1}$

$\kappa>0$ so that $\ell=1$ and $\bar{\ell}=0$ with $j=\ell-\frac{1}{2}, S_{\kappa} = \frac{1}{2}$.

The solutions for $\kappa = +1$ also represent a Dirac particle with $j=\frac{1}{2}$ which we call "$P_{\frac{1}{2}}$" where "P" refers to the fact that the upper components have $\ell=1$.

$$\psi_{+1}^{\mu} = A_1\begin{bmatrix} -j_1\ \underline{\sigma}\cdot\underline{\hat{r}}\ \chi^{\mu} \\ \\ \frac{ip}{E+M}\ j_o\ \chi^{\mu}\end{bmatrix}$$

3. $\underline{\kappa = -2}$

$\kappa<0$ so that $\ell=1$ and $\bar{\ell}=2$ with $j= 3/2$.

We refer to these solutions as "$P_{3/2}$" in the same spirit as the previous two examples.

$$\psi_{-2} = A_{-2}\begin{bmatrix} j_o\ \chi_{-2}^{\mu} \\ \frac{ip}{E+M}\ j_2\ \underline{\sigma}\cdot\underline{\hat{r}}\ \chi_{-2}^{\mu}\end{bmatrix}$$

For this case, $j=\ell+\frac{1}{2}$, the χ_{-2}^{μ} are identical to the $\phi_{j\mu}^{(+)}$ of Bjorken and Drell[11]: $\phi_{j\mu}^{(-)}$ differs by an overall phase from our conventions. We use the conventions of Rose in "Relativistic Electron Theory"[12] from which most of this appendix is derived.

APPENDIX B

1 Young Diagrams

To obtain the decomposition into irreducible representations of the product of two representations of SU(n), one starts with the corresponding Young tableaux. Since we are only concerned with the simplest products it suffices to know that

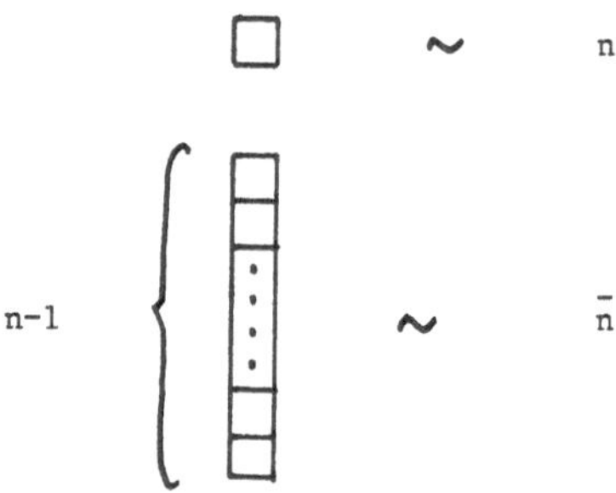

and a column of n boxes is the singlet. (More complete rules may be found elsewhere.[49] For the product of two tableaux,the allowed tableaux may be enumerated according to the Littlewood-Richardson rule. One must add to one of them taken as fixed, all the boxes of the other in all possible ways compatible with the following restrictions:-

i) The boxes of the multiplying diagram must be attached, to the fixed one, row by row starting from the first (upper) one.

ii) The resulting tableaux must be again a Young diagram. This means that the number of boxes is any row must not be increasing from the upper to the lower row.

iii) Two boxes from the same row of the multiplying diagram can never be placed in the same column.

iv) The maximum allowed number of rows for the resulting diagrams is n, with the convention that every column with n boxes must be dropped out of the final diagrams (just corresponding to a singlet).

v) In the multiplying diagram, put an index a into every box in the first row, b into every box of the second row, etc. For every diagram constructed according to the previous rules, one can read off the sequence of a's, b's, c's etc., starting from the right of the top row, the right of the second row and so on. The only allowed diagrams are the ones in which this sequence is a lattice permutation of the a's,b's, etc.

For example, a lattice permutation of $a^n b^m c^\ell$ is a sequence of n a's, m b's, ℓ c's like

aaabbac ...

such that to the left of any point in the sequence there are not less a's and b's, and not less b's than c's and so on.

For a^2bc, the lattice permutations are

aabc, abac, abca

Consider the example in SU(3) of $\underline{8} \otimes \underline{8}$ -- in terms of Young diagrams we have

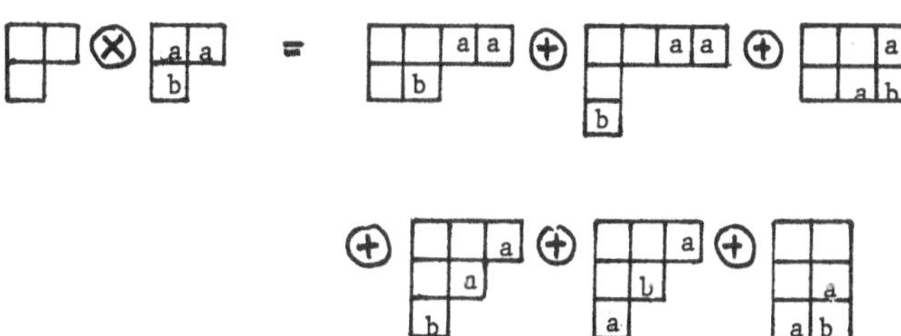

How does one calculate the dimension of the representation corresponding to a given Young diagram? The answer is either a long formula or a simple prescription - the so-called "Hook" Rule. This is best illustrated by an example. For a diagram in SU(n) assign numbers to each box as shown in the figure -- n's down the leading diagonal; n + 1, n + 2, etc., for diagonals to the right; n - 1,

n	n+1	n+2	n+3	n+4
n-1	n	n+1	n+2	
n-2	n-1			
n-3				
n-4				

n - 2, etc., for diagonals to the left.

The product of these numbers forms the numerator. The denominator is given by the "product of the hooks". The "hook" of each box is how many squares one must pass through, entering along the appropriate row from the right and leaving the diagram from the bottom of the appropriate column. For example, in SU(3) consider the $\underline{8}$ representation. The rule says that the dimension of

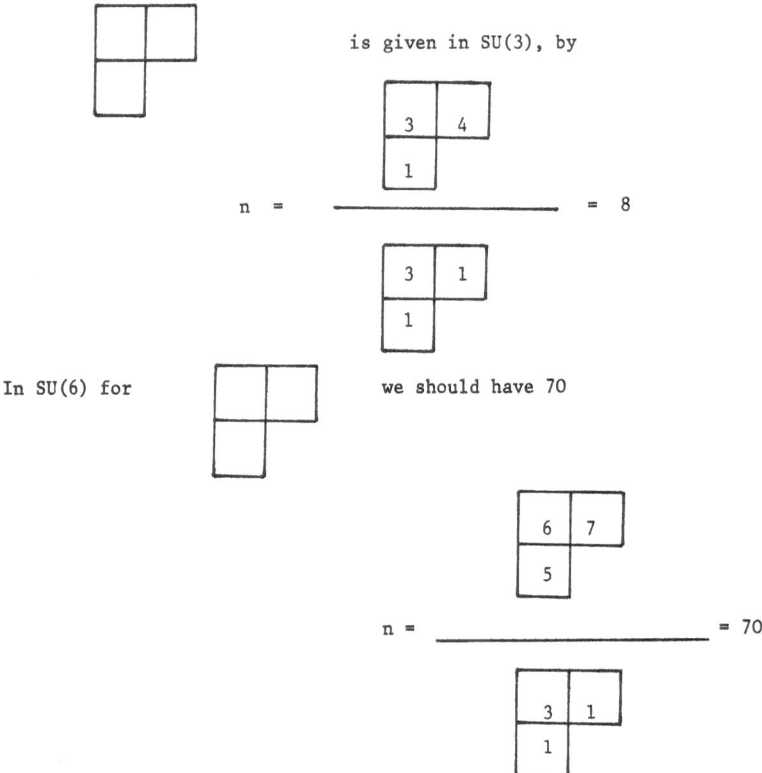

In this way we can easily find in our SU(3) example, the product $\underline{8} \otimes \underline{8}$

$$\underline{8} \otimes \underline{8} = \underline{27} \oplus \underline{10} \oplus \overline{\underline{10}} \oplus \underline{8} \oplus \underline{8} \oplus 1$$

(One must decide on whether or not it is the conjugate representation from the diagram).

These rules may easily be extended for calculations of the irreducible representations of O(n) and Sp(n).[50]

2 The Wave-Function Symmetries of Three Objects

If an object can be in one of a number of conditions x, y, z, ... we can, when we have three such objects, form states of four kinds of symmetry which we call S, α, β, A, symmetric (S), mixed-symmetric (α, β), and antisymmetric (A),

$$|S\rangle = |xyz\rangle_S = \frac{1}{\sqrt{6}}(|xyz\rangle + |xzy\rangle + |yxz\rangle + |yzx\rangle + |zxy\rangle + |zyx\rangle),$$

$$|\alpha\rangle = |xyz\rangle_\alpha = \frac{1}{2\sqrt{3}}(|xyz\rangle + |xzy\rangle + |yxz\rangle + |yzx\rangle - 2|zxy\rangle - 2|zyx\rangle),$$

$$|A\rangle = |xyz\rangle_A = \frac{1}{\sqrt{6}}(-|xyz\rangle + |xzy\rangle - |yzx\rangle + |yxz\rangle - |zxy\rangle + |zyx\rangle),$$

$$|\beta\rangle = |xyz\rangle_\beta = \frac{1}{2}(|xyz\rangle - |xzy\rangle + |yxz\rangle - |yzx\rangle),$$

where $|zxy\rangle$ means that the first object is in state z, the second in x, and the third in y. If, say, x, and y are the same state y = x, we must replace $|xyz\rangle + |yxz\rangle$ by $\sqrt{2}|xxz\rangle$. If x, y, z are all the same only the S state survives as $|xxx\rangle_s = |xxx\rangle$. The state α has been chosen to be symmetric in the last two quarks, the state β is antisymmetric. If we combine two states of these kinds, say $|1\rangle$, and $|2\rangle$, states of varying symmetry may be formed according to the rules:

$$|1\rangle_S|2\rangle_S = |\rangle_S,\quad |1\rangle_S|2\rangle_\alpha = |\rangle_\alpha,$$

$$|1\rangle_S|2\rangle_\beta = |\rangle_\beta,\quad |1\rangle_S|2\rangle_A = |\rangle_A,$$

$$|1\rangle_A|2\rangle_S = |\rangle_A,\quad |1\rangle_A|2\rangle_\alpha = |\rangle_\beta,$$

$$-|1\rangle_A|2\rangle_\beta = |\rangle_\alpha,\quad |1\rangle_A|2\rangle_A = |\rangle_S,$$

$$\sqrt{\tfrac{1}{2}}(+|1\rangle_\alpha|2\rangle_\alpha + |1\rangle_\beta|2\rangle_\beta) = |\rangle_S,$$

$$\sqrt{\tfrac{1}{2}}(-|1\rangle_\alpha|2\rangle_\alpha + |1\rangle_\beta|2\rangle_\beta) = |\rangle_\alpha,$$

$$\sqrt{\tfrac{1}{2}}(+|1\rangle_\alpha|2\rangle_\beta + |1\rangle_\beta|2\rangle_\alpha) = |\rangle_A,$$

$$\sqrt{\tfrac{1}{2}}(-|1\rangle_\alpha|2\rangle_\beta + |1\rangle_\beta|2\rangle_\alpha) = |\rangle_A.$$

REFERENCES AND FOOTNOTES

1. A. Chodos, R.L. Jaffe, K. Johnson, C.B. Thorn and V.F. Weisskopf, Phys. Rev. D9 (1974) 3471.

2. A. Chodos, R.L. Jaffe, K. Johnson and C.B. Thorn, Phys. Rev. D10 (1974) 2599.

3. T.A. DeGrand, R.L. Jaffe, K. Johnson and J. Kiskis, Phys. Rev. D12 (1975) 2060.

4. K. Johnson, Acta Physica Polonica B6, (1975) 865.

5. V. Weisskopf, Erice Summer School Lectures, 1974 and 1975.

6. K. Johnson, Scottish Universities Summer School Lectures, 1976.

7. R.L. Jaffe, invited talk at the XIth Rencontre de Moriond, Flaine, France, proceedings edited by J. Tran Than Vanh (1976).

8. R.L. Jaffe, invited talk at the Oxford Conference on Baryon Resonances, proceedings edited by R. Ross and D. Saxon (1976).

9. P.N. Bogolioubov, Ann. Inst. H. Poincare 8 (1967) 163.

10. For a discussion of the Klein paradox see for example the book "Advanced Quantum Mechanics" by J.J. Sakurai (Addison-Wesley, 1967).

11. J.D. Bjorken and S.D. Drell, Relativistic Quantum Mechanics (McGraw-Hill, 1964).

12. M.E. Rose, Relativistic Electron Theory (Wiley, 1961).

13. R. Dashen, invited talk at the 1975 International Symposium on Lepton and Photon Interactions at High Energies, edited by W.T. Kirk.

14. H.D. Politzer, Physics Reports 14C (1974) 129.

15. R.H. Dalitz, invited talk at the 13th International Conference on High Energy Physics, Berkeley, California, 1966, edited by M. Alston-Garnjost.

16. This terminology originates from K. Johnson, Rutherford Lecture, December, 1976.

17. R.L. Jaffe, Phys. Rev. D11 (1975) 1953; R.L. Jaffe and A. Patrascioiu, Phys. Rev. D12 (1975) 1314.

18. R.J. Hughes, MIT preprint CTP-615 (1977); to be published in Physical Review.

19. E. Golowich, Phys. Rev. D12 (1975) 2108.

20. T. Barnes, CalTech Report (1975).

21. E. Allen, MIT Report, CTP-471, (1975).

22. C. Bender and P. Hays, Phys. Rev. D14 (1976) 2622.

23. A. DeRújula, H. Georgi and S. Glashow, Phys. Rev. D12 (1975) 147.

24. A. Halprin and P. Sorba, Phys. Lett. 66B (1977) 177.

25. J.F. Donoghue, E. Golowich and B.R. Holstein, Phys. Rev. D12 (1975) 2875.

26. D.J. Broadhurst, Nucl. Phys. B105 (1976) 319; Open University preprint (1975).

27. P.J. Litchfield, rapporteur's report at the XVIIth International Conference on High Energy Physics, London 1974, edited by J.R. Smith.

28. Particle Data Group, Rev. Mod. Phys. 48, No. 2, Pt. II (1976).

29. For the spectrum of higher excitations, see the paper of G. Karl and E. Obryk, Nucl. Phys. B8 (1968) 609.

30. R. Horgan, invited talk at the Oxford Conference on Baryon Resonances, proceedings edited by R.Ross and D. Saxon (1976).

31. See for example, the talk by the present author at the Oxford Baryon Resonance Conference (reference above).

32. A.J.C. Hey, P.J. Litchfield and R.J. Cashmore, Nucl. Phys. B95 (1975) 516; R.J. Cashmore, A.J.G. Hey and P.J. Litchfield, Nucl. Phys. B98 (1975) 237; P.J. Litchfield, R.J. Cashmore and A.J.G. Hey, updated fit published in the proceedings of the Oxford Baryon Resonance Conference (see above).

33. M. Jones, R.H. Dalitz and R. Horgan, Oxford preprint 1977, to be published in Nuclear Physics.

34. See the talks by E. Pietarinen, D. Chew and R. Cutkosky at the Oxford Baryon Resonance Conference (see above).

35. R. Cutkosky, in his talk at the Oxford Baryon Resonance Conference (see above).

36. T.A. DeGrand and R.L. Jaffe, Ann. Phys. 100 (1976) 425.

37. T.A. DeGrand, Ann. Phys. 101 (1976) 496.

38. C. Rebbi, Phys. Rev. D14 (1976) 2362.

39. H.J. Melosh, Phys. Rev. D9 (1974) 1095.

40. A. DeRújula, tendentious suggestion from the audience.

41. D. Faiman and A. Hendry, Phys. Rev. 179 (1968) 1720; Phys. Rev. 180 (1969) 1572, 1609.

42. L.A. Copley, G. Karl and E. Obryk, Nucl. Phys. B13 (1969) 303.

43. R.P. Feynman, M. Kislinger and F. Ravndal, Phys. Rev. D3 (1971) 2706.

44. A. Love and D.V. Nanopoulos, Phys. Lett. 45B (1973) 507; F.J. Gilman and I. Karliner, Phys. Rev. D10 (1974) 2194; A.J.G. Hey and J. Weyers, Phys. Letts. 40B (1974) 69.

45. A.J.G. Hey, B.R. Holstein and D.P. Sidhu, Southampton University preprint THEP 76/7-9 (1977).

46. J. Babcock and J. Rosner, Ann. Phys. 96 (1976) 191.

47. K.C. Bowler and A.J.G. Hey, joint Edinburgh and Southampton preprint 1977, to be published in Physics Letters B; K.C. Bowler, A.J.G. Hey and P.J. Walters, in preparation.

48. R. Ayed and P. Bareye, reported at the 1973 Aix conference, Journal de Physique 34, Colloque C-1, suppl. to No. 10.

49. M. Hamermesh, 'Group Theory and its Applications to Physical Problems' (Addison-Wesley, 1962).

50. R.C. King, unpublished lectures at Southampton University, 1976.

51. A. Chodos and C.B. Thorn, Phys. Rev. D12 (1975) 2733.

52. R.L. Jaffe and K. Johnson, Phys. Lett. 60B (1976) 201.

53. R.L. Jaffe, Phys. Rev. D15 (1977) 267, 281.

54. For a review of present state of meson spectroscopy, see for example, A.J.G. Hey and D. Morgan, Rutherford report RL-77-060/A (1977).

55. R.L. Jaffe and J. Kiskis, Phys. Rev. D13 (1976) 1355.

56. J.F. Donoghue and E. Golowich, Phys. Rev. D14 (1976) 1386.

57. Work reported by K. Johnson in lecture at Rutherford Laboratory, December, 1976.

58. P. Gnädig, P. Hasenfratz, J. Kuti and A.S. Szalay, Phys. Lett. 64B (1976) 62.

59. G. Preparata and N. Craigie, Nucl. Phys. B102 (1976) 468; N. Craigie and G. Preparata, Nucl. Phys. B102 (1976) 497; G. Preparata, CERN Preprint TH-2271 (1977).

PART A: AN INTRODUCTION TO UNIFICATION

(Mirror-Symmetry, Confinement Versus Liberation, Unifying Mass-Scale etc.)

Jogesh C. Pati

Department of Physics and Astronomy
University of Maryland
College Park, Maryland

TABLE OF CONTENTS

---

*Supported in part by the National Science Foundation under Grant No. GP43662X.

TABLE OF CONTENTS (continued)

I. A PREVIEW

1.1 The Making and Breaking of Symmetries

Symmetries appear in Nature both in exact, so far as we know, and in approximate forms. For instance, the operations of rotation, space-time displacements, the joint transformation under (Time Reversal) x (Parity) x (Charge Conjugation), and electric charge gauge-transformation appear to be exact symmetries of Nature. On the other hand, the operations of CP, left-right symmetry, isospin, Yang-Mills local gauge-transformations,[1] weak and electromagnetic symmetry[2], quark-lepton-symmetry[3] and finally the postulated symmetry of the three basic forces[3,4] (weak, electromagnetic as well as strong) are broken in varying degrees.

A symmetry may break in two distinct ways: either because the basic lagrangian possesses an explicit asymmetry; or because the ground state of the system develops an asymmetry through nonperturbative solutions, even though the basic equations of motion are intrinsically symmetric. In quantum field theory the ground-state corresponds to the physical vacuum. This latter phenomenon of symmetry breakdown, induced through nonperturbative solutions is called "Spontaneous Symmetry Breaking". It is encountered in solid state physics (for example in superconductivity). Over the past decade this concept of spontaneous symmetry breaking has played an increasingly important role in the development of unified gauge field theories.

1.2 The Gauge-Hypothesis

The main idea underlying unified gauge theories[2-4] is that the observed asymmetries between basic particles (such as electrically neutral neutrinos versus charged electrons, or quarks versus leptons) and the distinctions between their basic forces[5] (i.e. weak, electromagnetic and strong) are not intrinsic to the basic equations of motion. These seemingly different particles and their forces are members of one set intimately linked to each other through the basic lagrangian; the observed asymmetries between them arise through the spontaneously induced asymmetry in the ground state. Such asymmetries, one can show, should prevail at low energies and must disappear at appropriately high energies. Such a view raises the attractive possibility[3] that perhaps all asymmetries may have their origin in this nonperturbative spontaneous breakdown of the symmetry.

To be specific, the "gauge-hypothesis" assumes that the observed interactions (weak, electromagnetic as well as strong) are generated through local gauge symmetries belonging to a gauge group G (not necessarily simple). G must contain at least one nonabelian factor, since weak-interactions involve charged currents. The symmetry G operates on a set of spin-1/2 multiplets $\psi(x)$ and spin-0 multiplets $\phi(x)$ with the transformation properties:[6]

$$\psi(x) \to e^{ig[\vec{F}_\psi]\cdot\vec{\Lambda}(x)}\psi(x)$$

$$\phi(x) \to e^{ig[\vec{F}_\phi]\cdot\vec{\Lambda}(x)}\phi(x) \tag{1}$$

where $[\vec{F}_\psi]$ and $[\vec{F}_\phi]$ are the matrix representations of G on $\psi(x)$ and $\phi(x)$ respectively and $\Lambda(x)$ is an arbitrary function of space and time variables. <u>Crucial to the hypothesis is the assumption that the local symmetry G is preserved as an exact symmetry of the basic lagrangian; it is broken only spontaneously</u>.

The invariance of the largrangian under <u>local</u> symmetry transformations (1) demands (as in the familiar case of electrodynamics) that we introduce a set of massless spin-1 gauge fields $\vec{A}_\mu(x)$ transforming as the adjoint representation of G, with the gauge-compensating transformation property:

$$A^i_\mu(x) \to A^i_\mu(x) + f^{ijk} A^j_\mu(x)\, \Lambda^k(x) + \partial_\mu \Lambda^i(x) \quad \text{(To first order in } \Lambda^i) \tag{2}$$

where f^{ijk} are the structure-constants of the group G. Simultaneously we must impose that $\partial_\mu\psi(x)$ in the lagrangian be replaced by the gauge-covariant derivative

$$D_\mu\psi(x) \equiv (\partial_\mu - ig[\vec{F}_\psi]\cdot A_\mu(x))\psi(x) \tag{3}$$

likewise for $\partial_\mu\phi(x)$; furthermore we must impose that the derivative of $A_\mu(x)$ appear in the gauge invariant combination:

$$G^i_{\mu\nu}(x) = \partial_\mu A^i_\nu(x) - \partial_\nu A^i_\mu(x) + gf^{ijk} A^j_\mu A^k_\nu \tag{4}$$

This finally leads to the gauge-invariant lagrangian of the form:

$$L = \bar{\psi}\,(i\not{D} - m)\psi - \frac{1}{4}\vec{G}_{\mu\nu}\cdot\vec{G}^{\mu\nu}$$

$$+ \left| D_\mu\phi \right|^2 - V(\phi) + h\,\bar{\psi}\Gamma\psi\phi \tag{5}$$

where $V(\phi)$ is a gauge invariant mass plus self-interaction term of the scalar fields ϕ, (subject to the condition of renormalisability $V(\phi)$ should contain in general only quadratic, cubic and quartic terms in ϕ). Note that gauge-invariance does not permit a mass-term $m^2_A(\vec{A}^2_\mu)$ for the spin-1 gauge-field. Depending upon the nature of the symmetry G, the fermi "mass" term $\bar{\psi}m\psi$ may or may not be permissible. In general the mass "m" and the Yukawa coupling $h\Gamma$ is a matrix in the space of the fermions. For most realistic theories, the bare fermi mass-term m needs to be zero. It has to be generated through spontaneous symmetry breaking like the gauge-mass.

Such a theory poses the dilemma of massless-spin-1-fields. If we did introduce spin-1 mass-term into the basic lagrangian (5), thereby destroying the local symmetry, the theory would immediately loose one desirable feature being sought for - renormalisability. On the other hand, if we leave the spin-1 fields massless, they cannot at least describe short range weak interactions.

The dilemma is neatly resolved through a mechanism commonly called the Higgs-phenomemon[7]. The mechanism assumes that the scalar-potential $V(\phi)$ is such that its minimum lies away from the symmetric point; in this case the true ground-state (or the physical vacuum) of the system, corresponding to the minimum of the energy eigenvalue, develops an asymmetry. So do the physical solutions. In other words the scalar potential $V(\phi)$ induces a spontaneous breakdown of the local (and in general the global) symmetry G into a lower symmetry G_o. Breaking of the local symmetry G into G_o makes the massless-spin-1 gauge fields associated with $(G-G_o)$ massive.

Spontaneous breakdown of the symmetry would in general have produced mass-less scalar particles - the Goldstone bosons[8]. Occurrence of such massless particles would have been undesirable (as much as massless weak gauge bosons), since none are found experimentally. In theories possessing local symmetries, however, this difficulty does not arise. The would be massless Goldstone bosons (ϕ) are "eaten up" by the respective two-component massless spin-1 gauge bosons A_μ. This produces the physical massive three component spin-1 gauge fields without any accompanying massless Goldstone-boson as desired. The re-maining members of the spin-0 multiplet ϕ (i.e. those which are are not "eaten up" by the spin-1 fields) emerge as massive spin-0 bosons.

The Case of the Abelian Local Symmetry: To illustrate the phenomenon described above consider the simple example of the theory possessing an abelian U(1)-local symmetry operating on a complex spin-0-field ϕ and the associated spin-1 gauge field A_μ. The gauge invariant lagrangian is:

$$L_1 = \left|(\partial_\mu - igA_\mu)\phi\right|^2 - (1/4)F_{\mu\nu}^2 - V(\phi) \tag{6}$$

Assume

$$V(\phi) = \mu^2 \phi\phi^\dagger + \lambda(\phi\phi^\dagger)^2 \tag{7}$$

The lagrangian is invariant under the local symmetry transformation:

$$\phi(x) \rightarrow e^{ig\Lambda(x)} \phi(x) \; ; \; A_\mu \rightarrow A_\mu + \partial_\mu \Lambda(x) \tag{8}$$

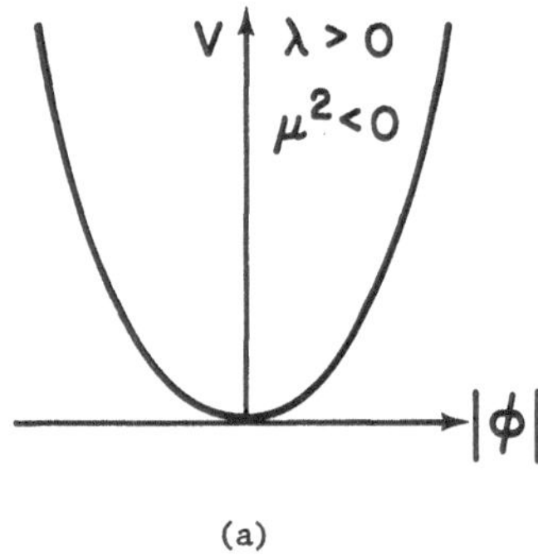

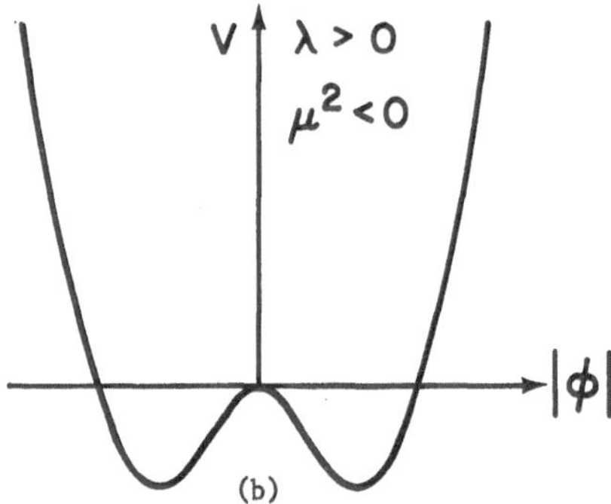

Fig. 1

Observe that if both μ^2 and λ are positive, then the classical minimum (treating ϕ as a classical field of the potential $V(\phi)$) lies at the origin $\phi = 0$ (Fig. 1a). In this case, the ground state of the system (the vacuum) corresponding to the minimum of the potential $V(\phi)$ retains[9] the symmetry (8). So do the physical solutions.

If on the other hand, μ^2 is negative, while λ is positive, the potential $V(\phi)$ as a function of ϕ (treated as a classical field) exhibits the behavior shown in Fig. 1b. In this case the origin $\phi = 0$ is a point of unstable maximum for the energy. The stable minimum of V lies at $|\phi| \equiv (v/\sqrt{2}) = (-\mu^2/2\lambda)^{1/2}$, which describes a circle of radius $v/\sqrt{2}$ in the space of (ϕ real, ϕ imaginary). The physical vacuum would thus correspond to ϕ taking a value at some particular point on this circle (rather than $\phi = 0$); all points on this circle being equivalent in so far as yielding a minimum of the energy. No matter, where is this point on the circle, the symmetry is broken. The ground state and therefore the excited physical states would exhibit the asymmetry despite the symmetry being exact in the basic lagrangian.

To see the manifestations of such a symmetry breaking, shift the field ϕ as follows:

$$\phi \to \phi = \frac{1}{\sqrt{2}} \left[v + \chi(x) + i\eta\,(x) \right] \tag{9}$$

where χ and η are quantum real fields defined such that their expectation values with respect to the physical vacuum vanish, so that the vacuum-expectation value of ϕ is $v/\sqrt{2}$ ($\langle\phi\rangle_{vac} = v/\sqrt{2}$). We have utilised the gauge-invariance of the basic lagrangian L, to choose $\langle\phi\rangle_{vac}$ to be real ($= v/\sqrt{2}$) without loss of generality.

Note that the shift in the field ϕ is only a convenient step towards developing a stable perturbation theory around the nonsymmetric but <u>stable</u> physical vacuum. The results thus obtained might have been obtained by doing perturbation theory around the symmetric vacuum; (this would correspond to no shift in ϕ; $\langle\phi\rangle = 0$); however knowing that small perturbations around the symmetric vacuum are unstable in the present case, one would need to sum the infinite perturbation series in order to obtain the same result - that the symmetry is broken. In this sense the symmetry breaking in this case is a nonperturbative phenomenon

Substituting (9) into (6) the lagrangian takes the form: -

$$L_1 \to L_2 = -\frac{1}{4} F_{\mu\nu}^2 + \frac{g^2 v^2}{2} A_\mu^2 + \frac{1}{2} (\partial_\mu \chi)^2 + \frac{1}{2} (\partial_\mu \eta)^2 + \mu^2 \chi^2 - gvA_\mu \partial^\mu \eta$$
$$+ \text{(cubic and quartic coupling terms)} \qquad (10)$$

The spin-1 field A_μ appears to have acquired a mass. In order to see the particle spectrum explicitly, we must eliminate the bilinear mixing term $A_\mu \partial^\mu \eta$. For this purpose, substitute polar variables for $\phi(x)$ (subject to $\langle\phi(x)\rangle = v/\sqrt{2}$):

$$\phi(x) = \frac{1}{\sqrt{2}} (v + \rho(x))\, e^{ig\xi(x)/v} \qquad (11)$$

$$A_\mu (x) = B_\mu(x) + \frac{1}{v} \partial_\mu \xi(x) \qquad (12)$$

Substituting (11) and (12) in (6) we obtain:

$$L_1 \to L_3 = -\frac{1}{4} B_{\mu\nu}^2 + \frac{g^2 v^2}{2} B_\mu^2 + \frac{1}{2} (\partial_\mu \rho)^2$$
$$-\frac{1}{2} (2\lambda v^2)\rho^2 - \frac{\lambda}{4} \rho^4 + \frac{1}{2} g^2 B_\mu^2 (2v\rho(x) + \rho^2 (x)) - \lambda v \rho^3 \qquad (13)$$

Note that the field $\xi(x)$ has disappeared. Eq. (13) describes a three component spin-1 field B_μ with mass $gv/\sqrt{2}$ and a leftover spin-0 Higgs-boson field of mass $\sqrt{2\lambda}v$. There are thus four independent components at the end (3 for B_μ and 1 for $\rho(x)$) - the same number that we started with for L_1 except that the two component complex ϕ present in L_1 has lost one member (the would be Goldstone boson) which has combined with the two component massless spin-1 field A_μ to yield the massive three component spin-1 field B_μ. As claimed earlier, there is no massless Goldstone boson, eventhough the symmetry is broken spontaneously. This desirable feature associated with the spontaneous breaking of local abelian guage symmetry is preserved, even if the gauge symmetry is a nonabelian one as long as it is local.[7]

In spite of the advantages of the Higgs-mechanism mentioned above, if all that it did for us is to produce massive spin-1 fields with no massless Goldstones, it might not have been so attractive; because after all one could have introduced mass-terms for the spin-1 fields explicitly into the lagrangian and thereby avoid the introduction of the elementary scalar fields altogether. Generating gauge boson masses spontaneously through the Higgs-mechanism offers, however, one distinct advantage - the resulting theory of massive charged and neutral spin-1 bosons become renormalisable[10] just like quantum electrodynamics. The tree diagrams of the resulting theory involving exchange of spin-1 and spin-1/2 as well as left over spin-0 particles (Higgs bosons) exhibit good high energy behaviour[11] consistent with unitarity; a feature which becomes instrumental in making the theory renormalisable. Thus, at last one obtains a theory in which higher order weak interactions are calculable unambiguously in terms of a finite number of parameters present in the "bare" lagrangian - thereby the very use of the lowest order weak-amplitude and in turn questions such as observed β-decay and μ-decay-universality receive an unambiguous theoretical status. One can understand why physical quantities or differences between quantities, such as β-decay versus μ-decay coupling constants and the (K_L-K_S)-mass-difference, that are known to be small, are small. Had we added a mass term for charged spin-1 particles explicitly into the bare lagrangian, we would have spoilt[12] this highly desirable feature - renormalisability - of the theory . In summary, the gauge-hypothesis - i.e. the hypothesis of massless Yang-Mills theory together with the hypothesis of spontaneous breaking of the local symmetry (presently implemented[13] through the Higgs-Kibble-mechanism) - appears to possess the best of all worlds:

(1) It generates massive gauge-particles (A massless photon corresponding to an unbroken local U(1)-symmetry can still emerge from the theory).

(2) In a realistic theory of fermions and bosons possessing Yukawa coupling $h\bar{\psi}\Gamma\psi\phi$ (Eq. (5)), fermions can also acquire mass through the vacuum expectation value of $\langle\phi\rangle$, even if the basic lagrangian does not permit such mass-terms.

(3) The theory generates no Goldstone-bosons.[14]

(4) It possesses good high energy behaviour. It is renormalisable.

(5) Extension[3,4] of the hypothesis to encompass weak, electromagnetic as well as strong interactions provides the scope for a unification of the basic particles and their forces under one underlying principle - The Gauge Principle. Additional accompanying bonus of the hypothesis is that strong interactions generated this time by nonabelian local symmetry[3] become "asymptotically free"[15] - a feature which provides an explanation[16] of approximate scaling and the pattern of its violation observed in deep inelastic lepton-hadron-scattering at SLAC and Fermilab.

I refer you to the excellent review articles by Abers and Lee, Bég and Sirlin, Bernstein, Llewellynsmith and Weinberg for a more detailed presentation of some of the points touched upon in the above preview.

II. THE STANDARD MODEL

2.1 $SU(2)_L \times U(1)$ - A Minimal Gauge Theory Linking Weak and Electromagnetic Interactions of Leptons

A minimal theory[2] describing the weak and electromagnetic interactions of electrons, muons and their neutrinos ν_e and ν_μ may be generated under the gauge-hypothesis by assuming a local symmetry $SU(2)_L \times U(1)$ with the left-handed components $(\nu_e, e^-)_L$ and $(\nu_\mu, \mu^-)_L$ transforming as doublets of $SU(2)_L$ and the right-handed e^-_R and μ^-_R transforming as singlets of $SU(2)_L$. The simplest scalar multiplets providing a breaking of the local symmetry is provided by a doublet $\phi = (\phi^+, \phi^\circ)$ of $SU(2)_L$. The transformation-property of the spinor and the scalar fields under $SU(2)_L \times U(1)$ and their $SU(2)_L \times U(1)$-quantum-numbers are listed below (with only e-multiplet exhibited, likewise for μ):

$$
\begin{aligned}
L_e &\equiv \frac{1}{2}(1+\gamma_5)\begin{pmatrix}\nu_e\\ e^-\end{pmatrix};\quad e^-_R \equiv \frac{1}{2}(1-\gamma_5)e^-;\quad \phi = \begin{pmatrix}\phi^+\\ \phi^\circ\end{pmatrix}\\
L_e &\rightarrow \left[e^{ig(\vec{\tau}/2)\cdot\vec{\Lambda}(x) - i g' \Lambda'(x)/2}\right] L_e\\
e^-_R &\rightarrow e^{-ig'\Lambda'(x)}\, e^-_R\\
\phi &\rightarrow \left[e^{ig(\vec{\tau}/2)\cdot\vec{\Lambda}(x) + ig'\Lambda'(x)/2}\right]\phi
\end{aligned}
\tag{14}
$$

	I_{3L}	Y
ν_{eL}	1/2	-1
e^-_L	1/2	-1
e^-_R	0	-2
ϕ^+	1/2	+1

Note that the abelian U(1)-quantum-number Y needs to be "adjusted" for each multiplet to yield the desired electric charge of the particles (which is defined by $Q \equiv I_{3L} + Y/2$, I_{3L} is the third component of $SU(2)_L$- generator. The only restriction in the choice of Y is that it must be the same for all members in an irreducible $SU(2)_L$ - multiplet, so that the group is $SU(2)_L \times U(1)$. (This unaesthetic freedom associated with abelian contribution to electric charge is avoided once the U(1)-factor is embedded within a unifying nonabelian symmetry discussed later). Note also that the theory manifestly violates left-right symmetry, since e^-_L and e^-_R are treated on very different grounds, as also ν_{eL} and ν_{eR}. In other words the theory is intrinsically parity nonconserving.

The corresponding lagrangian (respecting the local symmetry) is given by:

$$L = -\frac{1}{4}(\vec{W}_{\mu\nu}\cdot\vec{W}_{\mu\nu} + B^o_{\mu\nu}B^o_{\mu\nu})$$
$$+ \bar{L}_e i\gamma_\mu (\partial_\mu + ig\frac{\vec{\tau}}{2}\cdot\vec{W}_\mu - i\frac{g'}{2}B^o_\mu) L_e$$
$$+ \bar{e}_R i\gamma_\mu (\partial_\mu - ig'B^o_\mu) e_R$$
$$- h_e(\bar{L}_e \phi e_R + \bar{e}_R \phi^\dagger L_e) + (e\to\mu)$$
$$+ \left| (\partial_\mu - i\frac{g'}{2}B^o_\mu - ig\frac{\vec{\tau}}{2}\cdot\vec{W}_\mu)\phi^\dagger \right|^2 - V(\phi) \tag{15}$$

$$V(\phi) = \mu^2\phi^+\phi + \lambda(\phi^+\phi)^2 \tag{16}$$

Note that $SU(2)_L$-gauge-invariance forbids an electron mass term in the basic lagrangian (15). The symmetry group involves two basic gauge coupling constants g and g' since it is a direct product of $SU(2)_L$ and U(1).

As in the previous example, choose $\mu^2 < 0$ with $\lambda > 0$ to induce a spontaneous breakdown of the local symmetry $SU(2)_L \times U(1)$; choose the neutral component of ϕ to acquire nonzero vacuum expectation value:

$$\langle\phi\rangle = \frac{1}{\sqrt{2}}\binom{0}{v} \tag{17}$$

so that ϕ is shifted into $\langle\phi\rangle + \zeta(x)$. As in the previous example, one may choose without loss of generality v to be real:

$$v = \sqrt{-\mu^2/\lambda} \tag{18}$$

Substitution of $\phi \to \phi' = \langle\phi\rangle + \zeta(x)$ in the lagrangian generates masses for the gauge bosons as well as fermions. The mass-lagrangian is given by:

$$L_{mass} = \frac{g^2v^2}{8}\vec{W}^2_\mu + \frac{g'^2v^2}{8}(B^o_\mu)^2 + \frac{gg'}{4}B^0_\mu W^3_\mu - (h_e v/\sqrt{2})\bar{e}e - (h_\mu v/\sqrt{2})\bar{\mu}\mu \tag{19}$$

This implies

$$m^2_{W^+} = g^2v^2/4, \quad m_{e,\mu} = (h_{e,\mu})v/\sqrt{2} \tag{20}$$

$$m^2_{neutral} = \frac{v^2}{8} \begin{array}{c} \begin{matrix} B^0 & W^3 \end{matrix} \\ \begin{bmatrix} g'^2 & gg' \\ gg' & g^2 \end{bmatrix} \end{array} \tag{21}$$

It may be verified that three of the four real components of $\phi = (\phi^+, \phi^0)$ get absorbed to make $W^\pm$ and Z^0 massive; one neutral scalar field is left over with mass = $2\lambda v^2$. Note that the four component fermions e and μ have acquired mass $(h_e v/\sqrt{2})$ and $(h_\mu v/\sqrt{2})$ respectively. The Yukawa coupling constants h_e and h_μ presently are parameters, which need to be adjusted to fit the observed masses of the electron and the muon. The neutrino remains two-component and massless since its right-handed counterpart was never introduced into the theory.

The mass matrix $m^2_{neutral}$ has one massless eigenstate - the photon - and one massive eigenstate: the Z^0. Their compositions and masses are

$$A_\mu \equiv \frac{g' W^3_\mu + g\, B^0_\mu}{(g^2+g'^2)^{1/2}} \equiv \sin\theta_W\, W^3_\mu + \cos\theta_W\, B^0_\mu$$

$$Z^0_\mu = \frac{-g'\, B^0_\mu + g\, W^3_\mu}{(g^2+g'^2)^{1/2}} = \cos\theta_W\, W^3_\mu - \sin\theta_W\, B^0_\mu \qquad (22)$$

$$\tan\theta_W = g'/g$$

$$m_A = 0$$

$$m_Z = \frac{(g^2+g'^2)^{1/2} v}{2} = \frac{m_{W^+}}{\cos\theta_W} \geqslant m_{W^+} \qquad (23)$$

This special relationship $m_Z = (m_{W^+}/\cos\theta_W)$ is a joint consequence of generating gauge masses through Higgs-mechanism and equally important the special assumption that $SU(2)_L$-<u>doublet</u> scalar fields are used to break the symmetry [Experimentally, this relationship appears to be borne out, see later]. The electric charge e is given by

$$e = gg'/(g^2 + g'^2)^{1/2} = g\sin\theta_W = g'\cos\theta_W \leqslant (g,g') \qquad (24)$$

The interaction of the charged W^+ to fermions is:

$$L_{W^\pm} = [(g\sqrt{2})/4]\ \bar{\nu}_e \gamma_\mu (1+\gamma_5) e^-\ W^+_\mu + (e \rightarrow \mu) + \text{h.c.} \qquad (25)$$

This leads to an effective four fermion-interaction $[g^2/(8m^2_{W^+})](e^-\gamma_\mu(1+\gamma_5)\nu_e)$ $(\bar{\nu}_e\gamma_\mu(1+\gamma_5)e)$. Following familiar convention, the strength of this interaction, should be identified with $(G_F/\sqrt{2})$. Thus we obtain

$$g^2/(8m^2_{W^+}) = G_F/\sqrt{2} \qquad (26)$$

substituting $G_F M_N^2 \simeq 10^{-5}$ and $g^2 = e^2/\sin^2\theta_W$, we obtain

$$m_{W^+} \simeq 37.5\ \text{GeV}/\sin\theta_W \qquad (27)$$

(Recent neutrino experiments[18] show that $\sin^2\theta_W$ lie between $\simeq$ 0.25 to 0.30. Correspondingly, we would expect W^+ to lie between $\simeq$ 75 and 68 GeV. In turn using Eq. (23), this implies that the neutral Z^0 of the theory should lie between $\simeq$ 87 and 82 GeV.)

Using the composition of Z^0 in terms of the canonical fields W^3 and B^0 (Eq. (22)) and the interactions of W^3 and B^0, the Z^0-fermion interaction is given by:

$$L_{Z^o} = Z^o_\mu(g^2+g'^2)^{-1/2}[(g^2/2)(\bar{\nu}_{eL}\gamma_\mu\nu_{eL} - \bar{e}_L\gamma_\mu e_L) -(g'^2)(-\frac{1}{2}\bar{\nu}_{eL}\gamma_\mu\nu_{eL} - \frac{1}{2}\bar{e}_L\gamma_\mu e_L - \bar{e}_R\gamma_\mu e_R) + (e\to\mu)]$$

$$= Z^0_\mu(g^2+g'^2)^{1/2}(1/4)[\bar{\nu}_e\gamma_\mu(1+\gamma_5)\nu_e - \bar{e}\gamma_\mu(\frac{g^2-3g'^2}{g^2+g'^2} + \gamma_5)e + (e\to\mu)] \tag{28}$$

Such an interaction would induce $\nu_\mu e \to \nu_\mu e$ and $\bar{\nu}_\mu e - \bar{\nu}_\mu e$ scatterings. These processes can not be induced via charged $W^{\pm}$-exchange in the second order. Thus they are typical of neutral-current weak interactions. They have been seen to occur[19] with a characteristic amplitude of order G_F establishing the existence of this new class of weak-interactions. The Z^0-interaction would also contribute to $\bar{\nu}_e e \to \bar{\nu}_e e$ and $\nu_e e \to \nu_e e$-scatterings; this would be on top of W^+-contribution to these latter processes.

It is worth noting that first of all, the neutral current interaction of the theory is parity violating possessing both vector and axial vector couplings. (This is expected, since the basic lagrangian, prior to spontaneous symmetry breaking, is parity violating.) Second, unlike the charged case, the neutral current interaction is neither pure V-A, nor pure V+A (except, of course, for terms involving two component neutrinos).

Before entering into the extension of this theory to hadrons, two theoretical remarks are in order:

(1) High Energy Behaviour: As stated earlier, the fermion-gauge boson interactions of the theory (leading to exchages of massive $W^{\pm}$, Z^0 and fermions) plus the gauge and Yukawa interaction of the single leftover neutral Higgs-boson "ϕ^0" leads to good high energy behaviour of all tree amplitudes, consitent with unitarity bound. The fact that the Z^0 and the "ϕ^0" interactions are needed (in addition to those of $W^{\pm}$) may be seen by considering the tree amplitude for the process $e^-e^+ \to W^+W^-$ with e^- and e^+ having either the same or opposite helicities. There are four exchanges which contribute to this process: the exchanges of photon, Z^0 and "ϕ^0" in the s-channel and that of ν_e in the U-channel. Figs.2a,b,c,d):

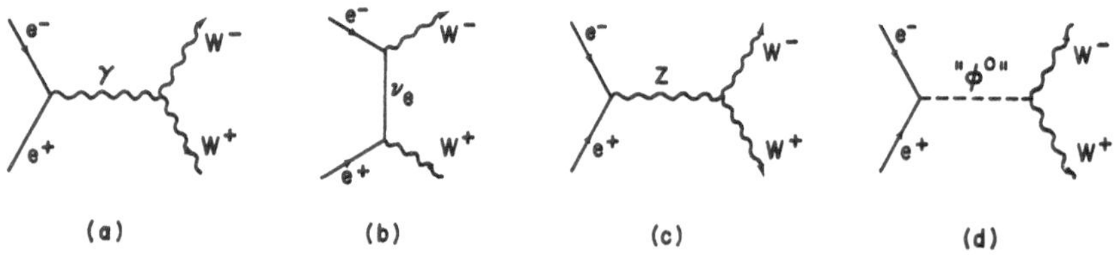

Fig. 2

It may be verified that the sum of the amplitudes ($f_a + f_b + f_c$) falls off like 1/E (for E >> m_W) with e^- ane e^+ having opposite helicities, where E denotes center of mass energy. Fig. 2(d) would contribute when e^- and e^+ have same helicity; in this case it may be verified that one needs to have the sum of all four amplitudes ($f_a + f_b + f_c + f_d$) to obtain the same good high energy behaviour. Without the neutral Z^0 and "ϕ^0", only 2(a) and (b) contribute, which lead to bad high energy behaviour for either helicity configuration. This demonstrates the role of Z^0 as well as the Higgs-meson "ϕ^0" in the realisation of good high energy behaviour of tree amplitudes.

(2) Adler-Bell-Jackiw Triangle anomalies: The spontaneously broken gauge theory described so far is not renormalisable (to all orders) due to the presence of axial couplings of the fermions. The anomalous divergences associated with axial vector currents spoil the symmetry apparent in the Feynman rules. These anomalies arise through divergences associated with triangle graphs as shown below.

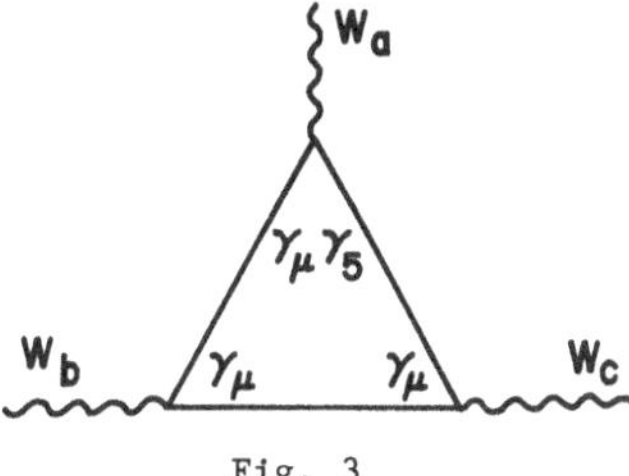

Fig. 3

One can show that the presence of such triangle anomalies would spoil[20] renormalisability of the theory. The demand that the theory (possessing axial vector couplings) be renormalisable to all orders can however be met if the anomalies mutually cancel among themselves. This can happen[21] if either the underlying group itself is safe (as for example SU(2)), or the representation content of the fermions turn out to be safe. Note that the cancellation of anomalies can occur due to contributions to trianglegraphs from fermions of differing masses, since the divergences associated with these graphs are independent of the masses of the contributing fermions.

For the $SU(2)_L \times U(1)$-theory, with only electron and muon-doublets, the anomalies do not cancel due to the axial interactions of the U(1)-gauge field (the B^0). This may explicitly be verified by drawing the contributions to the triangle graph with three external B^0 lines. Thus the $SU(2)_L \times U(1)$-theory, presented so far, needs to be supplemented by additional fermions (possessing appropriate couplings to the U(1)-field), in order that it may be anomaly free and thus be renormalisable. These additional fermions may even be hadrons (quarks).

The necessary and sufficient condition[22] for the theory to be anomaly free is:

$$\mathrm{Tr}\ [\gamma_5\{\Gamma_a,\Gamma_b\}\Gamma_c] = 0$$

for all a,b,c. $\Gamma_{a,b,c}$ denotes the vertices associated with the gauge fields $W_{a,b,c}$.

They are matrices in the space of all the fermions of the theory and, of course, they also involve γ-matrices. The trace is taken with respect to γ-matrices as well as fermion-incides.

For a general listing of safe groups, see Ref. 21. For the $SU(2)_L \times U(1)$-theory, the condition for the absence of anomalies reduces to

$$T\gamma\ [T_3^2\ Y] = 0$$

If only $SU(2)_L$ doublets and singlets are involved, the condition reduces to sum of fundamental charges being zero.

$$T\gamma\ Q^{em} = 0$$

It should be stressed that this is neither a necessary nor a sufficient condition for more general cases.

2.2 Extension To Hadrons: The Need for Charm

The minimal $SU(2)_L \times U(1)$-theory may be extended to include hadrons by postulating that of the three quark-fields (p,n_b,λ_b) the left-handed $(p,n_b)_L$ transform as a doublet of $SU(2)_L$; $(\lambda_b)_L$ and $(p,n_b,\lambda_b)_R$ transform as singlets of $SU(2)_L$ and that their U(1)-quantum numbers are adjusted to yield appropriate electric charges for these fields (as in the case of leptons). The subscript b here denotes "bare" and is used in anticipation that the spontaneously generated Fermi mass-matrix would in general mix the bare n_b and λ_b-fields; the physical fields would correspond to the eigenstates of this mass-matrix. Such a mixing is indeed needed, in order that the charged current of the theory in terms of the physical fields (p,n,λ) may have the familiar Cabibbo-form $\bar{p}(n\cos\theta + \lambda\sin\theta)$, where θ is the Cabibbo angle. It is straightforward to write down[23] gauge-invariant Yukawa coupling of the Higgs-doublets $\phi = (\phi^+,\phi^o)$ and $\tilde{\phi} = i\,\tau_2\,\phi^* = (\bar{\phi}^o, -\phi^-)$, introduced before, with the left and right-handed fermion-multiplets to generate a mass for the p-quark and diagonal as well as non-diagonal elements for the n-λ-mass matrix through $\langle\phi\rangle \neq 0$. In summary the $SU(2)_L$-gauge currents in terms of the physical fields have the form represented by the doublet:

$$\binom{p}{n_b}_L \xrightarrow[\text{mass-matrix}]{\text{Fermi}} \binom{p}{n\cos\theta + \lambda\sin\theta}_L$$

The orthogonal combination $(-n\sin\theta + \lambda\cos\theta)_L = (\lambda_b)_L$, as well as $(p,n_b,\lambda_b)_R$ are singlets of $SU(2)_L$. Thus the charged W^+ is coupled to the appropriate Cabibbo-current. By the same token W^3 coupling to quarks is given by:

$$\begin{aligned}
&\frac{g}{2} W_\mu^3\ [\bar{p}_L\gamma_\mu p_L - \bar{n}_{bL}\gamma_\mu n_{bL}] \\
&= \frac{g}{2} W_\mu^3\ [\bar{p}_L\gamma_\mu p_L - \{\cos^2\theta\ \bar{n}_L\gamma_\mu n_L + \sin^2\theta\ \bar{\lambda}_L\gamma_\mu\lambda_L \\
&\qquad + \cos\theta\sin\theta\ (\bar{n}_L\gamma_\mu\lambda_L + \bar{\lambda}_L\gamma_\mu n_L)\}]
\end{aligned} \tag{29}$$

Note the appearance of the <u>strangeness changing $\Delta s = \pm 1$ neutral current</u> with a coupling proportional to $\cos\theta \sin\theta$. Such a current appears in the interaction of the neutral eigenstate Z^o as well ($Z^o = (gW_\mu^3 - g' B_\mu^o)(g^2 + g'^2)^{-1/2}$, see Eq. (22)). Denoting the currents coupled to W_μ^3 and B_μ^o by $gJ_{L\mu}^3$ and $g'J_\mu^o$ respectively and noting that the electromagnetic current J_μ^{em} is the sum of the two (i.e. $J_\mu^{em} = J_{L\mu}^3 + J_\mu^o$), the Z^o-interaction is given by

$$\begin{aligned} L_{Z^o} &= (g^2 + g'^2)^{-1/2} (g^2 J_{L\mu}^3 - g'^2 J_\mu^o) Z_\mu^o \\ &= (g^2 + g'^2)^{-1/2} [(g^2 + g'^2) J_{L\mu}^3 - g'^2 J_\mu^{em}] Z_\mu^o \\ &= (g^2 + g'^2)^{1/2} (J_{L\mu}^3 - \sin^2\theta_W J_\mu^{em}) Z_\mu^o \end{aligned} \qquad (30)$$

where we have substituted $\sin^2\theta_W$ for $g'^2/(g^2+g'^2)$ (see Eq. (23)). Since $J_{L\mu}^3$ contains the strangeness changing neutral current (Eq. (29)), so does the current coupled to Z_μ^o. The same current contains neutral leptonic currents $(\bar\nu\nu)_L$, $(\bar\mu\mu)$ and $(\bar e e)$, as discussed in detail in sec. 2.1. Hence, second order Z^o-interaction can induce $|\Delta s| = 1$ as well as $\Delta s = 2$ strangeness changing neutral current processes with amplitudes as given below

$$\begin{array}{lll} & & \text{Amplitude} \\ \left.\begin{array}{l} K_L \to \mu^+\mu^- \\ K^\pm \to \pi^\pm \nu\bar\nu \\ K^\pm \to \pi^\pm e\bar e \end{array}\right\} & & O(G_F \cos\theta \sin\theta) \\ K^o \leftrightarrow \bar K^o & & O(G_F \cos^2\theta \sin^2\theta) \end{array} \qquad (31)$$

These amplitudes can not be suppressed by increasing the mass of Z^o. This is because, using Eq. (30), we observe that the amplitude for $K_L \to \mu\bar\mu$ for example is proportional to $(\cos\theta \sin\theta)(g^2+g'^2)/4m_Z^2 = (\cos\theta \sin\theta)(g^2+g'^2)\cos^2\theta_W/(4m_{W^+}^2) = (\cos\theta \sin\theta)g^2/(4m_{W^+}^2)$ $= (\cos\theta \sin\theta)(\sqrt{2}\, G_F)$, which is independent of m_Z.

Experimentally the revelant numbers are:

$$\frac{\Gamma(K_L \to \mu^+\mu^-)}{\Gamma(K_L \to \text{All})} \sim 10^{-8}, \qquad \frac{\Gamma(K^\pm \to \pi^\pm e^+e^-)}{\Gamma(K^\pm \to \text{All})} = (2.6 \pm 0.5) \times 10^{-7}$$

$$\frac{\Gamma(K^\pm \to \pi^\pm \nu\bar\nu)}{\Gamma(K^\pm \to \text{All})} < 0.6 \times 10^{-6}, \qquad \left|\frac{m_{K_L} - m_{K_S}}{m_K}\right| = 0.7 \times 10^{-14} \qquad (32)$$

These numbers and the limits suggest that the corresponding amplitudes are more like $O(G_F^2)$ rather than $O(G_F)$. Thus gauge theories, possessing neutral current interactions as in $SU(2)_L \times U(1)$ and introducing only (p,n,λ)-quarks, face a serious difficulty: They predict too large an amplitude for $\Delta s \neq 0$ neutral current processes.

The remedy is found neatly (and as it appears now rather uniquely) by postulating a fourth quark (c) - the charmed quark - with the same electric charge as the p-quark. This is the mechanism suggested by Glashow, Iliopoulos and Maiani.[24]

Assume two $SU(2)_L$-doublets

$$\binom{p}{n_b}_L \quad , \quad \binom{c}{\lambda_b}_L$$

Here p and c having the same charge and n_b and λ_b having the same charge have the same I_{3L} and therefore also same U(1)-quantum numbers Y. Assume that the spontaneously generated Fermi mass-matrix mixes n_b and λ_b; thus as before $n_b = n \cos\theta + \lambda \sin\theta$ and $\lambda_b = -n \sin\theta + \lambda \cos\theta$, where (n,λ) are the physical fields. The two $SU(2)_L$-doublets in terms of physical quark-fields are:

$$\begin{pmatrix} p \\ n\cos\theta + \lambda\sin\theta \end{pmatrix}_L \quad , \quad \begin{pmatrix} c \\ -n\sin\theta + \lambda\cos\theta \end{pmatrix}_L$$

Now W^3-coupling to quarks is given by:

$$(g/2)W_\mu^3[(\bar{p}_L\gamma_\mu p_L + \bar{c}_L\gamma_\mu c_L) - (\bar{n}_{bL}\gamma_\mu n_{bL} + \bar{\lambda}_{bL}\gamma_\mu\lambda_{bL})]$$

$$= (g/2)W_\mu^3[(\bar{p}_L\gamma_\mu p_L + \bar{c}_L\gamma_\mu c_L) - (\bar{n}_L\gamma_\mu n_L + \bar{\lambda}_L\gamma_\mu\lambda_L)] \qquad (33)$$

Thus the W^3-interaction no longer possesses the offending $|\Delta s| = 1$ neutral current piece regardless of the magnitude of the Cabibbo angle θ. The same is true of the Z^o-interaction. [In general, one may permit independent rotations θ_1, and θ_2 in the (p,c) and (n,λ)-spaces respectively through the Fermi mass-matrix. It may be verified that the resulting theory is the same as if there was a single Cabibbo rotation in the (n,λ)-space by an angle $\theta = \theta_1 - \theta_2$].

Since Z^o-current is devoid of $|\Delta s| = 1$-piece, the theory does not induce any of the $|\Delta s| = 1$ and $|\Delta s| = 2$ neutral-current processes listed in (31) to order g^2 and g'^2. Such processes would still be induced in the <u>fourth order</u> of the gauge interactions through a combination of ($W^\pm$ and Z^o) or double $W^\pm$ exchanges, as for example shown in Figs. 4a and b. One can show[25] that in a theory in which strong interactions are generated through a renormalisable gauge-principle utilising a symmetry, which <u>commutes</u> with the weak-symmetry (e.g. $SU(2)_L \times U(1)$), such fourth order diagrams involving loops are <u>damped</u> by one power of m_W^2 leading to amplitudes of order $\alpha^2/m_W^2 \sim G_F\alpha$. This is true without taking into account the <u>cancellation</u> between p and c quark-contributions, which are proportional to cosθ sinθ and -cosθ sinθ respectively. This cancellation would be exact if p and c quarks were mass-degenerate. Thus the $|\Delta s| = 1$ and $|\Delta s| = 2$ neutral current amplitudes are finally of order

$$G_F\alpha(\cos\theta\ \sin\theta)^n\ (m_c^2-m_p^2)/m_{W,Z}^2 \sim (G_F^2)(\cos\theta\ \sin\theta)^n(m_c^2-m_p^2)$$

where n = 1, 2 for $|\Delta s| = 1,2$. Such a magnitude would be compatible with the observed rates and upper limits shown in (32) provided $\Delta m^2 \equiv m_c^2 - m_p^2 \sim$ (1 to few) GeV^2. We conclude that the GIM mechanism provides a complete solution to the

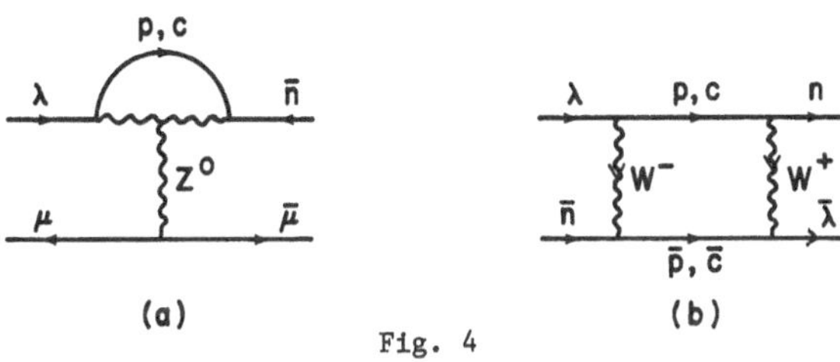

Fig. 4

problem of strangeness changing neutral-current processes provided the charm quark is not excessively heavy.

The best estimate of charmed quark mass is obtained from K_L-K_S-mass difference.[26] The diagram Fig. 4b is proportional to

$$\frac{g^4}{8} \sin^2\theta \cos^2\theta \int \frac{d^4K}{(2\pi)^4} \left(\frac{1}{K^2-m_c^2} - \frac{1}{K^2-m_p^2}\right)^2 \left(\frac{1}{K^2-m_W^2}\right)^2 \ (\gamma \text{ factors})$$

In the limit $m_p^2 << m_c^2 << m_W^2$, this leads to an effective lagrangian

$$L_{eff} \simeq -\frac{G_F}{\sqrt{2}} \frac{\alpha}{4\pi} \left(\frac{m_c}{38 \text{ GeV}}\right)^2 \cos^2\theta_c \sin^2\theta_c \ \{\bar{\lambda}\gamma_\mu \frac{(1+\gamma_5)}{2} n\} \{\bar{\lambda}\gamma_\mu \frac{(1+\gamma_5)}{2} n\}$$

Inserting this expression between K^o and $\bar{K}^o$ and retaining only the vacuum intermediate state between the two curly brackets yields

$$\Delta m_K \simeq \frac{1}{m_K} \langle \bar{K}^o | -L_{eff} | K^o \rangle \sim \frac{G_F}{\sqrt{2}} f_K^2 m_K \left(\frac{\alpha}{4\pi}\right) \left(\frac{m_c}{38 \text{ GeV}}\right)^2 \cos^2\theta_c \sin^2\theta_c$$

Using the observed value $\Delta m_K = 0.7 \times 10^{-14} m_K$, we obtain

$$m_c \sim 1.5 \text{ to } 2 \text{ GeV}$$

The mass thus estimated should be interpereted as the current and within a few hundred MeV the constituent charmed quark-mass, relveant for estimating masses of composites of charm quarks involving $c\bar{c}$, $c\bar{p}$, $c\bar{n}$ and $c\bar{\lambda}$-systems. (The physical mass of the charmed quark, if it can be liberated, may be somewhat higher, see remarks later.)

The naive estimate of the charmed quark-mass given above has turned out, rather surprisingly, to accord well with experiments.

The evidence for the observation of charm may be summarised as follows:

(1) The narrow resonances J/ψ and ψ' provide indirect evidence for charm in that their narrowness, decay modes and isospin properties are appropriately interpreted by postulating that, they are $^3S_1$ $q\bar{q}$ composites of quarks involving a new flavor (different from p, n, λ), with a constituent mass ≃ 1.5 GeV. This accords simultaneously with half the mass of the J/ψ, and with the masses of the D-particles.

(2) Direct evidence for charm comes from the discovery[27] of $(D^o, \bar{D}^o)$ with mass ≃ 1863 MeV and $D^\pm$ with mass ≃ 1868 MeV in $e^-e^+ \to D\bar{D} + X$ and in $e^-e^+ \to \psi(3772) \to D\bar{D}$. The $D^\pm$ are found to decay weakly predominantly into K-associated modes

($D^{\pm} \to K^{\mp}\pi^{\pm}\pi^{\pm}$; $D^{\pm} \to K_S\pi^{\pm}$) consistent with the GIM current. The ratio $\Gamma(D^o \to \pi^+\pi^-)/\Gamma(D^o \to K^-\pi^+)$, expected to be of order $\tan^2\theta \simeq 5\%$, is found to be less than[28] 8%. Thus one may identify the D^+ and D^o with ^1S_o $(c\bar{n})$ and $(c\bar{p})$ composites.

(3) Fair evidence for charm (to be precise for a new quantum number) arises from the observation[29] of like sign dimuons $\nu N \to \mu^-\mu^+X$, and in particular the preferential (K,Λ)-associated dilepton events[30] $\nu N \to \mu^-e^+ + (K,\Lambda)$.

2.3 The Need For Color

Prior to the need for charm, the need for a three valued "color" degree of freedom of quarks was realised as early as 1964. The color postulate is that quarks of each flavor $(p,n,\lambda, \ldots)$ come in three different colors (red, yellow and blue):

$$q = \begin{bmatrix} p_{red} & p_{yellow} & p_{blue} \\ n_{red} & n_{yellow} & n_{blue} \\ \lambda_{red} & \lambda_{yellow} & \lambda_{blue} \\ \vdots & \vdots & \vdots \end{bmatrix} \tag{34}$$

Motivations for postulating this new degree of freedom are the following:

(i) SU(6) and Pauli Principle: The spin $\frac{1}{2}$ baryon octet and the spin $\frac{3}{2}$ baryon decuplet can successfully be grouped to form a 56-plet of SU(6) containing internal SU(3) and spin SU(2). The 56-plet is however totally symmetric under SU(6). Such a classification would be consistent with the Pauli Principle, if one postulates[31] that quarks of each flavor come in three different colors and that the low lying baryons (this includes the 56-plet) are SU(3)'-color-singlet -- three quark composites and thus totally antisymmetric under SU(3)'-color. Simultaneously one needs to assume, consitent with the nonproliferation of low-lying states, that the low-lying mesons $(\pi,K,\eta,\rho,\omega,K^*,\phi)$ are also color singlet $q\bar{q}$-composites.

(ii) The $\pi^o \to 2\gamma$-Rate: The observed rate of $\pi^o \to 2\gamma$ turns out to be compatible with the theoretical value based on PCAC provided we postulate that quarks possess color (the rate is the same for either fractional or integer quark-charges). On the other hand it would be too low by a factor of 9 compared to the observed value, if we did not introduce color.

(iii) The $\Delta I = 1/2$-Rule: Current algebra and PCAC together with the color-hypothesis implies that, in the absence of $q\bar{q}$-pair excitations within the color-singlet baryons, only the $\Delta I = 1/2$-(or SU(3) octet)-piece of the effective nonleptonic interaction $(\bar{p}\gamma_\mu(1+\gamma_5)\lambda)(\bar{n}\gamma_\mu(1+\gamma_5)p)$ contibutes[32] to the amplitudes of nonleptonic hyperon decays $(\Lambda \to N+\pi,\ \Xi \to \Lambda+\pi$, etc.) thus providing an explanation for the observed $\Delta I = 1/2$-rule. Plausible analog of this argument for K-decays has been given by Nelson and Sebastian.[32] The above enhancement of the matrix element of the $\Delta I = 1/2$-piece of the interaction is above and beyond the short distance enhancement[33] of the octet

operator relative to the 27-plet operator, which may be calculated within the asymptotically free color-gauge theory. The latter is a factor $\simeq 4$ for the four flavor V-A theory. Both sources of enhancement utilise the color-hypothesis; both appear to be needed for an explanation of the $\Delta I = 1/2$-rule in hyperon and K-decays.
(iv) The Value of $R = \sigma(e^+e^- \to \text{hadrons})/\sigma(e^+e^- \to \mu^+\mu^-)$: A more recent motivation for the color-hypothesis arises from the observed value of $R \simeq 2$ below threshold for charm-excitation ($E_{cm} \lesssim 3.5$ GeV). Using the theoretically expected relationship $R = \Sigma Q_i^2$, one may infer that (p,n,λ)-flavors occur in three different colors.
(v) Chromodynamics: The existence of the color-degree of freedom permits the generation of strong interactions through color-gauges. The many advantages of chromodynamics generated through color-gauging, which in turn make the need for color compelling, are outlined in the next section.

2.4 $SU(2)_L \times U(1) \times SU(3)'_{L+R}$ - A Minimal Gauge Symmetry Generating Weak, Electromagnetic and Strong Interactions

The existence of the new SU(3)'-color degree of freedom commuting with flavor permits the generation of a renormalisable theory of weak, electromagnetic as well as strong interactions with weak interactions being associated with flavor and strong interactions with color-gauging.[34,35,36] Electromagnetism in general is a blend of flavor and color-gauges depending upon the nature of quark-charges (see below).

The advantages of generating effective strong interactions through SU(3)'-color-gauges (rather than e.g. through an abelian vector gluon or through scalar and pseudoscalar Yukawa interactions) are many fold. Most notably they are:

(a) Asymptotic freedom[15]

(b) Saturation properties[34,37,38] in quark-binding (qqq are strongly bound, but not qq and qqqq), and

(c) Unification of the Basic Forces[3,4] (Weak, electromagnetic as well as strong)

None of these features could be realised, had we generated effective strong interactions in any other way (e.g. through abelian U(1)-gluon or scalar and pseudoscalar Yukawa interactions). Given that the quarks do possess color, there are even further advantages of generating strong interactions through SU(3)'-color gauges, some of which are shared by the U(1)-gluon-interaction and some by the scalar or pseudoscalar Yukawa interactions, but not by both at the same time. They are:

(d) "Natural" Conservations of Parity and Strangeness to order G_F: Unified theories of weak and electromagnetic interactions in general pose the problem that they may lead to violations of parity and strangeness in pure hadronic processes in order α through weak W-loops. One needs to ensure that the offensive loops, after suitable mass and charge-renormalisations, are naturally damped and are of order $\alpha/m_W^2 \sim G_F$. One finds[25] that this is indeed the case if strong interactions are generated by a gauge-principle - either through the abelian U(1)-vector-gluon, or through the

SU(3)'-color-gauge interactions; both of which share the relevant common features that they have a gauge-origin and that they are invariant under the global $U(n)_L$ x $U(n)_R$-symmetry operating in the space of n quark-flavors. This desirable feature would be lost if the strong interactions are generated instead through scalar or pseudoscalar Yukawa interactions. In this case, one has to adjust counterterms for Yukawa interactions, so that their parity and strangenss violating parts have renormalised values of order G_F.

(e) $SU(3) \times SU(3)'_{color}$ rather than SU(9) as a Hadronic Symmetry: The SU(3)'-color-gauge interactions of (p,n,λ)-quarks are invariant under $U(3)_L \times U(3)_R \times SU(3)'$, which breaks down to the observed SU(3) x SU(3)'-symmetry through spontaneously generated quark-masses. By contrast, the U(1)-gluon interaction of (p,n,λ)-quarks possessing color is invariant under the big SU(9)-symmetry, which can not be broken badly (as needed experimentally) to the observed symmetries by quark-mass-terms without breaking SU(3) and/or SU(3)' equally badly.[39] This observation poses a constraint on the building of models of weak and electromagnetic interactions. Models of weak interactions[40] utilising the color-degree of freedom, thereby introducing weak-currents transforming as (1,8) and (8,8) under SU(3)x SU(3)', run into the problem that their basic strong interactions must be invariant under SU(9) in order that the weak currents be conserved, which in turn is necessitated in order that just the weak interactions be renormalisable. In other words, such models can not generate strong interactions through the SU(3)'-color-gauges without spoiling renormalisability of weak interactions. They may generate strong interactions for example through the U(1)-gluon or scalar Yukawa interactions; either of these however possess the big SU(9)-symmetry and thus pose the difficulty mentioned above. The moral of this is: one needs to introduce the color-degree of freedom, suggested on several grounds; but simultaneously one must reserve this degree of freedom to generate just strong interactions; the "basic" weak interactions[41] must be generated through the commuting flavor degree of freedom only. Let us now turn to a discussion of such a gauge-structure.

The minimal gauge-symmetry linking weak, electromagnetic and strong interactions is provided by the gauge-structure:[35]

$$G_o = SU(2)_L \times U(1) \times SU(3)'_{L+R} \tag{35}$$

operating on the minimal set of fermions

$$F_{L,R} = \begin{bmatrix} p_r & p_y & p_b \\ n_r & n_y & n_b \\ \lambda_r & \lambda_y & \lambda_b \\ c_r & c_y & c_b \\ \text{red} & \text{yellow} & \text{blue} \end{bmatrix}_{L,R} + \begin{bmatrix} \nu_{eL} \\ e^-_{L,R} \\ \mu^-_{L,R} \\ \nu_{\mu L} \end{bmatrix} \tag{36}$$

One needs the charm degree of freedom in accordance with the discussion presented earlier. Left-handed as well as right-handed quarks of each flavor $(q_r, q_y, q_b)_{q=p,n,\lambda,\text{or } c}$ transform as a $\underset{\sim}{3}^*$ of $SU(3)'_{color}$, while leptons are assumed to be singlets of $SU(3)'_{color}$. The demand of SU(3)'-local symmetry thus generates the fermion-color gauge boson interactions given by,

$$L_{color} = \sum_{q=p,n,\lambda,c} f_s(\bar{q}_r\bar{q}_y\bar{q}_b)\gamma_\mu \frac{\vec{\lambda}}{2} \begin{pmatrix} q_r \\ q_y \\ q_b \end{pmatrix} \cdot \vec{V}_\mu \tag{37}$$

f_s denotes the color-gauge coupling constant, $\vec{\lambda}$ the SU(3)'-matrices for the 3* representation and $\vec{V}_\mu$ are the octet of color-gluons.

Left-handed quarks of each color, divided into doublets, i.e. $(p_{iL}, n_{iL})_{i=r,y \text{ or } b}$ and $(c_{iL}, \lambda_{iL})_{i=r,y \text{ or } b}$, transform as doublets of $SU(2)_L$; so do (ν_{eL}, e^-_L) and $(\nu_{\mu L}, \mu^-_L)$. All right handed fields are assumed to be singlets of $SU(2)_L$. The U(1)-quantum numbers of the color and flavor-multiplets need to be adjusted such that (a) U(1) commutes with $SU(2)_L$ and $SU(3)'_{color}$ and, (b) the fermions possess the right electric charges. In short, the $SU(2)_L \times U(1)$-gauge-interactions are the same as before (discussed in secs. 2.1 and 2.2).

2.5 Spontaneous Breakdown of G_o: Quark-charges

There are two consistent solutions for the spontaneous breakdown of G_o and correspondingly two possible solutions[42] for quark-charges.

(i) Fractional Quark-charges: The color-degree of freedom is preserved as a local symmetry. The octet of color-gluons remain neutral and massless. Only the $SU(2)_L$-doublet scalar field ϕ introduced before needs to be introduced to break $SU(2)_L \times U(1)$. Photon is composed of only flavor $SU(2)_L \times U(1)$-gauge bosons; thus $Q = I_{3L} + \frac{Y}{2}$. Quarks acquire Gell-Mann-Zweig fractional charges. In this case, it has been conjectured[36] -- motivated by nonobservation of massless gluons (which could be emitted in pairs in π and K-decays, e.g. $\pi^+ \to \mu\nu + V\bar{V}$) and stable fractionally charged quarks -- that exact local color symmetry leads to absolute confinement of quarks, gluons and all color.

(ii) Integer Quark-charges: The SU(3)'-color-degree of freedom is broken spontaneously through a low mass-scale $\mu \sim 1$ to few GeV, such that the octet of color-gluons acquire mass and possess charged members; the photon is composed of flavor $SU(2)_L \times U(1)$ as well as SU(3)'-color-gauge bosons; the electric charge is[35]

$$Q = (I_{3L} + \frac{Y}{2})_{flav} + (F'_3 + F'_8/\sqrt{3})_{col} \tag{38}$$

The quarks acquire integer-charges like the leptons. The scalar-multiplets, which induce such a scheme of SSB, are the old $\phi = (2, Y = +1, 1)$ and a new multiplet $\sigma = (2+2, Y = -1, 3^*)$ with the vacuum expectation values:

$$\langle\phi\rangle = \frac{1}{\sqrt{2}}\binom{o}{v}\ , \quad \langle\sigma\rangle = \begin{bmatrix} \sigma & o & o \\ o & \sigma & o \\ o & o & \sigma \\ o & o & o \end{bmatrix} \tag{39}$$

A choice of <u>box-representation</u> for σ and the pattern for $\langle\sigma\rangle$ exhibited in (39) has the virtue that even though $\langle\sigma\rangle$ breaks local SU(3)'-color symmetry, it generates an effective $\widetilde{SU}(3)'_{global}$ color-symmetry, good to order α and equivalent to $SU(3)'_{col}$ in the space of quarks and gluons. Thus $\widetilde{SU}(3)'$ can serve[43] as a good color-classification symmetry.

The VEV $\langle\sigma\rangle$ breaks the SU(3)'-color and simultaneously the $SU(2)_L \times U(1)$-flavor-symmetry. This provides a mass ($\simeq f_s\, \sigma/\sqrt{2}$) to the octet of color-gluons, <u>but simultaneously generates a mixing between flavor and color-gauge particles encompassing neutral as well as charged members of each set</u>. This makes the massless photon a mixture of flavor (W^o) and color-gauge particle (U^o); this in turn inevitably generates a light mass "gluon" $\tilde{U}$, which is the orthogonal combination of W^o and U^o. Symbolically[44]

$$\begin{aligned} A &= \cos\delta\ W^o_{flav} + \sin\delta\ U^o_{col}\ ;\ m_A = 0 \\ \tilde{U} &= -\sin\delta\ W^o_{flav} + \cos\delta\ U^o_{col};\ m_{\tilde{U}} \simeq f_s\ \sigma/\sqrt{2} \sim 1 \text{ to few GeV} \end{aligned} \tag{40}$$

where $\tan\delta \sim (g/f_s) \ll 1$, and $U^o_{col} \equiv (1/2)(\sqrt{3}\ V_3 + V_8)$; V_3 and V_8 are the canonical color gauge particles coupled to the generators F'_3 and F'_8 respectively. Likewise the charged $W^\pm$ and charged gluons $V^\pm$ mix leading to eigenstates:

$$\begin{aligned} \tilde{V}^\pm &= \cos\beta\ V^\pm + \sin\beta\ V^\pm \\ \tilde{W}^\pm &= -\sin\beta\ V^\pm + \cos\beta\ W^\pm \end{aligned} \tag{41}$$

The mixing angle $\tan\beta \simeq (m_{V^\pm}/m_{W^\pm})^2(g/f) \lll 1$.

<u>Lepto-production of color</u>: The W^o-U^0-mixing has the subtle effect that photon as well as $\tilde{U}$-exchanges contribute to lepto-production of color ($eN \to e + X_{col}$) for the case of (gauge) integer-charge quarks. These two contributions are proportional to $\cos\delta\ \sin\delta$ and $-\cos\delta\ \sin\delta$ respectively, which mutually cancel each other except for the difference between their propagators ($m_{\tilde{U}} \neq m_A$). Taking this cancellation into account electroproduction-structure functions for the case of gauge-integer charge quarks are given by:[45]

$$F_i = F_i^{flav} + \left(\frac{m_U^2}{|q^2| + m_U^2}\right)^2 F_i^{col} \tag{42}$$

F_i^{flav} and F_i^{col} are defined in the standard manner by the Fourier transforms of matrix elements of ($J^{flav}\ J^{flav}$) and ($J_U^{col}\ J_U^{col}$) respectively. Note the damping factor $m_U^4/(|q|^2 + m_U^2)^2$ for color relative to flavor-production. This has the remarkable consequence that <u>asymptotically gauge integer-charge and fractionally</u>

charged quark-partons respond in one and the same manner towards leptonic probes. Note that not only electro and muon-production, but also neutrino-production-structure functions (arising through $\tilde{V}^{\pm}$ and $\tilde{W}^{\pm}$-exchanges) will have the form given by Eq. (42) with the substitution $m_U \to m_{V^\pm}$, where $m_{V^\pm} \simeq m_U$.

The two solutions (fractional versus integer-charge quarks) can still be distinguished from each other through spin-1 gluon-parton contributions to the structure functions. For the case of fractionally charged quarks, the gluon-partons are neutral and thus do not contribute; while for the case of integer-charge quarks, the gluon-octet possesses charged members, which do contribute; owing to the damping factor (Eq. (42)) their contributions scale. One needs[46] high q^2 low x-measurement of σ_L/σ_T to distinguish between these two solutions: gauge integer versus fractionally charged quarks.

The present electron, muon and neutrino-production experiments on face value appear to show[47] as though quark-charges are fractional. But precisely for the reasons mentioned above, this does not help decide whether the true quark-charges are fractional or integral[46] even if we assume that the effective physical color-threshold for the latter case is relatively low $\lesssim$ 3 to 5 GeV.

Theoretically the two solutions share one important common feature: they are both at least temporarily asymptotically free. Both solutions generate basic strong interactions through nonabelian SU(3)'-color-symmetry. Thus the strong gauge interactions in either case by themselves are asymptotically free. Asymptotic freedom is lost in both cases only due to the quartic couplings of the scalar fields (i.e. ϕ for the case of fractionally charged quarks and ϕ and σ for the case of integer-charge quarks). Since effective renormalised values of such quartic couplings are typically $\lesssim$ e, loss of asymptotic freedom due to such quartic couplings is not serious at presently available momentum-range.

The other similarities and dissimilarities between the two solutions for quark-charges within quark-lepton unified theories -- in particular the reasons why softly broken local color symmetry leading to liberated integer charge quarks becomes a viable alternative to the hypothesis of confinement -- are discussed in chapter IV.

2.6 Summary of the Standard Model

To summarise, the standard model is the hypothesis[2,35,36] that the effective weak interactions of quarks and leptons are generated through the local symmetry $SU(2)_L \times U(1)$ with the extension to hadrons encompassing GIM-charm as well as color; while effective strong interactions are generated via the commuting local symmetry SU(3)'-color. The model requires a minimum of twelve quarks and four leptons. The quarks and the charged leptons (e,μ) are four component objects, while the neutrinos (ν_e,ν_μ) can remain, by choice, as massless two component objects. The fermionic content of the model may be enlarged by adding new quark-flavors and new leptons.

Such enlargement appears arbitrary, unless one views that the symmetry G_o descends spontaneously from a higher unifying symmetry G (see discussions later).

The weak symmetry $SU(2)_L \times U(1)$ is assumed to be broken by the standard doublet ϕ. This hypothesis is commonly supplemented by the specific assumption that SU(3)'-color as a local symmetry is preserved; quarks are fractionally charged and are confined. For most applications of the model (including those based on asymptotic freedom, scaling behaviour and charmonium calculations) such a special assumption is however unnecessary[45] to the extent that the two solutions -- fractional versus integer charge quarks -- lead to the same consequences for reasons mentioned above. Distinctions between the two solutions could emerge only through tests outlined in the next chapter.

Salient features of the standard model, its experimental successes and possible drawbacks, are listed below:

(i) Neutral Currents: The model predicts the existence of neutral current-processes with all amplitudes determined in terms of a single parameter $\sin^2\theta_W$. Predictions of the model for neutrino-processes appear to be impressively brone out[18,19] by the recent data with a value of $\sin^2\theta_W$ lying between $\approx$ 0.25 to 0.30. Accurate measurements of elastic $\nu p \to \nu p$, $\bar{\nu} p \to \bar{\nu} p$-scatterings as well as leptonic scatterings $\nu_\mu e \to \nu_\mu e$, $\bar{\nu}_\mu e \to \bar{\nu}_\mu e$ and $\bar{\nu}_e e \to \bar{\nu}_e e$ would provide further valuable tests of the model.

(ii) Charm: The model requires the existence of charm. This too is impressively borne out by the data.

(iii) Atomic Parity: Neutral current interactions of the model contribute to neutrino, atomic and nuclear interactions. They are necessarily parity violating (see Eqs. (28) and (30)) with the degree of parity violation determined by the weak angle θ_W. Despite the successes of the model for neutrino-processes, its predictions for atomic parity violation appear to be in conflict with recent null or near null results of the search for parity violation in atomic Bismuth.[48] Theoretical predictions of the model, based on relativistic central field approximation together with an estimate of the shielding effect,[49] appear to be at least a factor of 3 lower than the corresponding experimental upper limits. Should continued experimental search for parity violation in Bismuth and other atoms (e.g. cesium, thalium and hydrogen) together with improved theoretical atomic calculations confirm this descrepancy, at least the weak/electromagnetic component $SU(2)_L \times U(1)$ of the standard symmetry G_o would have to be extended. (A simple solution incorporating such an extension is the postulate that Nature is fundamentally left-right symmetric rather than left-handed as in $SU(2)_L \times U(1)$. I discuss such extensions briefly in the next chapter and in more detail in part B.)

(iv) Chromodynamics: The model requires the existence of the color-degree of freedom. Whether color is physical and observable or it is permanently confined - equivalently[50] whether quarks are integer or fractionally charged - is still an experimentally open question.[46] Meanwhile, the successes of the hypothesis that the basic strong interactions are generated by the SU(3)'-color-gauge degree of freedom are many fold: asymptotic freedom, the associated bonus of scaling behaviour, charmonium calculations as well as the many advantages listed in Sec. 2.4. Even though the structure of the low energy weak and electromagnetic symmetry remains to be settled experimentally, it is thus conceivable that the correct theory for the basic low energy strong interactions has been found. It is the one given by the color-gauge interactions. More detailed predictions of the model pertaining for example to the nature of scaling violations in μp-scattering and emergence of jets in e^-e^+-annihilation await experimental tests.[51]

On the theoretical front, it is hoped that the model with exact or softly broken local SU(3)'-color-symmetry should provide the mechanism nonperturbatively to absolute[36] or partial[46] confinement. This is a burden[52] on the theory.

(v) Comments: The standard symmetry G_o, despite its successes, may not be regarded as a fundamental symmetry in as much as it does not relate the basic weak/electromagnetic and strong coupling constants g, g' and f_s to each other: the angle θ_W is a parameter in the theory. An uniconstant unifying symmetry[3,4] descending spontaneously to G_o (or other alternative forms), on the other hand, relates its effective low energy coupling constants to each other and thus determines[53] θ_W.

Furthermore, new flavors beyond charm are suggested recently experimentally through the discovery of the upsilon-structure.[54] Existence of such new flavors (as well as new heavy leptons) were motivated theoretically by considerations of higher unifying symmetries such as[3] $[SU(4)]^4$ and[55] E_7. I therefore now turn into a brief discussion of uniconstant unifying symmetries.

III. UNIFYING QUARKS AND LEPTONS AND THEIR FORCES

3.1 The Hypothesis

The idea of quark-lepton-unification is the postulate that fundamental baryons (quarks) and leptons are members of one multiplet.[3] All basic interactions - weak, electromagnetic and strong - derive their origin through a gauging of the symmetry group of this multiplet and are thus universal with respect to the members of this multiplet. The observed distinctions between quarks and leptons in the realm of strong interactions are then interpreted to be low energy phenomena, brought about by asymmetric boundary conditions (spontaneous symmetry breaking) rather than by asymmetries in the basic equations of motion. Such distinctions must disappear at appropriately high energies, where a new set of quark-lepton and lepton-lepton interactions, generated by the symmetry of the quark-lepton-multiplet, should begin to manifest.

Such a view opens the door to the next step: unification[3,4] of weak, electromagnetic as well as strong interactions[5] through a lagrangian governed by a single basic gauge coupling constant. These different forces (weak, electromagnetic and strong) are assumed to be components of a single set. Without any one of them, the set would be incomplete. The observed distinctions between weak/electromagnetic on the one hand and the strong interaction coupling constants on the other are attributed to differing finite renormalisation effects in different sectors, brought about through spontaneously induced differing gauge-masses. The true unity of the basic forces should manifest within this hypothesis at energies or momenta exceeding the masses of the heaviest gauge particles of the theory.

There are several appealing features, which emerge solely as a consequence of putting quarks and leptons together in one multiplet and generating their basic interactions in terms of one universal gauge coupling constant; these in turn provide the motivations for considering these ideas:

(a) First, because of the inclusion of the quark-lepton symmetry transformations, it becomes possible to generate electromagnetism without ever introducing an abelian U(1)-gauge group. This removes the unaesthetic freedom associated with the U(1)-contribution to electric charge and thereby provides the rationale for the quantisation of electric charge.

(b) Second, if quarks and leptons did not belong to one multiplet, there is no compelling reason why observed weak interactions would have picked positively charged proton but negatively charged electron to be left-handed (rather than positively charged proton and the positively charged positron).[3] On the other hand, if quarks and leptons are in one multiplet, it becomes possible in the first place to eliminate abelian U(1)-contributions to electric charge. This implies that the total charge of the quark-lepton-multiplet must add up to zero; thus there must exist a correlation between the charges of the quarks and the leptons carrying the same helicity. [To give the simplest example, which illustrates the point, consider a fictitious[56] 2x2 multiplet of quarks and leptons:

$$F = \begin{bmatrix} p & \nu_e \\ n & e^- \end{bmatrix}$$

Note with p and n having charges +1 and 0, it is (ν_e, e^-) and not $(\bar{\nu}_e, e^+)$, which can belong to the same multiplet as (p,n) in order that the total charge of the multiplet be zero. Thus $(\bar{\nu}_e, e^-)$ must have the same helicity as (p,n) in low energy weak interactions rather than $(\bar{\nu}_e, e^+)$. This feature carries over to realistic cases, in particular when lepton-number is treated as a fourth color[3] (see below).]

(c) If we assume that the lagrangian is generated by a unifying symmetry G, governed by a single basic gauge coupling constant g_G, and that G breaks spontaneously through a heavy mass-scale M ($\gg m_W$) into the symmetry of the low-energy interactions (G_{low}); it becomes possible through renormalisation group equations:

(i) to relate[53,57,58] the effective weak/electromagnetic and strong coupling constants to each other in terms of the unifying mass M;

(ii) For M sufficiently large, one obtains, for realistic examples of G_{low}, (e.g. $SU(2)_L \times U(1) \times SU(3)'$ or other similar examples), a qualitative understanding[59] of why low-energy effective SU(3)' color-gauge interactions are strong compared to the flavor weak/electromagnetic interactions ($\alpha_{strong}/\alpha \gg 1$); and

(iii) from the observed value of (α_{strong}/α), one can predict, for a given G and the low-energy symmetry G_{low}, the unifying mass-scale M and therefore simultaneously the weak angle[60] θ_W.

Several alternative models of unifying symmetries have been proposed. While they all share the common hypotheses of quark-lepton-unification as well as the postulate of unification of the basic forces,[3,4] they differ from each other:

(a) in respect of their fermion contents;

(b) in respect of the new class of quark-lepton and lepton-lepton-interactions;

(c) equally important, in respect of the origins of nonconservations of parity as well as of baryon and lepton-numbers, consequently in respect of the mechanism for (possible) proton-decay;

(d) in respect of the allowed solutions for quark-charges;

(e) in respect of the possible solutions for low-energy symmetries (G_{low}) and correspondingly the allowed solutions for the weak angle θ_W (or its analogs); and last but not the least

(f) in respect of the unifying mass-scale M, where the underlying unification of the basic particles and their forces should manifest.

I shall illustrate these features by presenting mainly one class of unifying models -- having the structure:[3]

$$G = G_{flavor} \times G_{color} \qquad (44)$$

The two crucial characteristics of this class of models are:

(a) Lepton number is a distinct color,

(b) Discrete flavor-color and left-right symmetries[3,61] together with the chosen gauge-structure ensure one single basic gauge coupling constant despite the semi-simple character of the symmetry G.

I shall at the end point to some of the crucial differences between this class of models on the one hand and the alternative unifying symmetries[4] -- e.g. SU(5), SO(10) and E_7, based on simple groups -- on the other.

3.2 Symmetry Structures $G_{flavor} \times G_{color}$: The Need for Mirror Fermions

The main ingredients of this class of symmetries are the following:

(A) Assume that the quark-lepton-multiplet defines two commuting symmetries G_{flavor} and G_{color}. Quarks possess flavors and three colors (red, yellow and blue); leptons possess the same flavors as the quarks, but differ from them only in having a different color. In short, this is the hypothesis[3] that "lepton-number is the fourth color." The symmetry G_{color} operates on quark as well as leptonic colors. Since quarks need to be introduced into the basic lagrangian with left- and right-components, this necessitates that leptons (including neutrinos) must be introduced with left and right components as well.

(B) The assumption of one fundamental gauge coupling constant in turn requires that we must impose a discrete

$$\text{flavor} \leftrightarrow \text{color} \qquad (45)$$

symmetry. This is fulfilled if the number of flavors in the basic fermionic multiplet F equals the number of colors, which minimally is four, as mentioned above. Hence the minimal basic quark-lepton multiplet possesses[3] four flavors (p,n,λ,c) and four colors (r,y,b,ℓ).

$$F_{L,R} = \begin{bmatrix} p_r & p_y & p_b & p_\ell = \nu_e \\ n_r & n_y & n_b & n_\ell = e^- \\ \lambda_r & \lambda_y & \lambda_b & \lambda_\ell = \mu^- \\ c_r & c_y & c_b & c_\ell = \nu_\mu \end{bmatrix}_{L,R} \qquad (46)$$

Note the necessity for the right-handed neutrinos $(\nu_{e,\mu})_R$ within this hypothesis, as mentioned above. The symmetry at this stage is $SU(4)_{flavor} \times SU(4)'_{color}$.

(B) In order to generate realistic weak interactions, we must introduce chiral flavor gauging, and therefore following flavor-color-discrete symmetry, also chiral color-gauging (for an alternative possibility of doubly vectorial color-gauging see later). The gauge symmetry now[62,63] is $[SU(4)_L \times SU(4)_R]^{flav} \times SU(4)'_L \times SU(4)'_R]^{col}$. A single gauge coupling constant is ensured by imposing flavor ↔ color as well as left ↔ right discrete symmetries:

$$\begin{array}{c} \text{flavor} \leftrightarrow \text{color} \\ \text{left} \leftrightarrow \text{right} \end{array} \tag{47}$$

However an important new ingredient must now be added to the theory. With chiral gauges, the SU(4)-symmetry of each sort generates triangle anomalies, which must be cancelled by introducing a new mirror[62] set of fermions $F^m_{L,R}$ (analogous to $F_{L,R}$):

$$F^m_{L,R} = \begin{pmatrix} p'_r & p'_y & p'_b & E^\circ \\ n'_r & n'_y & n'_b & E^- \\ \lambda'_r & \lambda'_y & \lambda'_b & M^- \\ c'_r & c'_y & c'_b & M^\circ \end{pmatrix}_{L,R} \tag{48}$$

This set possesses four new quark-flavors (p',n',λ',c') and four new leptons $(E^\circ,E^-,M^-,M^\circ)$, of which at least the charged ones need to be heavy $\gtrsim$ 1.8 GeV. Apriori there are two possible ways, consistent with the absence of anomalies, to gauge the fermions under the local symmetry:

$$G = [SU(4)_A \times SU(4)_B]^{flavor} \times [SU(4)'_C \times SU(4)'_D]^{color} \tag{49}$$

which we denote for short as $[SU(4)]^4$.

(I) Chiral Color: In this case the symmetries (A,B) and (C,D) correspond to chiral flavor and chiral color-gauging respectively, with the fermions transforming as:

$$F_L \sim F^m_R = (4,1,\bar{4},1);\ F_R \sim F^m_L = (1,4,1,\bar{4}) \tag{50}$$

Symbolically, the gauge interactions have the form

$$\begin{aligned} \mathcal{L}^I_{gauge} &= g_G[W_A(\bar{F}_L F_L + \bar{F}^m_R F^m_R)_{flav} + (A \to B;\ L \leftrightarrow R)] \\ &+ g_G\ [V_C(\bar{F}_L F_L + \bar{F}^m_R F^m_R)_{col} + (C \to D;\ L \leftrightarrow R)] \end{aligned} \tag{51}$$

where each of $W_{A,B}$ and $V_{C,D}$ represents a 15-fold. Note that the above gauge-lagrangian respects the discrete symmetries denoted by (47). Indeed it is invariant under the mirror-symmetry-transformation:

$$F_{L,R} \leftrightarrow F^m_{R,L} \tag{52}$$

In this case, spontaneous symmetry breaking may in general permit the emergence of either the diagonal sum $SU(3)'_{C+D}$, or the split chiral three color-symmetry $SU(3)'_C \times SU(3)'_D$, as good low energy symmetry, with the latter breaking softly (by a mass-scale ~1 to 2 GeV) into the diagonal sum $SU(3)'_{C+D}$ (see discussions later). The second pattern -- i.e. soft breaking of chiral $SU(3)'_C \times SU(3)'_D$ into $SU(3)'_{C+D}$ -- would generate an octet of light axial color gluons[3] in addition to the QCD octet of still lighter or massless color-gluons. In turn, this generates the novel possibility that vector chromodynamics may be supplemented by an axial chromodynamics mediated by an octet of light axial color gluons.

(II) Doubly Vectorial Color: In this case (A,B) still refer to chiral flavor, but (C,D) to vectorial color-gauging[58] of F and F^m respectively with the fermions transforming as:

$$F_L = (4,1,\bar{4},1),\; F_R = (1,4,\bar{4},1),\; F^m_L = (1,4,1,\bar{4}),\; F^m_R = (4,1,1,\bar{4}). \tag{53}$$

The corresponding gauge-interactions have the form:[64]

$$\begin{aligned} \mathcal{L}^{II}_{gauge} &= g_G[W_A(\bar{F}_L F_L + \bar{F}^m_R F^m_R)_{flav} + (A\rightarrow B;\; L\leftrightarrow R)] \\ &+ g'_G[V_C(\bar{F}F)^{col}_{L+R} + (C\rightarrow D;\; F_{L,R}\rightarrow F^m_{L,R})] \end{aligned} \tag{54}$$

Note that the basic color-gauge interactions here are vectorial in contrast to case I, where they are chiral. This case generates the possibility, if "split color" $SU(3)'_C \times SU(3)'_D$ (rather than the diagonal sum $SU(3)'_{C+D}$) emerges as a good low-energy symmetry, that the "basic" quarks contained in F and the mirror quarks contained in F^m may couple to two different octets of color-gluons ($V_C(\underset{\sim}{8})$ and $V_D(\underset{\sim}{8})$), with perhaps a small mixing[65] between the two sets.

For the case of chiral color-gauging, the fermions can acquire mass via their Yukawa interactions to the flavor↔color left↔right symmetric multiplet $\phi=(4,\bar{4},\bar{4},4)$ consistent with the hypothesis of a single gauge coupling constant. For the case of doubly vectorial color-gauging, there is however a notable reservation: the relevant scalar multiplet capable of having allowed invariant Yukawa coupling with fermions $\bar{F}_L F_R$ and $\bar{F}^m_L F^m_R$ etc. is $\phi'=(4,\bar{4},1,1)$, which is not flavor-color symmetric. Thus the case of doubly vectorial color-gauging would not permit the hypothesis of a single basic coupling constant (i.e. $g'_G = g_G$), unless the fermions acquired their masses through a dynamical spontaneous breakdown of the gauge-symmetry rather than via Yukawa coupling of ϕ'. This case is still worthy of consideration, however (a) because of its novelty, and (b) because our understanding of the origin of spontaneous symmetry breaking is still in its early stages. For concreteness, I shall from now on base my discussions pertaining to unification, on the case of chiral color-gauging only as it permits the generation of fermion masses consistent with the hypothesis of a single gauge coupling constant.

3.3 Salient Features of the Mirror Theory

The following salient features of the mirror-theory[3] based on the unifying symmetry structures $G_{flavor} \times G_{color}$ may be noted. (These features with the exception of feature (iii), are common to either pattern of color-gauging: chiral or doubly vectorial.)

(i) Parity and CP: The mirror-theory is intrinsically left↔right symmetric[3,61] and thus conserves parity. Parity nonconservation (as also CP-violation, see later) may arise within this theory through spontaneous symmetry breaking leading to a splitting of W_A and W_B-masses for example. (Advantages of left-right symmetry are noted later.)

(ii) Baryon and Lepton-Number Conservations: The mirror-theory, based on the symmetry-structure $G_{flavor} \times G_{color}$, intrinsically conserves basic quantum numbers such as baryon number (B) and lepton-number (L), as well as fermion number F = B+L. Violations of these quentum numbers in general arise within this theory only via spontaneous symmetry breaking,[3] e.g. via spontaneously induced mixing of gauge-particles carrying different B and L. (Violation of fermion number arises spontaneously only provided one gauges fermion-number putting fermions F_L and anti-fermions (F_L^C) in the same multiplet under a symmetry structure such as[66] SU(8) x SU(8)'.) This should be contrasted from alternative unifying symmetries[4] e.g. SU(5), SO(10) or E_7, in which violations of these quantum numbers (B, L and F) are intrinsic to the basic equations of motion.

(iii) Axial Chromodynamics: The mirror theory, with chiral color-gauging, inevitably generates an octet of axial color-gluons[62] in addition to the QCD-octet. Such axial gluons can in general be rather light (~1 to few GeV). One distinct "advantage" of the existence of light axial gluons is that in turn it permits[58] unification of the basic forces to be manifest at a relatively low mass-scale (~10^5 GeV, see remarks below). Such light axial gluons would generate, as mentioned before, an effective low-energy axial chromodynamics supplementing the familiar vector QCD. The consistency of this new possibility, motivated by the hypothesis of low-mass-unification (see remarks below), with the observed phenomena, in particular as to whether this may help remove some of the lingering discrepancies between vector QCD and observed charmonium physics -- as regards for example the rates of radiative transitions[67] -- remains to be examined. It is conceivable that hyperfine splitting of the type exhibited by the separation between $^3S_1$ (J/ψ) and ^1S_o (η_c) is a consequence[68] of such low energy axial QCD.

(iv) Unifying Mass-Scale: The chiral version of the symmetry-structure $G_{flavor} \times G_{color}$, e.g. $[SU(4)]^4$, permits a descent into low energy symmetry of the form $G_{weak} \times SU(3)'_C \times SU(3)'_D$ containing chiral three-color symmetry $SU(3)'_C \times SU(3)'_D$ as a good low energy symmetry. Such a descent in turn permits[58] (as shown later) unification of the basic forces to be manifest at a relatively "low" mass-scale (~10^5 GeV). This possibility of low mass-unification appears to be a special feature of the mirror theory based on the symmetry-structure $G_{flavor} \times G_{color}$. It is not available within alternative unifying symmetries e.g. SU(5), SO(10) or E_7 (see discussions later). [It should be noted that the emergence of chiral three color symmetry $SU(3)'_C \times SU(3)'_D$ as a good "low energy" symmetry would imply the existence of octets of vector and light axial vector gluons ($m_A \le m_W$).]

(v) Global Flavor Symmetry: For the minimal mirror-theory $[SU(4)]^4$, there are altogether 8 quark-flavors (p,n,λ,c) + (p',n',λ',c'). The strong interactions generated by either the diagonal 3-color symmetry $SU(3)'_{C+D}$ or the split-symmetry $SU(3)'_C \times SU(3)'_D$ are invariant under the global flavor-symmetry $U(8)_L \times U(8)_R$, which is broken to observed flavor-symmetries only by spontaneously induced quark-masses and mass-splittings plus electromagnetic and weak interactions. Of the four mirror flavors, it is likely that at least one flavor (n') has manifested itself[69] already through the discovery of the upsilon (Υ)-structure[54] at about 10 GeV.

(vi) Number of Flavors versus Number of Colors: Symmetries of the type $G_{flavor} \times G_{color}$ need the number of flavors (including those of the mirror) to be twice the number of colors (including the color(s) of the leptons). Minimally, if leptons (including e and μ) define only one color, the symmetry is $[SU(4)]^4$ needing four colors and eight flavors, i.e. a total of 32 four component fermions. More generally, the symmetry[70] is $[SU(n)]^4$ based on n colors (n ≥ 4) and 2n flavors needing a total of $2n^2$ four component fermions. Within such an extended structure (n > 4), electron and muon need not have the same color; for example with $[SU(6)]^4$, the basic fermions can have the structure:

$$F_{L,R} = \begin{bmatrix} p_r & p_y & p_b & \nu_e & \nu_\mu & p_I^+ \\ n_r & n_y & n_b & e^- & \mu^- & n_I^\circ \\ \lambda_r & \lambda_y & \lambda_b & E^- & M^- & \lambda_I^\circ \\ c_r & c_y & c_b & E^\circ & M^\circ & c_I^+ \\ b_r & b_y & b_b & L^- & N^- & b_I^\circ \\ t_r & t_y & t_b & L^\circ & N^\circ & t_I^+ \end{bmatrix} \qquad (55)$$

with an analogous composition for the mirror fermions. The sixth color may or may not correspond to a leptonic color. It is worth noting that if the observed[71] τ^- turns out to be a heavy lepton with (V-A) interaction (rather than an integer charge quark (see remarks later), or a heavy lepton with V+A-interaction), then within the mirror-theory, it may have to be assigned[69,72] to the basic multiplet F rather than to the mirror-multiplet F^m. For example the τ-particle, in this case, may be identified with E^- and E°. This in turn would necessitate an extension of $[SU(4)]^4$ beyond n = 4 and correspondingly an enlargement in the size of the basic and mirror fermionic multiplets as shown above. There is no experimental need yet for such an extension in the basic degree of freedom; whether such an extension will be needed, it should be stressed, appears to depend crucially on the nature of τ^-. Should such an extension be called for, however, one may begin to wonder seriously as to whether quarks are elementary; the idea that they are composites of more fundamental substructures -- the prequarks[73] -- would become still more compelling.

(vii) Fermi Mass-Matrix: The physical weak interactions depend upon the gauging pattern, as well as upon the Fermi-mass-matrix. The latter apriori may induce Cabibbo-like ($n\leftrightarrow\lambda$, $p\leftrightarrow c$, $n'\leftrightarrow\lambda'$, $p'\leftrightarrow c'$) as well "skew" F-F' mixings ($p\leftrightarrow p'$, $n\leftrightarrow n'$, $\lambda\leftrightarrow\lambda'$, $c\leftrightarrow c'$). Within a unified theory (in contrast to the SU(2) x U(1)-theory possessing an abelian factor), the Fermi-mass matrix turns out to be quite restrictive, as indicated before. It does not appear to permit[74] the sort of skew-physical couplings (e.g. $\vec{W}_A$ coupled to the physical doublets $\binom{p}{n_\theta}_L + \binom{c}{\lambda_\theta}_L + \binom{p}{n'}_R + \binom{c}{\lambda_\phi}_R$), which were actively considered in the literature, motivated by the announcement of the y-anomaly in $\bar{\nu}$-interactions. With the recent disappearance[75] of such anomalous effects, there is no experimental motivation at present for such skew physical couplings either.(See remarks later about semistable or stable mirror composites).

(viii) Mirror Versus Standard Vector-Like Theories: At a stage, where the fermions are massless, the mirror theory is vector-like for either chiral or doubly vectorial color-gauging. This is because of the presence of the basic (F) as well as the mirror fermions (F^m) (see Eqs. (51) and (54)). To see how the "weak" gauge lagrangian looks like, it is useful to write down the coupling of the gauge bosons belonging to the GIM[24] $SU(2)_A \times SU(2)_B$ subgroup of $SU(4)_A \times SU(4)_B$.

$$\begin{aligned} L = g \sum_{4\,\text{colors}} & \vec{W}_A\cdot[\binom{p}{n_c}_L + \binom{c}{\lambda_c}_L + \binom{p'}{n'}_R + \binom{c'}{\lambda'}_R] \\ + g \sum_{4\text{colors}} & \vec{W}_B\cdot[\binom{p}{n_c}_R + \binom{c}{\lambda_c}_R + \binom{p'}{n'}_L + \binom{c'}{\lambda'}_L] \end{aligned} \tag{56}$$

where $(n_c,\lambda_c)_{L,R}$ are related to the physical $(n,\lambda)_{L,R}$ fields in terms of the Cabibbo-angles. In general other Cabibbo-like rotations, for example rotation in the space of (n',λ')-fields, can be induced through the Fermi mass-matrix. In the context of a unified theory, skewness angles, brought about through the mixing of canonical basic and mirror fermions, (e.g. the mixing of (p,p') or (n,n') etc.), are assumed to be small (as remarked above).

Note two distinctive features of this coupling:

(a) For every left current, there is an associated right current, both coupled to the same gauge particle (W_A), likewise for W_B. (For example, corresponding to the left current $(\bar{p}n)_L$, there is the associated mirror right current $(\bar{p}'n')_R$, both coupled to W_A+). This is what makes the theory vector-like. Second,

(b) for every left (right) current coupled to $\vec{W}_A$, there is a parallel right (left) current coupled to a distinct set of gauge-particles $\vec{W}_B$ (e.g. $(\bar{p}n)_L$ is coupled to W_A+, while the parallel right current $(\bar{p}n)_R$ is coupled to W_B+). This is what makes the theory, within our terminology, "left-right symmetric." (A meaningful distinction between the two words -- i.e. "associated" and "parallel" -- is derived under the supposition that spontaneously induced Fermi mass-matrix would combine ($\bar{F}_L$ and F_R) and ($\bar{F}_L^m$ and F_R^m) to make massive four component particles with little or no mass-mixing between F and F^m). Note that the two features -- (a) and (b) -- are simultaneously realised, because of the presence of the mirror set F^m supplementing the basic set F. Left-right symmetry (feature (b)), by itself, can be realised[3,61] without the presence of the mirror.

While the mirror-theory is vector-like, there are important differences between this class of vector-like theories and the standard "left-right conjugated" vector-like theories based for example on the symmetry group[76] $SU(2)_{L+R} \times U(1)_{L+R}$. A typical example of the weak gauge interactions of such a theory, requiring a minimum of 6 flavors, is provided by the following pattern of interactions:

$$g\vec{W}\cdot[\binom{p}{n_c}_L + \binom{c}{\lambda_c}_L + \binom{t}{b}_L + \binom{p}{b}_R + \binom{c}{\lambda_c}_R + \binom{t}{n_c}_R] \tag{57}$$

Such a theory is intrinsically parity conserving like the mirror-theory. We call such a theory "left-right conjugated" because left and right handed fermions (e.g. p_L and p_R), which eventually combine to make four component particles, couple to the same gauge particles within this latter class of theories, unlike the mirror-theory (compare Eq. (57) with (56)).

There are important differences between the left-right symmetric mirror-vector-like theories and the standard -- what we call "left-right conjugaged" -- vector-like theories:

(a) The left-right conjugated theories depend on Fermi mass-matrix to provide the skew orientations between left and right handed doublets in terms of physical fermion fields and thereby to induce low-energy parity violation (e.g. compare $(\bar{p}n_c)_L$ versus $(\bar{p}b)_R$); while the left-right symmetric theories depend primarily[77] upon inequality of spontaneously induced W_A and W_B gauge-masses[3] for low-energy parity violation. Both classes of theories share the common feature that they are asymptotically parity conserving.

(b) The neutral current-interactions of the standard vector-like theories are necessarily pure vectorial (see Eq. (57)), which in turn implies that neutrino and antineutrino neutral current-cross-sections for such theories must be equal ($\sigma(\nu N)_{NC} = \sigma(\bar{\nu}N)_{NC}$); this special prediction of the standard vector-like theories is excluded experimentally[78]. By contrast the mirror vector-like theories imply in general a mixture of vector and axial vector-interactions involving VV, AA as well as VA-pieces. Even in the absence of the parity violating VA-piece, the parity conserving (VV+AA)-interaction distinguishes between neutrinos and antineutrinos, as observed experimentally. In other words, neutrino-experiments[78] rule out the standard vector-like theories, but are in full accord with the (left-right symmetric) mirror vector-like theories. (This aspect will be discussed in more detail in Part B).

3.4 The Gauge Interactions and Low-Energy Restrictions

The symmetry $[SU(4)]^4$ generates four sets of gauge particles W_A, W_B, V_C and V_D -- each set a 15-fold. Let us represent the triplets of GIM - $SU(2)_{A,B}$ gauge particles $\vec{W}_A$ and $\vec{W}_B$, relevant for low-energy interactions (see Eq. (56)), and the 15-folds of V_C and V_D by the matrices:

$$\vec{W}_{A,B} = \begin{bmatrix} \frac{\vec{\tau}}{2}\cdot\vec{W} & 0 \\ 0 & \tau_1\frac{(\vec{\tau}\cdot\vec{W})}{2}\tau_1 \end{bmatrix}_{A,B}, \quad V_{C,D} = \left[\begin{array}{c|c} V(\underset{\sim}{8}) & \begin{matrix}\bar{X}_1\\ \bar{X}_2\\ \bar{X}_3\end{matrix} \\ \hline X_1\ X_2\ X_3 & \sqrt{3/4}\ S^\circ \end{array}\right]_{C,D} \tag{58}$$

The interactions of $\vec{W}_{A,B}$ are given before (Eq. (56)). $V(\underset{\sim}{8})_{C,D}$ are the (chiral) octets of color-gluons coupled to the quark-colors as well as the mirror quark-colors; X's are coupled to chiral $(\bar{q}\ell)$ and $(\bar{q}'\ell')$-currents, while $S^\circ_{C,D}$ are coupled to diagonal chiral $\bar{q}q$ and $\bar{\ell}\ell$ and likewise to mirror-currents. Explicitly, their gauge-interactions with the fermions are given by:

$$L_{color} = \sum_{flavors} g_G[(\bar{q}_r q_y q_b)_L \gamma_\mu \frac{\vec{\lambda}}{2}\begin{pmatrix} q_r \\ q_y \\ q_b \end{pmatrix}_L + (L\to R; q\to q')]\vec{V}_{C\mu}(\underset{\sim}{8}) + (C\to D; L\leftrightarrow R) \tag{59}$$

$$L_x = \sum_{i=r,y,b}\ \sum_{flavors} g_G[(\bar{q}_i\gamma_\mu\ell)_L + (\bar{q}'_i\gamma_\mu\ell')_R]X^i_{C\mu} + (C\to D, L\leftrightarrow R) + h.c. \tag{60}$$

$$L_{S^\circ} = \sum_{flavors} - (\frac{g_G}{2\sqrt{6}})[(\sum_{i=r,y,b} \bar{q}_i\gamma_\mu q_i - 3\bar{\ell}\gamma_\mu\ell)_L + (\sum_{i=r,y,b} \bar{q}'\gamma^\mu q' - 3\bar{\ell}'\gamma_\mu\ell')_R]$$

$$S^\circ_{C\mu} + (C\to D; L\leftrightarrow R) \tag{61}$$

The above notation utilises the feature that leptons and quarks carry matching flavors (e.g. ν_e and $p_{r,y,b}$ have the same flavor; e^- and $n_{r,y,b}$ have the same flavor etc.; see Eqs. (46) and (48)). The $\vec{\lambda}$-matrices in eq. (59) represent SU(3)-matrices for the 3*-representation.

Spontaneous breakdown of the unifying symmetry must be such as to respect the following low-energy-restrictions:

(i) The charged ($W_B^{\pm}$ and $W_A^{\pm}$) generate right-handed (V+A and V-A) β-decay interactions respectively (see Eq. (56)). The ratio of V+A to V-A amplitude is given by $(m_{W_A^+}/m_{W_B^+})^2$ $(\cos\theta_R/\cos\theta_L)$, where $\theta_{L,R}$ are the respective Cabibbo-angles defined in the $(n,\lambda)_{L,R}$-spaces. From the experimental observation that the longitudinal polarisation of electrons emitted in β-decays is (-v/c) within about 1%, we may deduce[3]

$$m_{W_B^+} \gtrsim 3(\cos\theta_R/\cos\theta_L)^{1/2}\, m_{W_A^+} \tag{62}$$

Thus for[74] $\cos\theta_R \approx \cos\theta_L$, the charged W_B^+ need to be at least about three times heavier than the charged W_A^+. (This is barring[77] large mixing between F and F'.)

(ii) Given the absence or suppression of ($\Delta s = 1$)-neutral current processes, and the $\mathcal{O}(G_F)$-strength for left-handed V-A weak interactions, $SU(4)_A$ operating on the <u>left-handed</u> quarks and leptons must descend through a heavy mass-scale $M \gtrsim 10^4$ GeV into the GIM-$SU(2)_A^{I+II}$-subgroup, which treats $(p,n,)_L$ as well as $(c,\lambda)_L$ as doublets (likewise for the mirrors). $SU(2)_A^{I+II}$ should break subsequently with a <u>"low"-mass-scale</u> ($\lesssim m_{W_A^+}$). The symmetry $SU(4)_B$ must also break likewise, except that in this case, even the GIM $SU(2)_B^{I+II}$-subgroup may in general be broken through a heavy mass-scale. (Whether it does, or does not, is relevant for the question of atomic parity conservation, see later.) Thus at the very least, the following breaking pattern is needed:

$$SU(4)_A \times SU(4)_B \xrightarrow[M \gtrsim 10^4 \text{ GeV}]{\text{Heavy mass-scale}} [SU(2)_A^{I+II} \times SU(2)_B^{I+II}]_{GIM} \tag{63}$$

(iii) The exotic X-particles induce the unobserved decay $K_L \to \bar{\mu}e$, as they couple simultaneously to $(\bar{n}e)$ and $(\bar{\lambda}\mu)$-currents. From the known upper limit on the rate of such decays, one may deduce the lower limit[79]

$$m_X \gtrsim 10^4 \text{ GeV} \tag{64}$$

(iv) Finally, noting that the strong-interactions generated by SU(3)'-color-subgroups are parity conserving, the four color symmetry $SU(4)'_C \times SU(4)'_D$ must break so as to conserve left-right symmetry in the three color-sector. This may happen in two alternative ways:

Either

$$\text{(A)}\quad SU(4)'_C \times SU(4)'_D \xrightarrow[M \geq 10^4 \text{ GeV}]{\text{Heavy Mass}} SU(3)'_{C+D} \times \begin{cases} U(1)_C \times U(1)_D \\ \text{or } U(1)_{C+D} \end{cases}$$

or,

$$\text{(B)}\quad SU(4)'_C \times SU(4)'_D \xrightarrow[M \geq 10^4 \text{ GeV}]{\text{Heavy Mass}} \underbrace{SU(3)'_C \times SU(3)'_D}_{\downarrow \text{ Light Mass} \atop SU(3)'_{C+D}} \times \begin{cases} U(1)_C \times U(1)_D \\ \text{or } U(1)_{C+D} \end{cases} \qquad (65)$$

Both cases (A) and (B) generate pure vector and axial vector color-gluon eigenstates. The two cases differ from each other in that the axial color-gluons are superheavy for case (A) and that they are light ($\lesssim m_{W_L}$) for case (B). The symmetries $U(1)_{C,D}$ correspond to the 15th generators of $SU(4)'_{C,D}$; they generate the gauge particles $S^\circ_{C,D}$ (see Eqs. (58) and (61)); the symmetry $U(1)_{C+D}$ is their diagonal sum. Both the breaking patterns (A) and (B) would make the exotic X-gauge particles massive in accordance with the requirement (64); <u>this is what causes a low-energy breakdown of quark-lepton unification</u>.

3.5 Spontaneous Breakdown of $[SU(4)]^4$

To generate spontaneous symmetry breaking, one must introduce scalar multiplets, their gauge and self-interactions and their mass-terms[80] consistent with the gauge symmetry <u>as well as</u> the flavor-color left-right discrete symmetries;[3,61,81] the preservation of the discrete symmetries is needed for the preservation of the hypothesis of a single gauge coupling constant.

In general one needs scalar multiplets transforming as the direct products of the fundamental representations of two of the four SU(4)'s and also as adjoint representation of each of the four SU(4)'s. A possible set of scalar multiplets respecting the A↔B↔C↔D discrete symmetries and capable of inducing the desired pattern of spontaneous breakdown of $[SU(4)]^4$, consistent with the requirements (62)-(65), is exhibited below:[82]

<u>Fundamental Set</u> (192 real fields)

$$A = (4,\bar{4},1,1),\ B = (1,4,1,\bar{4}),\ C = (4,1,\bar{4},1)$$
$$H = (1,1,4,\bar{4}),\ J = (1,4,\bar{4},1),\ K = (4,1,1,\bar{4}) \qquad (66)$$

<u>Adjoint Set</u> (four sets of 15 folds of each SU(4) = 240 real fields)

$$(D,E,L,M)_A = (15,1,1,1);\ (D,E,L,M)_B = (1,15,1,1)$$
$$(D,E,L,M)_C = (1,1,15,1);\ (D,E,L,M)_D = (1,1,1,15) \qquad (67)$$

(The $M_{A,B,C,D}$ can be 6 folds of each SU(4) rather than 15-folds.)

The subscripts A,B,C,D on the fields listed under the adjoint set have the following meaning ($D_{A,B,C,D}$ are adjoint representations of $SU(4)_{A,B,C,D}$). The fields A,B,C,D_i on the other hand, it should be noted, have no correspondence to the groups $SU(4)_{A,B,C,D}$. The sets D_i, E_i, L_i and also M_i (for n = 15), while transforming similarly, differ from each other in respect of their pattern of vacuum expectation values (see below). The breaking of color and flavor-sectors are exhibited separately.

Breakdown of Color-Symmetry:

Symmetry	Multiplet	VEV	Mass To	Residual Symmetry
$SU(4)'_C$	$D_C=(1,1,15,1)$	$\begin{bmatrix} d & & & \\ & d & & \\ & & d & \\ & & & -3d \end{bmatrix}_C$	X_C	$SU(3)'_C \times U(1)_C$
$SU(4)'_D$	$D_D=(1,1,1,15)$	$\begin{bmatrix} d & & & \\ & d & & \\ & & d & \\ & & & -3d \end{bmatrix}_D$	X_D	$SU(3)'_D \times U(1)_D$
$SU(4)'_C \times SU(4)'_D$	$H=(1,1,4,\bar{4})$	$\begin{bmatrix} h_1 & & & \\ & h_1 & & \\ & & h_1 & \\ & & & h_4 \end{bmatrix}$	$V_C(\underset{\sim}{15}) - V_D(\underset{\sim}{15})$	$SU(4)'_{C+D}$

h_1 gives mass to $V_C(\underset{\sim}{8})-V_D(\underset{\sim}{8})$ as well as to $S^\circ_C-S^\circ_D$, while h_4 gives mass to $X^\circ_C-X^\circ_D$ and to $S^\circ_C-S^\circ_D$. Depending upon the magnitudes of the VEV of the multiplets $D_{C,D}$ and H, there are different alternative hierarchies possible, as shown below:

$$\text{(I)}\quad SU(4)'_C \times SU(4)'_D \xrightarrow[\langle H\rangle]{M_I \gg 10^4\text{ GeV}} SU(4)'_{C+D} \xrightarrow[\langle D\rangle_{C,D}]{\bar{M}_I \gtrsim 10^4\text{ GeV}} SU(3)'_{C+D} \times U(1)_{C+D} \tag{68}$$

(The vectorial $SU(4)'_{C+D}$ still retains quark-lepton-unification and is the one considered for illustration of the quark-lepton unification-hypothesis in the second and third papers of Ref. 3).

$$\text{(II)}\quad SU(4)'_C \times SU(4)'_D \xrightarrow[\langle D\rangle_{C,D}]{M_I \gtrsim 10^4\text{ GeV}} SU(3)'_C \times U(1)_C \times SU(3)'_D \times U(1)_D$$

Eff. Residual Symmetry

$$\oplus \xrightarrow[\langle H\rangle]{} \begin{cases} \text{(a)}\ h_1 \sim 0; 0 \leq h_4 \lesssim m_W \rightarrow SU(3)'_C \times U(1)_C \times SU(3)'_D \times U(1)_D \\ \text{(b)}\ h_1 \sim 0; h_4 \gg m_W \rightarrow SU(3)'_C \times SU(3)'_D \times U(1)_{C+D} \\ \text{(c)}\ (h_1, h_4) \gg m_W \rightarrow SU(3)'_{C+D} \times U(1)_{C+D} \end{cases} \tag{69}$$

Here M_I and $\bar{M}_I$, denoting mass-scales >> m_W, are responsible for the primary stage(s) of symmetry-breaking; m_W denotes the mass-scale of $W^{\pm}_L$ ($\simeq$ 70 GeV). Note that case II(c), for most practical purposes, is identical to case I. These different alternatives I, II(a), (b) and (c), it turns out, lead to drastically different unifying mass-scales M_I and, of course, also to different complexions of low-lying gauge particles (see discussions later).

Breakdown of Flavor Symmetry:

Assume patterns of VEV as follows:[83]

$$\langle L_{A,B}\rangle = \begin{bmatrix} \ell & & & \\ & \ell & & \\ & & -\ell & \\ & & & -\ell \end{bmatrix}_{A,B} ; \quad E_{A,B} = \begin{bmatrix} \varepsilon & & & \\ & -\varepsilon & & \\ & & -\varepsilon & \\ & & & \varepsilon \end{bmatrix}_{A,B} , \quad \langle M_{A,B}\rangle = \begin{bmatrix} & & & m \\ & & m & \\ & -m & & \\ -m & & & \end{bmatrix}_{A,B} \tag{70}$$

A possible hierarchical chain is as follows:

$$SU(4)_A \times SU(4)_B \xrightarrow[\langle L_{A,B}\rangle]{M_I \gtrsim 10^4\ \text{GeV}} [SU(2)^I \times SU(2)^{II}]_A \times [SU(2)^I \times SU(2)^{II}]_B$$

$$\oplus \xrightarrow[\langle M_{A,B}\rangle]{M_I \gtrsim 10^4\ \text{GeV}} SU(2)_A^{I+II} \times SU(2)_B^{I+II} \quad \text{(GIM)}$$

$$\oplus \xrightarrow[\langle E_B\rangle]{M_{II} \sim m_{W_R^+}} SU(2)_A^{I+II} \times U(1)_B^{I+II}$$

Neutral part of $SU(2)_B^{I+II}$ survives; $L \leftrightarrow R$ symmetry is preserved in the neutral sector (71)

(Perhaps $M_{II} = O(\sqrt{\alpha}) M_I$)

The superscripts I and II refer to SU(2) subgroups operating on the (p,n) and (c,λ)-doublets respectively, while I+II is their diagonal sum representing the GIM-SU(2). The relevant low-energy flavor-symmetry G_W, responsible for weak/electromagnetic interactions, is obtained by combining the symmetry noted above with the surviving color-singlet piece $U(1)_C \times U(1)_D$ or $U(1)_{C+D}$) of the four color-symmetry. Thus

$$G_W = SU(2)_A^{I+II} \times U(1)_B^{I+II} \times U(1)_{C+D}, \text{ or } U(1)_C \times U(1)_D \tag{72}$$

The subsequent breaking of G_W is realized via a mass-scale $\leq m_{W_L^+}$ through VEV of fields belonging to the fundamental set:[3]

$$\langle B\rangle = \begin{bmatrix} 0 & & & \\ & 0 & & \\ & & 0 & \\ & & & b_4 \end{bmatrix} , \qquad \langle C\rangle = \begin{bmatrix} c_1 & & & \\ & c_1 & & \\ & & c_1 & \\ & & & c_4 \end{bmatrix} \tag{73}$$

with $c_1 = 0$ or ~ 1 GeV (the crucial distinction between the alternative possibilities $c_1 = 0$ and $c_1 \sim 1$ GeV is noted later). The parameters b_4 and c_4 are primarily responsible for the masses of $W_L^{\pm}$ (same as $W_A^{\pm}$) and the neutral weak gauge-particles of the theory. Thus, for example,

$$G_W = SU(2)_A^{I+II} \times U(1)_B^{I+II} \times U(1)_{C+D}$$

$$\downarrow \langle B \rangle$$

$$SU(2)_A^{I+II} \times U(1)_{"B+C+D"} \quad (\text{standard } SU(2)_L \times U(1)) \tag{74}$$

$$\downarrow \langle C \rangle \sim m_{W_L^+}$$

$$U(1)_{flav}^{EM} \tag{75}$$

With $\langle B \rangle \simeq \langle C \rangle$, the $L \leftrightarrow R$-symmetry of the theory in the neutral sector would be preserved (approximately), even though that in the charged-sector is broken through $\langle E_B \rangle \gg \langle E_A \rangle$.

If on the other hand, $b_4 \gg c_4$, the charged as well as neutral sectors would violate $L \leftrightarrow R$-symmetry; the relevant low-energy weak/EM symmetry would become

$$G_W = SU(2)_A^{I+II} \times U(1)_{"B+C+D"} = SU(2)_L \times U(1) \tag{76}$$

which is the standard left-handed $SU(2)_L \times U(1)$-symmetry (in the space of the basic fermions).

In exhibiting the above patterns of spontaneous-breakdown, we have assumed VEV fields such as $(4,\bar{4},\bar{4},4)$, needed to give mass to fermions to be small ($\ll c_4$). Similarly, VEV of fields such as $D_{A,B}, L_{C,D}, M_{C,D}, E_{A,C,D}$, J and K (see Eqs. (66) and (67) for definitions) are assumed to be zero[84] or small (i.e. $O(\sqrt{\alpha})$ compared to VEV of respective multiplets related by discrete symmetries).

In summary, ignoring the mass-splittings of $W_L^{\pm}$ and $W_R^{\pm}$ (in comparison with the primary versus secondary mass-splittings) we see that:

(A) There are two alternative possibilities for the relevant low energy three color-symmetry. They are

(i) The vector $SU(3)'_{L+R}$ (8 massless vector color-gluons if $c_1=0$); or

(ii) The chiral $SU(3)'_L \times SU(3)'_R$ (8 massless vector color-gluons (if $c_1=0$) + 8 light axial color gluons).

The subscripts L,R refer to the gauging pattern of the basic fermions.

(B) Simultaneously, there are three possible low-energy manifestations (G_W) of the same parent symmetry $[SU(4)]^4$, relevant for low-energy weak/electromagnetic interactions. They are:

(i) The left handed $SU(2)_L \times U(1)$ ($gb_4 \gg 2m_{W_L^+}$; $gh_i \gg 2m_{W_L^+}$ only one "light" weak boson Z°)

(ii) The L$\leftrightarrow$R symmetry $\bar{G}_{LR}$=$SU(2)_L \times SU(2)_R \times U(1)_{L+R}$ ($b_4 \simeq c_4$; $g(h_i) >> 2m_{W_L^+}$, Two light weak bosons N_1, N_2 (Ref. 3,61,85)

(iii) The extended L$\leftrightarrow$R symmetric $G_{LR} = SU(2)_L \times SU(2)_R \times U(1)_L \times U(1)_R$ — $b_4 \simeq c_4$, $g(h_i) \lesssim m_{W_L^+}$, Three light weak bosons $Z_{1,2,3}$. (Ref. 62,86)

We have listed the number of relatively light weak bosons, which are expected to emerge in each case, and have used the subscripts L,R to refer to the coupling of the basic fermions. (Thus $SU(2)_{L,R}$ correspond to $SU(2)_{A,B}$: $U(1)_{L,R}$ to $U(1)_{C,D}$ and $U(1)_{L+R}$ to $U(1)_{C+D}$.)

Borrowing from recent work,[87,86] and the more detailed discussions in Part B, I summarize below the allowed mass-range for the lightest neutral weak-boson in each case, listing the relevant experiments which either set a lower limit on the mass, or can help improve the lower limit on the mass (or will determine the mass) of the lightest weak gauge particle. [The values given are for $\sin^2\theta_W \simeq .3$.]

Low Energy Symmetry (G_W)	Lightest Weak Boson	Mass-Range (GeV)	Relevant Experiments
$SU(2)_L \times U(1)$	$Z°$	82	Neutrino Experiments
$SU(2)_L \times SU(2)_R \times U(1)_{L+R}$	N_1	58-70	(i) Lower limit set[87] by present atomic parity experiments (ii) e^-e^+ F-B asymmetry measurements at PETRA & PEP[87] (iii) Polarized ep-scattering[88] (iv) High energy pp & $\bar{p}p \to \mu\bar{\mu}+X$[87]
$SU(2)_L \times SU(2)_R \times U(1)_L \times U(1)_R$	Z_1	35-70	(i) hfs effect[82] $m_{Z_1}>15$ GeV (ii) $(g-2)_\mu$ $m_{Z_1} \gtrsim 35$ GeV[89] (iii) Preliminary unpublished F-B asymmetry measurements at SPEAR $m_{Z_1} \gtrsim 39$ GeV[82] (iv) e^-e^+ F-B asymmetry measurements at PETRA & PEP[86] (v) High energy pp & $\bar{p}p \to \mu\bar{\mu}+X$

As to which of the three low energy manifestations -- i.e. $SU(2)_L \times U(1)$, or $\bar{G}_{LR}$, or G_{LR} -- is chosen by nature (if any), depends crucially on the mass of the lightest neutral weak gauge-particle; this in turn can be decided by the experiments listed above. The prospect of discovering the lightest neutral weak gauge particle (e.g. via dilepton-production in high energy pp & $\bar{p}$p-collisions) over the next two to three years is high, if G_{LR} happens to be the correct choice.

Advantages of Left-Right Symmetry: The symmetry $[SU(4)]^4$, permits the emergence of the left-right symmetric substructures $\bar{G}_{LR}$ and G_{LR}, as relevant symmetries for low energy weak and electromagnetic interactions. Some of the advantages of such $L\leftrightarrow R$ symmetric-substructures are:

(a) On the one hand, they retain[90] all the successes of the left-handed $SU(2)_L \times U(1)$-theory for neutral current interactions involving neutrinos (this is primarily because laboratory neutrinos, produced by charged current interactions, are dominantly left-handed and are "blind" to right-handed V+A-interactions present in G_{LR} and $\bar{G}_{LR}$)

(b) On the other hand, they can accommodate[85] the null or near null results of recent atomic parity experiments.[48] (This is because, to begin with they are $L\leftrightarrow R$-symmetric, and they permit the possibility that the degree of $L\leftrightarrow R$-symmetry-breaking in the neutral mass-sector may be much smaller than that in the charged mass-sector. Such a possibility can not arise within the $L\leftrightarrow R$ nonsymmetric substructure $SU(2)_L \times U(1)$ or within unifying symmetrics such as[4] SU(5) which are intrinsically $L\leftrightarrow R$ non-symmetric.)

(c) The left-right symmetric substructures such as G_{LR} or $\bar{G}_{LR}$ permit the emergence of a desirable milliweak theory of CP-violation,[61] its origin is tied to "complex" Cabibbo-rotations in the (n,λ)-space and whose magnitude is linked to the observed suppression of V+A relative to V-A charged current interactions.

(These features are discussed in more detail in part B.)

Semistable or Stable Mirror Matter: As remarked in section 3.3, consistent with natural conservation quantum numbers, the mirror theory does not permit large mixing between the basic (F) and the mirror fermions (F^m). This raises an interesting possibility: if the mising angles are really tiny $<<10^{-5}$ (rather than $O(\alpha)$), the quantum number associated with the lightest mirror quark would be preserved to an extent much better than strangeness and charm; consequently at last some mirror-matter (e.g. composites of the lightest mirror quark and the lightest basic anti-quark $(\bar{p})$) would be relatively stable possesing life times $>>10^{-10}$ sec. So also would be the lightest mirror-lepton. In the extreme limit of no F-F^m mixing, such matter would be absolutely stable. Such semistable or stable objects, if they exist, would provide a new vista for experimental physics.

IV. THE TWO BASIC ALTERNATIVES: FRACTIONAL VERSUS INTEGER QUARK-CHARGES; CONFINEMENT VERSUS LIBERATION

4.1 General Results

Depending upon the nature of spontaneous symmetry breaking, $c_1=0$ or $c_1\neq0$, (see Eq. (73)), unifying symmetry structures of the form $G_{flavor} \times G_{color}$ give rise to two possible compositions for the massless photon in terms of the canonical flavor and color gauge fields. Correspondingly there are two possible solutions for quark-charges:

Either (i) quarks are fractionally charged ($c_1=0$). In this case SU(3)'-color defined by the symmetry of the three quark-colors does not contribute to electric charge; quarks of all three colors with a given flavor carry the same charge. SU(3)'-color, as a local symmetry, is preserved. The octet of color gluons remain neutral and massless. Since emission of such massless gluons in pairs has not been -- e.g. in weak decays such as $\pi^+\to\mu\nu+V\bar{V}$ -- not to mention of the nonobservation of fractionally charged stable quarks -- one is obliged to assume on experimental grounds that massless QCD confines (nonperturbatively) quarks and gluons and all color.

Or (ii) quarks are integer-charged ($c_1\neq0$). SU(3)'-color-symmetry does contribute to electric charge on par with flavor. The octet of gluons acquire a symmetrical mass (barring $\mathcal{O}(\alpha)$ corrections) and possess four charged and four neutral members. An effective global $\widetilde{SU}(3)$'-color-symmetry is preserved[43] (to $\mathcal{O}(\alpha)$), which serves as a good classification symmetry. Neutral as well as charged gluons $V(\underset{\sim}{8})$ mix with corresponding flavor W-gauge-particles; thus the gluons become unstable against weak and electromagnetic decays into leptons and hadrons.

These features, as well as the consistency of the integer-charge solution with lepto-production-experiments, were aldready outlined for the nonunifying (low-energy) symmetry $SU(2)_L \times U(1) \times SU(3)'_{L+R}$ in Chapter II. There are two important new features, which emerge solely (1) because of putting quarks and leptons into one multiplet, and (2) because of the elimination of abelian contributions to electric charge. They are listed in the next two sections:

4.2 Relationship Between Charge, Gluon-Masses and Symmetry-Structures The physical requirement that a good global SU(3)'-color-symmetry must be preserved (even if the local symmetry is broken), in order that it may serve as a classification-symmetry, turns out to be a nontrivial requirement within unifying symmetries. The reasons briefly are as follows: The only known mechanism[91,43] -- allowing the emergence of a good global symmetry despite the breakdown of the corresponding local symmetry -- involves the introduction of a scalar multiplet, carrying a "box"-representation, with nonzero equal VEV along the diagonal. (Only the first three elements should be equal if global SU(3)'- should emerge.) Furthermore the "box" must transform as the fundamental representation of the relevant local symmetry being broken. This is the case for $<\sigma>$ introduced in Ch. II (see Eq. (39)) and $<C>$ introduced in the present context (see Eq. (73)).

The diagonal elements of the "box" referred to above need to have electrically neutral fields in order that the global-symmetry-strategy may be implemented. These elements would be electrically neutral, however, provided the symmetries associated with the row and the column-indices of the "box" contribute symmetrically to electric charge. In the context of a <u>nonabelian quark-lepton-unifying symmetry, this turns out to be possible, provided</u>[92]

(i) The unifying symmetry is semisimple of the form $G_{flavor} \times G_{color}$ with the row and the column indices of the box transforming like $(n,\bar{n})$, ["n" being the fundamental representation of G_{color} or G_{flavor}], and provided

(ii) quarks acquire integer-charges rather than fractional charges. (This is possible, if flavor and color-generators of the unifying group contribute symmetrically to electric charge.)

There are many important corollaries[92] of this result:

(I) Simple unifying groups[4] -- such as SU(5), SO(10) and E_7 -- can not permit the implementation of the global symmetry-strategy. Thus within such symmetries, SU(3)'-color as a local symmetry must be left unbroken: quarks must acquire fractional charges.

(II) If the unifying symmetry is of the form $G_{flavor} \times G_{color}$, <u>there are two and only two allowed solutions for quark-charges</u> -- the familiar fractional and integral-- with the accompanying consequences listed before. The two solutions need to be distinguished from each other experimentally.

(III) In the context of a renormalisable nonabelian quark-lepton <u>unifying symmetry</u>, with quarks and leptons sharing same multiplet, <u>color gluons can acquire mass, preserving a good global color-symmetry, provided the symmetry is of the form $G_{flavor} \times G_{color}$ and quarks acquire integer-charges.</u>[93]

The expressions for the electric charge for these two solutions are given below:[3]

$$Q = (F_3 + F_8/\sqrt{3} - \sqrt{2/3}\, F_{15})_{A+B} - \sqrt{2/3}\, (F'_{15})_{C+D} \quad \text{(Fractional quark-charges)}$$

$$Q = (F_3 + F_8/\sqrt{3} - \sqrt{2/3}\, F_{15})_{A+B} + (F'_3 + F'_8/\sqrt{3} - \sqrt{2/3}\, F'_{15})_{C+D} \quad \text{(Integer-charge quarks)} \qquad (77)$$

where $F_{3,8,15}$ are the diagonal generators of the respective SU(4)-group. It is easy to verify that either solution <u>predicts</u> the same charge for the leptons (0,-1,-1,0) for the basic <u>and</u> for the mirror multiplet. (In other words, even if one wanted to, one could not put $(\bar{\nu}_e, e^+, \mu^+, \bar{\nu}_\mu)$ in the same multiplet as the quarks.) This in turn explains the point raised before: Negatively charged electron (rather than e^+) and positively charged proton must exhibit the same helicity in low energy weak interactions. It appears that this is a special virtue of the hypothesis that lepton-number defines a distinct color.[3]

I now turn into the second important consequence of putting quarks and leptons into one multiplet.

4.3 Violation of Baryon and Lepton-Numbers: The Unconfined Unstable Integer-Charge Quark Within quark-lepton unifying symmetries, if quarks acquire integer-charges, the exotic X-gauge particles (carrying B = L = ±1) mix spontaneously with the flavor W-gauge particles (carrying B=L=0). This induces baryon and lepton-number violations (fermion-number is still conserved). Integer-charge quarks become unstable against decay into leptons. The relevant diagrams,[94] exhibiting the dominant decay mechanisms of quarks are shown in Figures 5a,b,c.

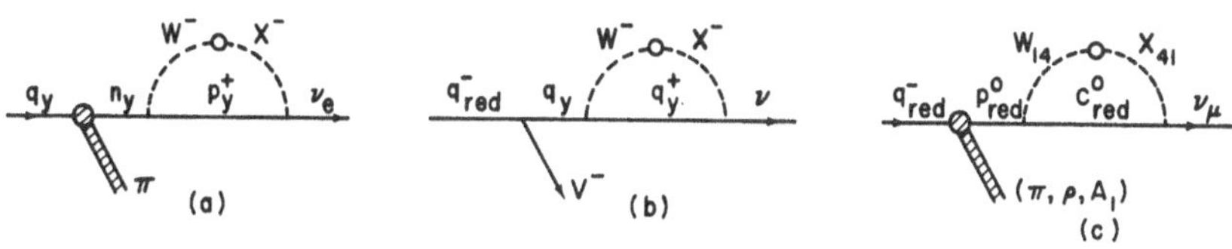

Fig. 5

The corresponding dominant decay modes of quarks turn out to be:

$$\begin{aligned} q_{y,b} &\to \nu + \text{mesons } (\pi, K, \eta) \\ q^-_{red} &\to \nu + \text{gluon } (V^-) \\ q^-_{red} &\to \nu + (\pi, \rho, A_1) \end{aligned} \tag{78}$$

Other decay modes -- e.g. those involving emission of 3 leptons $(q \to \ell+\ell+\bar{\ell})$ or emission of a single charged lepton (e.g. $q^o_y \to e^- + \pi^+$) turn out to be strongly damped[95] compared to the decay modes listed above. (It should be noted that the decay modes $q^-_{red_4} \to \nu$ + mesons (π, ρ, A_1) etc. are expected[96] within the unifying symmetry $[SU(4)]^4$, but not within the restricted basic model $SU(2)_L \times SU(2)_R \times SU(4)'_{L+R}$, for which W_{14} coupling to $(\bar{p}c)$-currents is nonexistent.)

Lifetimes of quarks with physical masses in the range of 2-3 GeV are found to be[94] $\simeq 10^{-11} - 10^{-13}$ sec. Such short lived integer-charge quarks would have escaped direct detection by all bubble-chamber as well as emulsion searches made so far, even if they may have been pair-produced in hadronic collisions with reasonable cross-section[97] $\sim 10^{-30} - 10^{-31}$ cm^2 (say) at ISR and Fermilab energies. (Emulsion-searches are sensitive to the life-time range $\sim 10^{-11} - 10^{-13}$ sec; however, with production cross-section in the range indicated above, the background-effect in searches made so far would be serious.) This, as well as several associated considerations, based on lepto-production experiments (duscussed earlier) and proton-stability (see discussions below) obviate the need for the hypothesis of fractionally charged quarks and confinement. Experimental tests of the alternative of unconfined integer-charge quarks are given in the next sections.

4.4 Unstable Proton With quarks being unstable, the proton becomes unstable as well against decay into leptons. However, assuming that the physical masses of quarks and diquarks are heavier than those of the proton, the proton - a three quark composite - can decay only provided all three quarks within the proton "simultaneously" convert into leptons.[3] This makes the proton a rather stable object, and accounts for our prolonged existence. Yet the proton must decay. Depending upon the details of quark-decay mechanism their physical masses and the probability of all three quarks within the proton being near the origin ($|\psi(o)|^2$), one obtains[94] a proton lifetime in the range:[98]

$$\tau_{proton} \sim 10^{29} - 10^{32} \text{ years} \tag{79}$$

with the dominant decay modes of the proton being of the type

$$p \to 3\nu + (\text{mesons}) \tag{80}$$

It is important to note that within the unconfined unstable integer-charge quark-hypothesis, quarks cannot be too longlived ($\tau_{quark} \lesssim 10^{-11}$ sec.), or else they would have been discovered (assuming physical $m_q < 5$ GeV); this in turn implies that the proton, within this hypothesis, can not be too longlived either. A life time of $\simeq 10^{32}$ years is in fact a reasonable upper limit for the proton to live within this hypothesis, and thus can be used to test the hypothesis. [This is unlike alternative models[4] of proton decay, based on confined fractionally charged quarks, in which proton-decay is induced directly by the gauge interactions in the second order. In such models, proton-lifetime depends directly upon the mass of the exotic superheavy gauge particles. Estimating the mass ($\simeq 10^{16}$ GeV) from the unification hypothesis, the proton-lifetime in such models is estimated[99] to be $\simeq 10^{38}$ years; in other words the proton is too stable in those models for its decay-possibility to be tested in the forseeable future. This is not so for the case of unconfined unstable integer-charge quarks as mentioned above.]

In summary, based on these considerations as well as discussions in Chapter II (involving scaling phenomena, lepto-production experiments and spectroscopy of quark-composites), we see that the hypothesis of liberated integer-charge quarks is a theoretically allowed and at present experimentally <u>viable alternative</u> to the one of confined fractionally charged quarks. Endowed with liberation, the former offers several intriguing observational possibilities for experiments in planning.[100] Some of these are presented below with the assumption of relatively low-mass (~ 1 to few GeV) physical color. A more complete discussion, including those of possible high mass color-threshold ($\gtrsim$ 8 GeV), may be found in Ref. 100.

4.5 <u>Signatures of Physical Color</u> In addition to quarks being observable, the octet of spin-1 color-gluons (involving four charged[3] $(V_\rho^\pm, V_{K*}^\pm)$ and four neutral members $(\tilde{U}, \tilde{V}, V_{K*}^o, \bar{V}_{K*}^o))$, color-octet $q\bar{q}$-composites in various spin-parity configurations, as well as color-nonsinglet qqq-baryons would be observable entities in the theory, if quarks carry integer-charges. The decay-modes of the lowest colored octet mesons give some of the distinctive signatures for physical color. These are listed below under the assumption that the 1^--gluon octet is the lightest among the color-octet states.

<u>Decay Modes Of Neutral Gluons</u>: There are four neutral members. First consider the ones coupled to diagonal-currents ($\tilde{U}$ and $\tilde{V}$). In the limit of ideal mixing $\tilde{V} = V^o = \frac{1}{2}(\sqrt{3}\ V_8 - V_3)$ and $U \simeq [U^o + O(g/f)W^o_{flavor}]$ are the eigenstates, with $U^o = \frac{1}{2}(\sqrt{3}\ V_3 + V_8)$; in this limit $\tilde{V}$ is decoupled from e^-e^+ and $\mu^-\mu^+$. In general, the physical particles ($\tilde{U}$ and $\tilde{V}$) are mixtures of U and V^o, and both are coupled to e^-e^+ with amplitudes proportional to $\cos\xi$ and $\sin\xi$ respectively, ξ being the mixing angle. The expected decay modes and partial widths are:[101,100]

	Expected Partial Width (For $m_U \simeq$ 1-2 GeV)	
$(\tilde{U},\tilde{V}) \to e^-e^+$ or $\mu^-\mu^+$	6 to 30 KeV $(\cos^2\xi,\ \sin^2\xi)$	
$\to \eta'\gamma$	40 to 1000 KeV $(\cos^2\xi,\ \sin^2\xi)$	
$\to 2\pi\gamma, 4\pi\gamma, K\bar{K}\gamma$	50 to 500 KeV $(\cos^2\xi,\ \sin^2\xi)$	
$\to 3\pi, \rho\pi, 5\pi, K\bar{K}$	10 to 1000 KeV	
$\to 2\pi, 4\pi, K\bar{K}$	10 to 1000 KeV	(81)

The existence of such <u>narrow</u> ($\Gamma \sim \frac{1}{10}$ to few MeV) gluons in the 1.1 to about 1.8 GeV-region is at present compatible with the data involving e^-e^+ and photo-production experiments.[101] There are at present likely candidates for such gluons: the narrow 1498 MeV-resonance, observed at Frascati and reported at the Hamburg Conference, appears to be a promising candidate for the $\tilde{V}$-gluon (with $\sin^2\xi \simeq 10^{-2}$). A study of its decay modes, as well as a search for the accompanying <u>narrow</u> $\tilde{U}$-gluon, expected to possess a leptonic partial width larger than that of $\tilde{V}$ (see Eq. (81)), in the mass-region within ± 100 MeV of $\tilde{V}$, should shed further light on such an identification. In general, a search for narrow objects in the region $\simeq$ 1.1 to 1.8 GeV at Frascati, Orsay and Novosibirsk would be of interest in this regard.

The remaining two neutral members (V^o_{K*} and $\bar{V}^o_{K*}$) possess essentially the same decay-modes and branching ratios as the neutral $\tilde{U}$; their absolute widths are related to that of $\tilde{U}$ by the factor $(m_V^4/m_W^4)(m_V^2/(m_V^2-m_U^2))^2$, where m_V and m_U are the masses of V^o_{K*} and $\tilde{U}$ respectively. Thus, if m_V and m_U differ by 10 to 100 MeV, we expect $\Gamma(V^o_{K*}) \sim (\frac{1}{3.6} \text{ to } \frac{1}{36})\Gamma(\tilde{U})$.

Decay Modes of the Charged Gluons:

	$m_V \simeq$ 1-2 GeV	
$(V^{\pm}_{\rho}, V^{\pm}_{K*}) \to e^{+}\nu_e$	$(30 \pm 5)\%$	
$\to \mu^{+}\nu_{\mu}$	$(30 \pm 5)\%$	
$\to 2\pi, 3\pi, K\bar{K}, \ldots$	$(30 \mp 10)\%$	
$\to \pi\pi e\nu, K\bar{K}e\nu, \ldots$	(1 to 5)%	
$\to \pi e\nu, Ke\nu,$	forbidden by I-spin	
$\to \eta e\nu$	forbidden by SU(3) and SU(3)'	(82)

Note the distinction from charm, for which semileptonic modes involving single Kaon-emission are allowed and frequent. Using $(V^{\pm}_{\rho,K*}-W-)$-mixings (see Ch. II), their lifetimes are found to be:

$$\tau(V^{\pm}_{\rho}, V^{\pm}_{K*}) \simeq 10^{-13} \text{ sec. } (m_V/ 1 \text{ GeV})^{-5} (1/2) \tag{83}$$

Color, in general may be expected to be produced[100] in high energy hadronic collisions, photo-production,[101] lepto-production and e^-e^+-annihilation:

$p+p \to V^{+}_{\rho} + V^{-}_{\rho} + X$ $\to V^{\pm}_{\rho} + X^{\mp}_{col} + X$	Expect $\sigma_{col} \sim 10^{-29} - 10^{-31}$ cm^2 at Fermilab and ISR energies
$e^-e^+ \to V^{\pm}_{\rho,K*} + \bar{V}^{\mp}_{\rho,K*}$ $\mu\nu \qquad e\nu$	$R_{V\bar{V}} = 1/8$ A partial source of μe-events, if gluons are light. In this case red quarks would be a more dominant source, see later.
$\nu_{\mu}N \to \mu^- + V^{+}_{\rho} + X$ $\mu^{+}\nu_{\mu}$ $\to \mu^- + X^{+}_{col} + X$ $\mu^{+} + \ldots$	$(d\sigma_{col}/d\sigma_{flav})$ depends on glue content of the nucleon and the effective-gluon-parton mass $\bar{\mu}$ (see 3rd paper of Ref. 45). With a recent estimate of glue-content being nearly 25%, $m_V \simeq 1.2$ to 1.5 GeV, and $\bar{\mu}^2 \simeq .8$ GeV^2, one expects $(d\sigma_{col}/d\sigma_{flav}) \simeq 2\text{-}3\%$, and thus a dimuon-rate from inclusive color-production $\simeq$(1/3 to 1/2)% (inclusive color-muonic branching ratio $\simeq$ 15 to 20%).
$\mu^-N \to \mu^- + V^{+}_{\rho} + V^{-}_{\rho} + X$ $\mu^{+}\nu \qquad \mu^{-}\bar{\nu}$	(Trimuons) (83)

I now turn into a possible interpretation of the SPEAR-DORIS $\bar{\mu}e$-events[102] in terms of production and decay of integer-charge quarks.

4.6 SPEAR-DORIS-$\bar{\mu}$e-Events

If (p,n,λ) quarks have a physical mass $\approx$ 1.8 to 2 GeV, and if gluons are lighter than quarks with $m(q_{red})-m(q_{yel,blue}) < m_\pi$ -- a perfectly feasible possibility -- then one can attribute the $\bar{\mu}e$-events observed[71] at SPEAR and DORIS to pair production and decay of charged red quarks:

$$\begin{array}{lll} e^-e^+ \to q^-_{red} & + & q^+_{red} \\ \quad\quad\;\; \downarrow & & \;\downarrow \\ \quad\quad\;\; V^- + \nu & & V^+ + \nu \\ \quad\quad\;\; \hookrightarrow e^- + \bar{\nu}_e & & \hookrightarrow \mu^+ + \nu_\mu \end{array} \tag{84}$$

(i) Spectrum: The momentum-spectrum[102] of the muon (or the electron) thus arising via a two-step three-body decay of quarks, differs from that arising via a single step 3-body decay of a heavy lepton in the low and medium momentum region ($p_{e,\mu}$ ~200 to 500 MeV/c). But the two spectra coincide in the high-momentum region ($p_{e,\mu}$ > 600 MeV/c), especially if charged gluons are relatively light < 1.4 GeV. Experimentall measurements are made in the high-momentum-region (p_μ > 650 MeV/c and p_e > 400 MeV/c); the data within the errors are compatible with the quark as well as the heavy-lepton hypothesis for the $\bar{\mu}$e-events.

(ii) "Point" Coupling: Quark-electromagnetic form factor, determined by the "temporarily" asymptotically free color-gauge theory,[3] is expected to be nearly unity and slowly varying in the SPEAR energy range. Correspondingly the rate of production of ($\bar{\mu}$e)-events is expected[94] to follow (neglecting small variation) the (1/s)-behaviour, characteristic of a pointlike heavy lepton. This is observed experimentally.[71] [Taking the branching ratios of $\mu\nu$ and $e\nu$ decays of $V^\pm$ to be each $\approx$ 25 to 35% (see Eq. (82)), and the asymptotic contribution of pair-production of charged red quarks $(n_r^- n_r^+)$ and $(\lambda_r^- \lambda_r^+)$ to the R-parameter[103] to be (2/9), the net contribution to R of the μ^+e^--signal arising from $q\bar{q}$-production and decays is given by $R_{q\bar{q}}(\mu^+e^-) \approx (2/9)(1/4 \text{ to } 1/3)^2 \approx (1.6 \text{ to } 2.5)\%$, compatible with the data.]

(iii) Semileptonic signals (e.g. $e^-e^+ \to \bar{\mu}e + \pi^+\pi^-$ + missing momentum): Direct decays of red quarks into $V^\pm + \nu$ + (pions) are inhibited by phase-space since gluon is relatively heavy (~ 1.3 to 1.4 GeV) compared to the mass of the quark (~ 1.8 GeV). Semileptonic decays of charged gluons themselves must involve at least a pair of pions, and are also relatively suppressed (see Eq. (82)). Combining both sources, the semileptonic ($\bar{\mu}$e)-events are predicted[94] to be $\approx$ 1 to 5% compared to the pure leptonic ($\bar{\mu}$e)-events. This too is borne out by the data.

(iv) Jets: Missing Momentum: The yellow and blue quarks, as well as the neutral red quarks (p_r^o and c_r^o), eventually decay with the emission of neutrinos plus mesons (π,K,η, etc.), but not charged leptons (see Eq. (78)). Thus pair-production of these quarks, contributing a value $\approx$ (10/3) - (2/9) = 28/9 to the R-parameter, would lead to jet-like distribution of the decay hadrons (π,K,η) with missing energy and momentum carried away by neutrinos. [The gluonic contribution[104] (1/8) (m_U^2/μ^2) to R may reflect itself in direct production of hadrons through $q\bar{q}$ and gluon-antigluon-recombination. Such a contribution could lead to an increase in average pion multiplicity $\langle n_\pi \rangle$ with increasing energy, as observed experimentally.]

(v) $e^-e^+ \to \bar{\mu} + (\rho$, or A_1, or $\pi)$ + missing momentum: Within $[SU(4)]^4$-symmetry charged red quarks may decay into $V^- + \nu$ (Fig. 5b) as well as to mesons $(\rho, A_1, \pi, \text{etc.}) + \nu$ (Fig. 5c). Thus one would expect semileptonic events of the type ($\bar{\mu}$ + meson + missing momentum) in addition to pure leptonic events. For the quark-hypothesis, the relative frequency of $\rho : A_1 : \pi$ semileptonic events depends on the ratios of $\bar{q}q\rho : \bar{q}qA_1 : \bar{q}q\pi$ couplings (see Fig. 5c). Thus the suppression of π relative to pure leptonic or semileptonic ρ and A_1-modes, indicated by the recent PLUTO data,[105] can be accommodated under the quark-hypothesis for the $\bar{\mu}e$-events. Such an observation, if confirmed, would, however be incompatible with the heavy lepton-interpretation.

We conclude that the present data on $\bar{\mu}e$-leptonic and (μ + meson)-semileptonic events are equally compatible with the quark as well as the heavy lepton-origin of these events (with a possible difficulty for the heavy lepton-interpretation in respect of the suppression of $(\bar{\mu}\pi)$-events). One needs refined lepton-momentum measurement in the 200-500 MeV/c as well as measurements of rates of semileptonic $(\bar{\mu}e + \pi^+\pi^-)$-events to help distinguish between these alternatives.

4.7 Distinction Between Low Mass Physical Color and Hidden Color

To summarize, for either alternative - absolute versus partial confinement (or fractional versus integer-charge quarks) - one must essentially assume that the effective masses of the valence quarks and gluons inside the nucleonic environment are small ($\lesssim$ 300 MeV, getting even smaller for high momentum probes) compared to their physical outside masses. This is the so-called "Archimedes Effect." The main physical distinction between the two is that the physical masses of quarks and gluons outside of the environment are (assumed to be) infinite for the case of the former, while they are large (~1 to few GeV), but finite, for the case of the latter. Correspondingly, the transmission probability for quarks and gluons to be liberated out of the "bags" is zero for the case of absolute, but finite though small for the case of partial confinement. One expects that asymptotic freedom, relevant at short distances ($\lesssim 10^{-14}$ cm), together with nonperturbative solution of the nonabelian quark-gluon-interactions, relevant at large distances ($\gtrsim 10^{-13}$ cm), should produce this peculiar "Archimedes Effect."[46]

An experimental distinction between the two alternatives would be of major importance inasmuch as that would shed light on a fundamental issue. I list a few crucial tests, which should help provide such a distinction.

Phenomena	Low mass gauge physical color	Confined color
1) $\left(\frac{\sigma_L}{\sigma_T}\right) \xrightarrow{\text{Asymp}} f(x) \neq 0$	Yes	No
2) Two-step, three-body decay spectrum for the μe events: $e^- e^+ \to \bar{\mu} e + X$, which can be distinguished from genuine three-body decay by a measurement of the lepton momentum spectrum in the region of 200-500 MeV.	Yes $e^- e^+ \to \bar{q}_{red} + q^+_{red}$ $\bar{q}_{red} \to V^- + \nu$ $\quad \hookrightarrow e^- + \nu_e$ $q^+_{red} \to V^+ + \nu$ $\quad \hookrightarrow \mu^+ + \nu_\mu$ (see text)	No Heavy lepton decaying to charged leptons ($L^- \to \mu + \nu + \bar{\nu}$) would give rise to genuine three-body decay spectrum
3) Observation of high-momentum ($\bar{\mu}e$) signals accompanied by pair of pions $e^- e^+ \to \bar{\mu} e + \pi^+ + \pi^- + \ldots$	Yes Expected rate for such semileptonic signal ~1 to 5% compared with leptonic ($\bar{\mu}e$) events	No Heavy lepton origin of the $\bar{\mu}e$ events cannot give rise to semileptonic signals
4) Missing energy and momentum carried away by neutrinos in the <u>hadronic jets</u> produced by $e^- e^+$ collision	Yes Such missing energy and momentum must exist for the quark interpretation of the $\bar{\mu}e$ events (see text)	No No simple explanation would arise for <u>such</u> missing energy and momentum with confined quark
5) Observation of mono-energetic photons in $e^- e^+$ collision in the vicinity of threshold for $\bar{\mu}e$ production	Yes $e^- e^+ \to p^o_{red} + \bar{p}^o_{red}$ $p^o_{red} \to n^o_{yel} + \gamma$ $\quad \hookrightarrow \nu$ + mesons (see text)	Not expected
6) Observation of short-lived charged particles ($\tau \sim 10^{-13}$ to 10^{-15} sec.), whose semileptonic decay modes <u>always</u> involve a <u>pair</u> of pions, a pair of kaons but never a single π or K	Yes A clear signal for charged gluons $V^\pm_\rho \to \mu^\pm + \nu + \pi^+ + \pi^-$ $\quad \not\to \pi + \mu + \nu$ (see text for branching ratio)	No All flavor objects (like charm particles) will have semi-leptonic decay modes involving single K or π

Phenomena	Low mass gauge physical color	Confined color
7) Observation of narrow states ($\Gamma \sim \frac{1}{5}$ to 5 MeV) in the Frascati, Novosibirsk and Orsay region ($E_{cm}(e^-e^+)$ $\simeq$ 1.1 - 1.8 GeV) with at least one of them having a significant radiative decay branching ratio and leptonic partial width $\gtrsim$ 1 keV	Yes Clear signal for the neutral gluons $(\tilde{U},\tilde{V})$	No
8) Observation of particles in emulsion-bubble chamber studies with lifetimes $\sim 10^{-12}$-10^{-13} sec., which decay either into (mesons + missing neutrinos), or into (charged leptons + missing neutrinos) like a heavy lepton, but which can scatter relatively strongly against nuclei ($\sigma \sim$ mb)	Yes Clear signal for decaying integer-charge quarks $q^+_{yel,blue} \to \nu + (\pi, K, \ldots)$ $q^-_{red} \to V^- + \nu$ $\quad \hookrightarrow e^- + \bar{\nu}_e$ Expected quark pair production cross-section in hadronic collisions $\sim 10^{-30}$-10^{-31} cm^2 at Fermilab ISR-energies for $m_q \sim$ 2 to 3 GeV.	No

As pointed out before, a crucial feature of low-mass physical color, in addition to the tests mentioned above, is that the lightest color-octet vector mesons (whether these are gluons or $q\bar{q}$ color-octet mesons) must lie in the 1.1 to 1.8 GeV region. This is required on the basis of existing experimental searches. The same mass region for the gluons is also required so far as the simple basic model is concerned for a consistent interpretation of the SPEAR $\bar{\mu}$e events in terms of pair production and decays of quarks. This should serve to emphasize the importance of a search for the light gluon in this low-mass region. Such a search, especially via a scan in e^-e^+ annihilation (at Frascati, Novosibirsk and Orsay) and photoproduction deserves priority so that it may clearly eliminate or establish this intriguing hypothesis of low-mass physical color.

V. UNIFICATION OF THE BASIC PARTICLE FORCES AT 10^3 or 10^{13} m_W?

5.1. The Role of Embedding

The hypothesis[3,4] that the fundamental particles and their interactions are unified at a basic level, through a lagrangian characterised by a single gauge coupling constant, raises two important questions: (1) At what mass-scale (M) would this complete unification (lost at low energies through spontaneous breaking of the symmetry), be observable? (2) What is the value of the renormalised weak angle $\sin^2\theta_W$?

Both these questions can be answered in the context of a unified gauge theory through renormalisation group equations.

It has generally been claimed[53,57] over the past that the unifying mass-scale M needs to be superheavy ($M \gtrsim 10^{15}$ GeV) in order that "strong" interactions may be strong at low-energies. If such a superheavy mass-scale was indeed unavoidable, there would appear to be no hope that even some faint traces of the unification-hypothesis might be visible in the foreseeable future. (Fortunately), it has recently been realised,[58] that such a superheavy unifying mass-scale M is a consequence of a special assumption: the embedding of the low-energy weak interaction-symmetry (G_W) and the chromodynamic strong interaction-symmetry ($SU(3)'_{Col}$) within the unifying symmetry G is such that the gauge coupling constants associated with G_W and SU(3)'-color are equal to each other in the (bare) symmetric limit.

The symmetry group $[SU(4)]^4$-more generally $[SU(n)]^4 (n \geqslant 4)$ - permit a descent departing from this assumption. For example, corresponding to the descent from SU(4) to GIM-SU(2) for flavor, together with the direct descent of $SU(4)'_{C,D}$ to $SU(3)'_{C,D}$ for Color (see Ch. III for notations), the coupling constant associated with the GIM-SU(2)-group is smaller (by a factor $1/\sqrt{2}$) than that associated with the SU(3)'-color group.

Such a "small" departure (involving a factor of only $1/\sqrt{2}$ in the ratio of the bare coupling constants) has the dramatic consequence that the unifying mass-scale M is lowered by at least ten orders of magnitude below all previous estimates. Such a relatively low unifying mass-scale in turn raises the attractive possibility that the unification hypothesis may in fact be testable through ongoing cosmic ray-experiments, and perhaps also through the next generation of accelerators.

5.2. Renormalisation

To see the underlying reason behind this dramatic effect, assume (for simplicity) that the symmetry $[SU(4)]^4$ descends spontaneously at the primary stage of symmetry breaking through a single superheavy mass-scale $M(\gg m_W)$

to a "low-energy" symmetry having the form $G_W \times G_{Color}$:

$$[SU(4)]^4 \xrightarrow{M >> m_W} G_W \times G_{Color} \tag{84}$$

$$\text{where, } G_W = \begin{cases} \text{either} & G_L = SU(2)_A \times U(1) \quad \text{(See Eq. (76))} \\ \text{or} & \bar{G}_{LR} = SU(2)_A \times SU(2)_B \times U(1)_{C+D} \quad \text{(See Eqs. (71), (72))} \end{cases} \tag{85}$$

$$\text{and } G_{col} = \begin{cases} \text{either } SU(3)'_{C+D} & \text{(conventional: Case I)} \\ \text{or } SU(3)'_C \times SU(3)'_D & \text{(Split Color: Case II)} \end{cases} \tag{86}$$

Recall (in accordance with the notations in Ch. III) that $SU(2)_{A,B}$ are the GIM-subgroups of $SU(4)_{A,B}$; G_L is the standard left handed, and $\bar{G}_{LR}$ the left-right symmetric weak/electromagnetic subgroups. The split color-symmetry $SU(3)'_C \times SU(3)'_D$ refers to chiral color (if color-gauging is chiral), while $SU(3)'_{C+D}$ is their vector diagonal sum. Case II differs from Case I in that for Case II, the split chiral color $SU(3)'_C \times SU(3)'_D$ is preserved as a good low-energy symmetry, it is broken softly by a mass-scale ($\lesssim m_W$) at the secondary stage of symmetry breaking into $SU(3)'_{C+D}$; while for Case I, split color breaks down "strongly" by a heavy mass-scale ($>>m_W$) into the diagonal sum $SU(3)'_{C+D}$ at the primary stage of symmetry breaking (compare Eqs. (69 (a) and (b)) versus (69 (c)).

Let g_1, g_2, g_3 $\bar{g}_3$ and g_G denote the coupling constants associated with the abelian U(1), "weak" $SU(2)_{A,B}$, "strong" $SU(3)'_{C,D}$, the diagonal sum $SU(3)'_{C+D}$ and the unifying group $[SU(4)]^4$ respectively. Noting that diagonal summing (e.g. $SU(2)^I_A \times SU(2)^{II}_A \to SU(2)^{I+II}_A$ or $SU(3)'_C \times SU(3)'_D \to SU(3)'_{C+D}$) reduces the associated coupling by a factor $1/\sqrt{2}$, we observe the following relationships between the above coupling constants in their large momentum, or (bare)symmetric limits.

Case I: $$g_1 = g_2 = \bar{g}_3 = g_{G}/\sqrt{2} \tag{86}$$

Case II $$g_1 = g_2 = g_{G}/\sqrt{2}$$
$$g_3 = g_G = \sqrt{2}\, g_{1,2} \tag{87}$$

Case II thus provides the crucial departure as regards the symmetric limit relationship between $g_{1,2}$ versus g_3, mentioned above.

Let us now write down the renormalisation group equations for the coupling constants, assuming that they are all small ($|g_i| << 1$) in the domain of interest.

Renormalisation In the Weak Sector: Using decoupling theorem[106] arguments and the normalisation conditions (Eq. (86) or (87)), the running coupling constants $g_{1,2}(\mu)$ for momenta $\mu \leq M$ are given by:

$$g_i^{-2}(\mu) = 2g_G^{-2}(M) + 2b_i \ln (M/\mu) \quad (i = 1,2) \tag{88}$$

where $b_2 = -22/(3(4\pi)^2) + b_1$. For algebraic simplicity, assume first that $G_W = G_L = SU(2)_A \times U(1)$ and that quarks are fractionally charged. In this case,

$$e^{-2} = g_2^{-2} + 5\, g_1^{-2}/3; \; \sin^2\theta_W = 3\, g_1^2/(3\, g_1^2 + 5\, g_2^2) \tag{89}$$

Combining (88) with (89), we obtain

$$g_2^{-2} - g_1^{-2} = (8 \sin^2\theta_W - 3)/5e^2 = -(11/12\pi^2) \ln M/\mu \tag{90}$$

<u>Renormalisation In The Color Sector: Emergence of Chromodynamics</u>:

<u>Case I: (The Conventional Case)</u>: In this case QCD-coupling is given by $\overline{g_3}$ ($\alpha_s(QCD) = (\bar{g}_3)^2/4\pi$). Using the normalisation condition (86), we obtain

$$(\bar{g}_3^{(\mu)})^{-2} = 2g_G^{-2}(M) + 2b_3 \ln (M/\mu) \tag{91}$$

where $b_3 = -11/(4\pi)^2 + b_1$. Using Eqs. (90) and (91) and writing $u_S \equiv \alpha_s(QCD)/\alpha = (\bar{g}_3)^2/e^2$, we obtain

$$u_S = (8/3)\, [1 - (11\alpha/\pi) \ln (M/\mu)]^{-1} \tag{92}$$

$$\sin^2\theta_W = 1/6 + 5/(9\, u_S) \tag{93}$$

<u>Case II: (Split Color)</u>: Here we will assume (for simplicity)[107] that the split color $SU(3)'_C \times SU(3)'_D$ is broken by a mass-scale $M' \sim 1$ GeV, <u>low</u> compared to electroproduction vertex momenta. In this case, electroproduction (as also charmonium physics) would still be described by an effective coupling constant $\alpha_s(QCD) = g_3^2(\mu)/4\pi$, where g_3 denotes the coupling constant associated with $SU(3)'_{C,D}$. Using the symmetric limit normalisation condition $g_3 = g_G$ (see Eq. (87)), we obtain:

$$g_3^{-2}(\mu) = g_G^{-2}(M) + 2b'_3 \ln (M/\mu) \tag{94}$$

with $b'_3 = -11/(4\pi)^2 + b_1/2$. Using (90) and (94), we now obtain:

$$u_S = \frac{\alpha_s(QCD)}{\alpha} = 2(8/3)\, [1 - (77\alpha/(3\pi)) \ln M/\mu]^{-1} \tag{95}$$

$$\sin^2\theta_W = 2/7 + 10/(21\, u_S) \tag{96}$$

Apart from the overall factor of 2, note the <u>crucial distinction</u> between the coefficients of ($\ln M/\mu$) in Eqs. (92) versus (95). The bigger this coefficient, the smaller the unifying mass M (for a given value of u_S).

Putting $\mu \approx 10$ GeV and requiring that $u_S = \alpha_s/\alpha \approx 15$ (i.e. $\alpha_s \approx 0.1$ at $\mu \approx 10$ GeV), we obtain:

$$M \approx 10^{15} \text{ GeV} \qquad \text{Case I}$$
$$\sin^2\theta_W \approx 0.20 \text{ (For } G_W = SU(2)_A \times U(1) \qquad (97)$$

This is the ultraheavy unifying mass, obtained traditionally. By contrast,

$$M \approx 10^5 \text{ GeV} \qquad \text{Case II}$$
$$\sin^2\theta_W \approx 0.30 \qquad (98)$$

Note the lowering of the unifying mass-scale M for Case II compared to Case I.

The origin of this enormous difference between the unifying mass-scales for Cases I and II is easy to trace. The coupling constant renormalisations depend upon the relevant residual symmetries; the bigger the residual symmetry, the stronger is the renormalisation effect. Thus, the bigger the residual symmetry in the color-sector relative to the weak flavor-sector, the stronger is the growth of the effective color-coupling constant relative to the flavor SU(2)-Coupling Constant with decreasing momentum; correspondingly the lower is the unifying mass-scale M. The residual color-symmetry relative to flavor-symmetry is clearly bigger for Case II, than it is for Case I. This leads to the enormous lowering in the mass-scale M (see Fig. 6 for a graphic demonstration).

Analogous considerations apply[58] to the case of $G_W = \bar{G}_{LR}$.

For either choice of $G_W (= G_L$ or $\bar{G}_{LR})$, the same qualitative results hold for the case of integer-charge quarks. The main change is that the unifying mass M turns out to be lower by about a factor of 10 compared to that for the case of fractionally charged quarks. This permits unification at a mass-scale as low as about 10^4-10^5 GeV. Such a mass-scale coincides with the estimate of the exotic X gauge-masses obtained from independent considerations[3] (See Ch. III). This in turn strengthens the compatibility of the hypothesis of unconfined decaying integer charge-quarks within a unified theory.

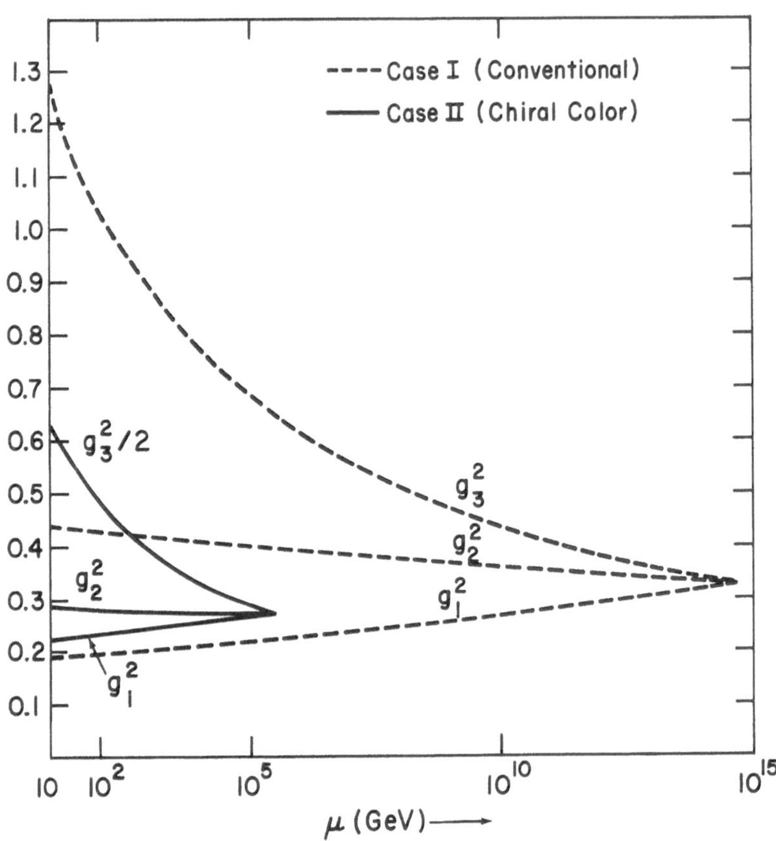

Fig. 6. Effect of the Hierarchy of Symmetry Breaking on the Unifying Mass-scale M. The variation of coupling constants $g_{1,2,3}$ for the two cases of symmetry breakdown (I and II) are shown corresponding to the specific choice that the low energy weak/electromagnetic symmetry is $G_W = G_L = SU(2)_L \times U(1)$ and that the quarks are fractionally charged. The qualitative features of the result that the unifying mass-scale M is superheavy ($\gtrsim 10^{15}$ GeV) for Case I (conventional color) and that it is relatively low ($\sim 10^5$ GeV) for case II (chiral color) is not altered by choosing $G_W = G_{LR} = SU(2)_L \times SU(2)_R \times U(1)_{L+R}$ and/or quark-charges to be integral. (For cases I and II, it is g_3^2 and $g_3^2/2$ respectively, which coincide with $g_{1,2}^2$ at large momenta. However, for either case, it is g_3, which denotes the relevant "strong" interaction coupling constant at low momenta.)

5.3 Remarks and a Summary

(i) Low-mass-unification-hypothesis suggests that chiral color $SU(3)'_L \times SU(3)'_R$ may be a good low energy symmetry. The QCD-octet of vector color gluons must then be accompanied by a low mass octet of axial color-gluons. These axial color-gluons would be neutral and confined (like the vector color gluons) if quarks are fractionally charged, and possess charged members and would be liberated, if quarks are integer-charged. The existence of such light axial gluons (mass ~ 1 to few GeV) seems to be compatible with all known phenomenology such as that encountered in lepto-production.

(ii) The mechanism for low-mass unification involving a symmetric limit dichotomy between "weak" SU(2) and "strong" SU(3)'-color coupling constants, as discussed above, appears to be a special feature of symmetries of the form $[SU(n)]^4$ $(n \geq 4)$. It is not available within unifying symmetries such as[4] SU(5) and SO(10). Correspondingly, the unifying mass-mass-scale M needs to exceed $\approx 10^{14}$ GeV within symmetries of the latter type. A comparison of the salient features of the unifying symmetries $[SU(4)]^4$, SU(5), SO(10) and E_7 is presented as a table at the end (p.290).

(iii) Had the unifying mass-scale M been necessarily as large as 10^{14} GeV or higher, the question of a direct test of the hypothesis of unification of the basic forces (weak, electromagnetic, as well as strong) would not even arise, at least within one's own generation. The realization of the low unifying mass-scale $M \approx 10^4 - 10^6$ GeV serves to remove this gloomy view: it raises the attractive hope that the hypothesis may after all be tested directly through ongoing cosmic ray experiments and also through the next generation of accelerators. Nothing could be more dramatic than the observation that quarks and leptons begin to interact universally in all respects and that simultaneously the so-called "strong" interactions acquire their basic $\mathcal{O}(\alpha)$-strength. That such a possibility could arise at a momentum-scale as low as 10^4-10^5 GeV should provide a strong new incentive to build high energy accelerators in the 10^3-10^4 GeV range, which is being discussed at present.

(iv) The major challenge presently confronting unified theories is a natural understanding of the phenomenologically imposed mass-scales of fermions and gauge-mesons. Mass differences such as that between n-p quarks, ν_e-e, μ-e, or the mass-ratios such as (e/μ), (p/c) and (quark/lepton), as well as the many Cabibbo-like angles and CP-violating phases should hopefully be predictions of the theory.[108] As regards gauge-masses, one would need a natural explanation for the ratio of the superheavy unifying mass-scale M versus the masses of the lighter gauge particles W's and Z's. In this respect, a low unifying mass-scale $M \approx 10^4$-10^6 GeV generates the attractive possibility that perhaps only one superheavy mass-scale triggers spontaneous symmetry breaking, with radiative corrections[109] of order $\sqrt{\alpha}$ and α providing masses to the lighter gauge mesons (the familiar W's and Z's). This remains to be actually implemented within a realistic model such as $[SU(4)]^4$.

REFERENCES AND FOOTNOTES

1. C. N. Yang and R. C. Mills, Phys. Rev. D96, 191 (1954).
2. J. Schwinger, Ann. Phys. (N.Y.) 2, 407 (1957); S. L. Glashow, Nucl. Phys. 22, 57a (1961); Abdus Salam and J. C. Ward, Physics Letters 13, 168 (1964). In the present context, the model based on spontaneous breakdown of the local symmetry $SU(2)_L \times U(1)$ is due to S. Weinberg, Phys. Rev. Lett. 19, 1264 (1967) and Abdus Salam, in Elementary Particle Theory, Nobel Symposium, Ed. by N. Svartholm (Almqvist, Stockholm, 1968), p. 367.
3. J. C. Pati and Abdus Salam, Phys. Rev. D8, 1240 (1973); Phys. Rev. Lett. 31, 661 (1973); Phys. Rev. D10, 275 (1974), and Physics Letters 58B, 333 (1975).
4. H. Georgi and S. L. Glashow, Phys. Rev. Lett. 32, 438 (1974); H. Fritzsch and P. Minkowski, Ann. Phys. (NY) 93, 193 (1975); F. Gürsey and P. Sikivie, Phys. Rev. Lett. 36, 775 (1976); P. Ramond, Nucl. Phys. B110, 214 (1976).
5. These attempts are admittedly incomplete, until gravity is in the picture.
6. In general, if G is a direct product of several groups, correspondingly several coupling constants would appear.
7. P. W. Higgs, Phys. Rev. Lett. 12, 132 (1964); P. Englert and R. Brout, Phys. Rev. Lett. 13, 321 (1964); G. S. Guralnik, C. R. Hagen and T. W. B. Kibble, Phys. Rev. Lett. 13, 585 (1965); P. W. Higgs, Phys. Rev. 145, 1156 (1966); T. W. B. Kibble, Phys. Rev. 155, 1554 (1967).
8. J. Goldstone, Nuovo Cim. 19, 15 (1961); Y. Nambu, G. Jona-Lasinio, Phys. Rev. 122, 345 (1961); 124, 246 (1961); J. Goldstone, A. Salam and S. Weinberg, Phys. Rev. 127, 965 (1962).
9. One still needs to examine if quantum-corrections could disturb the character of the ground state of the theory (see e.g. S. Coleman and E. Weinberg, Phys. Rev. D7, 1888 (1973)).
10. G' t Hooft, Nucl. Phys. B33, 173 (1971); B35, 167 (1971); B. W. Lee and J. Zinn-Justin, Phys. Rev. D5, 3121, 3137 (1972).
11. See for example, J. M. Cornwall, D. N. Levin and G. Tiktopoulos, Phys. Rev. Letters, 30, 1268 (1973); C. H. Llewellyn Smith, University of Hawaii Summer School Lecture Notes (1973) and references therein.
12. See for example R. N. Mohapatra, S. Sakakibara and J. Sucher, Phys. Rev. D10, 1844 (1974).
13. As to whether spontaneous symmetry breaking is entirely due to the presence of elementary Higgs-scalars, or whether it is partly or entirely induced dynamically is not yet clear.
14. If potential $V(\phi)$ has a global symmetry higher than the local symmetry, one needs to examine whether light pseudo goldstone bosons are generated in the theory.

15. D. Gross and F. Wilczek, Phys. Rev. Lett. 30, 1343 (1973); H. D. Politzer, Phys. Rev. Lett. 30, 1346 (1973).

16. See for example H. D. Politzer "Testing of Scaling and QCD with muon Beams", HUTP-77/A038 and references therein; G. Altarelli, G. Parisi and R. Petronzio, Physics Letters 63B, 183 (1976).

17. S. Abers and B. W. Lee, Physics Reports 9C, (1973); M.A.B. Bég and A. Sirlin, Ann. Rev. Nucl. Science, 24, 379 (1974). C. Llewellynsmith, Hawaii University Lecture Notes (1973). S. Weinberg, Rev. Mod. Phys. 46, No. 2, (1976).

18. Recent CDHS measurements yield $\sin^2\theta_W = 0.24 \pm 0.02$ (M. Holder et. al. CERN Preprint, 1977). Earlier measurements, gave somewhat higher values for $\sin^2\theta_W$, see for example, J. Blietschau et. al., Nucl. Phys. B118, 218 (1977); B. Barish et. al., Proc. Internat. Neutrino Conf. Aachen, 1976 (p. 289); A. Benvenuti et. al., Proc. Internat. Neutrino Conf. Aachen, 1976 (P. 296).

19. J. Blietschau et. al., Nucl. Phys. B118, 218 (1977); H. Faissner et. al., reports of Aachen-Padova results at Aachen, Tbilisi and Ben-Lee Memorial Conference (1977).

20. C. Bouchiat, J. Iliopoulos and P. Meyer, Phys. Lett. 38B, 519 (1972).

21. H. Georgi and S. L. Glashow, Phys. Rev. D6, 429 (1972).

22. D. J. Gross and R. Jackiw, Phys. Rev. D6, 477 (1972).

23. See for example the first paper of Ref. 3 where the relevant Yukawa interaction is written down.

24. S. Glashow, J. Iliopoulos and L. Maiani, Phys. Rev. D2, 1285 (1970).

25. R. N. Mohapatra, J. C. Pati and P. Vinciarelli, Phys. Rev. D8, 3652 (1973); S. Weinberg, Phys. Rev. Lett. 31, 494 (1973).

26. A. I. Vainshtein and I. B. Khriplovich, JETP Letters 18, 141 (1973); M. G. Gaillard and B. W. Lee, Phys. Rev. D10, 897 (1974).

27. G. Goldhaber et. al., Phys. Rev. Lett. 37, 569 (1976). For a recent review, see the report by A. M. Litke, Invited Talk, European Physical Society Meeting, held at Budapest (July, 1977), SLAC-PUB-2024.

28. B. Wiik, Lectures presented at Cargese summer school, 1977.

29. See e.g. C. Baltay et. al., Phys. Rev. Lett. 39, 62 (1977) and references therein.

30. E. G. Cazzoli et. al., Phys. Rev. Lett. 34, 1125 (1975); G. Blietschau et. al., Phys. Lett. 60B, 207 (1975); J. Von Krogh et. al., Phys. Rev. Lett. 36, 710 (1976).

31. "Color" was first introduced implicitly through paraquarks of order three by O. W. Greenberg, Phys. Rev. Letters, 13, 598 (1964), and subsequently explicitly through three triplets of quarks, by M. Han

and Y. Nambu, Phys. Rev. 139, B1006 (1965) and P.G.O. Freund, Phys. Letters 15, 352 (1965). "Explicit" Color, which in general permits either solution for quark-charges (integral or fractional), appears to be needed for the generation of local SU(3)'-color-gauge interactions. [A straightforward formulation to generate Yang-Mills gauge interactions in the parafield-theory generates SO(3) rather than local color SU(3)-symmetry: see P. G. O. Freund, Phys. Rev. D13, 2322 (1976), and also remarks by O. W. Greenberg and C. A. Nelson, Physics Reports 32C, Page 88 (1977)].

32. J. C. Pati and C. H. Woo, Phys. Rev. D3, 2920 (1971); K. Miura, T. Minamikawa, Prog. Theor. Phys. 38, 954 (1967). Plausible arguments for K-decays were given by C. A. Nelson and K. J. Sebastian, Phys. Rev. D8, 3144 (1973).

33. M. K. Gaillard and B. W. Lee, Phys. Rev. Letters 33, 108 (1974); G. Altarelli and L. Maiani, Phys. Letters 52B, 351 (1974).

34. The first suggestion of using SU(3)'-Color-gauges to generate superstrong interactions, with additional interactions generating strong (or medium strong) interactions, is due to Han and Nambu (Ref. 31).

35. The first suggestion of a renormalisable gauge theory of weak, electromagnetic and strong interaction using flavor to generate basic weak and SU(3)'-Color to generate basic strong interactions (via the gauge structure $SU(2)_L \times U(1) \times SU(3)'_{Col}$ was made by J. C. Pati and A. Salam (Aug, 1972, unpublished), work reported by J. D. Bjorken, Proceedings of the 1972 Batavia Conference (Vol. 2, Page 304). The same local symmetry permits either solution for quark-charges depending upon the nature of spontaneous symmetry breaking (see text). This work is elaborated in the first paper of Ref. 3.

36. The hypothesis that local color symmetry remains unbroken, and that color is confined was proposed in stages by H. Fritzsch , M. Gell-Mann and H.Leutwyler, Physics Letters B47 ,365 (1973). S. Weinberg, Phys. Rev. Letters 31, 494 (1973) and D. J. Gross and F. Wilczek, Phys. Rev. D8, 3633 (1973). See also, C. Itoh, et. al. (Preprint, 1973, unpublished).

37. O. W. Greenberg and D. Zwanziger, Phys. Rev. 150, 1177 (1966); H. J. Lipkin, Phys. Lett. 45B, 267 (1973).

38. For a good review of the color-hypothesis and its various versions, see O. W. Greenberg and C. A. Nelson. Physics Reports 32C (1977).

39. J.C. Pati, Phys. Rev. D4, 2143 (1971); J. C. Pati and A. Salam, Trieste Report IC/73/81, A. D. Dolgov, L. B. Okun, and V. I. Zacharov)

Phys. Lett. 47B, 258 (1973). Breaking global SU(9) via Higgs mechanism would require a rather elaborate set of Higgs-scalars, such a large global symmetry would inevitably generate a large number of light pseudo-Goldstone bosons. Thus spontaneous breakdown of global SU(9) is unlikely to remove the SU(9)-difficulty discussed in the text.

40. A number of such models were proposed in the early stages of model building. See the last two references in Ref. 39 for a listing of references on these models.

41. In such a theory, spontaneous symmetry breaking may still induce a mixing between flavor and color-gauge particles; such mixing in turn may generate a color-nonsinglet component in the weak-current. Such a theory, in which color is "brought" into weak interactions only through spontaneous symmetry breaking, differs drastically from the one in which the basic weak interactions use the color-symmetry, prior to spontaneous symmetry breaking. In particular, the former does not suffer from the SU(9)-difficulty. See discussions later for other characteristic features of such a theory pertaining to leptoproduction experiments.

42. In general, if there is an abelian U(1)-factor contributing to electric charge, SU(3)'-generators (F_3' and F_8') may contribute to electric charge with arbitrary coefficients associated with them. This arbitrariness is removed, once the symmetry (e.g. G_0) is embedded within a nonabelian local symmetry; and the constraint that an effective good global SU(3)-Color Symmetry must be preserved (despite the breaking of the local SU(3)-Color Symmetry) is imposed. (See discussions in Ref. 43).

43. R. N. Mohapatra, J. C. Pati and A. Salam, Phys. Rev. D13, 1733 (1976).

44. See Ref. 35 for details.

45. J. C. Pati and A. Salam, Phys. Rev. Lett. 36, 11 (1976); G. Rajasekharan and P. Roy, Pramana 6, 303 (1975); V. Elias, J. C. Pati, A. Salam and J. Strathdee, Pramana 8, 303 (1977).

46. See Ref. 45 and J. C. Pati, "Unification-Liberation of Quarks and Leptons", Univ. of Md. Tech. Rep. 77-073; Proceedings of the 1976 Scotthish University Summer School Lecture Notes ed. by I. M. Barbour and A. T. Davies (Pages 89-145, 1976). An account of the "Philosophical" implications of partial versus absolute confinement may be found in J. C. Pati and Abdus Salam, "Lepton-Hadron-Unification", Proceedings of the 1976 Aachen Neutrino Conference, ed. by H. Faissner, H. Reithler and P. Zerwas (Pages 589-629, see especially Sec. V).

47. See for example, B. Barish, Particles and Fields, 1974 (APS/DPF-Williamsburg), ed. C. E. Carlson (AIP, New York, 1975).

48. L. L. Lewis, et. al., Phys. Rev. Lett. 39, 795 (1977); P. E. G. Baird, et. al., Phys. Rev. Lett. 39, 798 (1977).

49. For a review on present experimental and theoretical status, including calculations of shielding effect, see P. Sandars, Invited Talk at Ben Lee Memorial Conference, held at Batavia, Illinois (Oct., 1977), To appear in the Proceedings.

50. The possibility of observable fractionally charged quarks, though logically permissible, is disfavored within quark-lepton unifying symmetries (without abelian U(1)-factors), unless massless gluons are seen; as for such symmetries, color-gluons can acquire mass and yet an effective good SU(3)-color global symmetry can emerge, provided quarks acquire integer-charges (See Ref. 43 and discussions in Ch. IV).

51. See for example H. D. Politzer (Ref. 16); H. Georgi and H. D. Politzer "Clean Tests of QCD in μp-Scattering (HUTP-77/A063). For tests pertaining to jet-structure, see H. Georgi and M. Machacek, Phys. Rev. Lett. 39, 1237 (1977); G. Steerman and S. Weinberg, Phys. Rev. Lett. 39, 1436 (1977), and E. Farhi, Phys. Rev. Lett. Vol. 39, 1587 (1977).

52. Abdus Salam and J. Strathdee propose that spin-2 multiplet should play a vital role in generating partial or absolute confinement (Trieste Preprints, 1977). The link between Yang-Mills spin-1 and Einstein-Weyl spin-2 gauge-theories remains to be understood, quite independently of the question of whether spin-2 exchange produces confinement or not.

53. The first such calculation utilising the Georgi-Glashow SU(5)-model (Ref. 4) for illustration, was carried out by H. Georgi, H. R. Quinn and S. Weinberg, Phys. Rev. Lett. 33, 451 (1974).

54. S. W. Herb et. al., Phys. Rev. Lett. 39, 252 (1977); W. R. Innis, et. al., Phys. Rev. Lett. 39, 1249 (1977).

55. F. Gürsey and P. Sikivie; and P. Ramond (Ref. 4).

56. The example is fictitious, because with p and n-quarks having charges +1 and 0, the proton acquires a charge +2. This feature is remedied yet the essential content of the illustration leading to a correlation between quark and lepton-charges is retained - once quarks are assigned color as well as flavor (see discussions in Ch. IV).

57. V. Elias, Phys. Rev. D14, 1896 (1976); ibid Phys. Rev. D16, 1586 (1977); ibid Phys. Rev. D (To be published); F. J. Yndurdin, Nucl. Phys. B115, 293 (1976). M. Chanowitz, J. Ellis and M. K. Gaillard, CERN preprint TH-2312 (1977), To appear in Nuclear Physics.

58. V. Elias, J. C. Pati and Abdus Salam, "Unification of The Basic Particle Forces at a Mass-Scale of order 1000 m_W", Univ. of Md. Tech. Rep. 78-041 (Sept., 1977).

59. A compulsive explanation would require an opriori understanding of the ratio of the unifying mass-scale M and themass of the weak boson m_W (~100 GeV).

60. For weak/electromagnetic symmetries other than $SU(2)_L \times U(1)$ (e.g. $SU(2)_L \times SU(2)_R \times U(1)_{L+R}$, see later), one can still define a parameter analogous to θ_W.

61. R. N. Mohapatra and J. C. Pati, Phys. Rev. D11, 566, 2558 (1975).

62. J. C. Pati and Abdus Salam, Phys. Letters, 58B, 333 (1975); J. C. Pati, Proc. Second Orbis Scientae, Coral Gables, Florida, Jan., 1975 (p. 253-256), ed. by A. Perlmutter andS. Widmayer.

63. H. Fritzsch and P. Minkowski, Ann. Phys. (NY) 93, 193 (1975).

64. The basic distinction between the patterns I and II arises under the presumption that Fermi mass-matrix, generated by spontaneous symmetry breaking would mix $\bar{F}_L$ with F_R and $\bar{F}^m_L$ with F^m_R with little or (in the extreme case) no mixing between F and F^m. [See also remarks later about the difficulty of generating Fermi masses for Case II].

65. The decays of mirror $\bar{F}^m F^m$-composites to normal hadrons made of $\bar{F}F$-components would in this case be governed by a super Zweig rule.

66. J. C. Pati and Abdus Salam: "SU(8) x SU(8)"-unpublished notes; J. C. Pati, Abdus Salam andJ. Strathdee, Nuovo Cimento 26, 72 (1975).

67. See for example G. Feinberg and J. Sucher, Phys. Rev. Letters, 13, 1740 (1975); J. D. Jackson, invited talk at the American Physical Society Meeting, Budapest (July, 1977), CERN-Preprint (1977), To appear in the Proceedings.

68. The consistency of this hypothesis and other general consequences of low-mass axial color-gluons are being examined in collaboration with J. Sucher.

69. V. Elias, J. C. Pati and Abdus Salam, In preparation (1977).

70. The symmetries $[SU(5)]^4$ and $[SU(6)]^4$ belonging to this class have been considered (J. C. Pati and Abdus Salam, unpublished), see also Ref. 46. The $[SU(5)]^4$ symmetry has been studied in some detail by S. Rajpoot (Ph.D. Thesis, Imperial College, 1977).

71. M. L. Perl et. al., Phys. Rev. Lett. 35, 1489 (1975). For a good review of the recent experimental status covering SPEAR and DORIS-experiments, see M. L. Perl, invited talk presented at the Photon-Lepton-Symposium, Hamburg, W. Germany (August, 1977), SLAC-PUB 2022 (1977). The review emphasises the compatibility of the data with the heavy lepton-interpretation; the equally viable quark-interpretation of the data is not considered in this review. (The quark versus heavy lepton origins for the $(\bar{\mu}e)$-events are considered in Ch. IV).

72. Barring large mixing of F and F^m via Fermi mass-matrix, which would, in general, lead to "unnatural" violations of quantum numbers as well as of β-decay versus μ-decay and μ versus e universalities. Small mixing - e.g. of ν_e^o and E^o - can be present. Such mixing may indeed account naturally for the masslessness or small mass of the physical neutrinos (ν_e and ν_μ). [Consider the possibility that ν_e^o - diagonal mass element is $\alpha^2 M_{E^o}$ and the non-diagonal $(\nu_e^o E^o)$-mixing mass is αm_{E^o}. Such a mass-matrix would generate a massless eigenstate; the mixing angle would however be tiny ($\mathcal{O}(\alpha)$) and thus compatible with the observed universalities. These considerations will be elaborated in a note under preparation.]

73. The prequark-hypothesis has been considered in different forms. For gauge motivated considerations, see J. C. Pati and A. Salam, Phys. Rev. D10, 275 (1974), Footnote 7; 1975 Palermo Conference Proceedings, ed. by A. Zichichi (page); J. C. Pati, A. Salam and J. Strathdee, Phys. Lett. 59B, 265 (1975) and Supplement Trieste Preprint IC/75/139/ (unpublished). The hypothesis under other motivations has been considered by W. Krolikowski, Nuovo Cim. 72, 645 (1972); J. D. Bjorken (private communications, 1973, unpublished); C. H. Woo (private communications 1974, unpublished); K. Matumoto, Prog. Theor. Phys. 52, 1973 (1974); and O. W. Greenberg, Phys. Rev. Lett. 35, 1120 (1975). Certain Perturbative dynamical calcualtions testing the compataibility of the hypothesis has recently been carried out by E. Nowak, J. Sucher and C. H. Woo (Phys. Rev. D. To be published).

74. J. C. Pati (Ref. 46, Page 109 of the Proceedings). The same constraint follows by demanding that the physical weak currents satisfy left-right symmetry, which would require that the left and right Cabibbo-like angles be equal up to $\mathcal{O}(\alpha)$-corrections (M.A.B. Bég, R. Budny, R. N. Mohapatra and A. Sirlin, Phys. Rev. Lett. 38, 1252 (1977)).

75. M. Holder et al., Phys. Rev. Letters, 39, 433 (1977).

76. M.A.B. Bég, and A. Zee, Phys. Rev. Letters 30, 675 (1973). For a list of references on other vector-like models see R. M. Barnett, Review talk, Proceedings of the Brookhaven APS meeting (Sept. 1976).

77. Skewness angles, arising due to F-F' mixing, are assumed to be small for reasons mentioned before (see also Ref. 74).

78. A Benvenutti et al., Phys. Rev. Letters 37., 1039 (1976); J. Blietschau et al., Preprint CERN/EP/Phys 76-55; B. C. Barish et al. (CALT-68-544). The clearest distinction is shown by elastic scattering data (i.e. measurements of $\sigma(\bar{\nu}p \to \bar{\nu}p)$ versus $\sigma(\nu p \to \nu p)$; D. Cline et al., Phys. Rev. Lett. 37, 252 648 (1976) and W. Lee et al., Phys. Rev. Lett. 37, 186 (1976).

79. Precise limit would depend upon low-momentum effective X-coupling constant. This is being studied by V. Elias.

80. Technically, the scalar mass terms need not respect the discrete symmetries; since such terms, having dimension two, induce only finite corrections to coupling constants. However, aesthetically they should.

81. G. Senjanovic and R. N. Mohapatra, Phys. Rev. D12, 1502 (1975).

82. The discussion on the spontaneous breaking of $[SU(4)]^4$ follows a recent paper by V. Elias, J. C. Pati and A. Salam, "Light Neutral Z°-Boson In the PETRA and PEP Energy Range Within the Unifying Symmetry $[SU(4)]^4$", Univ. of Md. Tech. Rep. No. 78-043 (Nov. 1977).

83. As to whether an allowed minimum of the potential (subject to the parameters being in a given range) will permit such different patterns of VEV for fields transforming similarly (e.g. L_i, E_i and M_i), remains to be examined.

84. Two multiplets, related by discrete symmetries, may well have unequal vacuum expectation values (see e.g. Ref. 81).

85. H. Fritzsch and P. Minkowski, Nucl. Phys. B103, 61 (1976); R. N. Mohapatra and D. P. Sidhu, Phys. Rev. Letters 38, 667 (1977); A. DeRujula, H. Georgi and S. L. Glashow, Preprint (1977); J. C. Pati, S. Rajpoot and Abdus Salam, Imperial College Preprint ICTP/76/11 (Phys. Rev. D To be published).

86. The possible importance of the group $SU(2)_L \times SU(2)_R \times U(1)_L \times U(1)_R$ has been emphasized by J. C. Pati (Ref. 62), J. C. Pati, S. Rajpoot and Abdus Salam (Ref. 85, Footnote 15). Recently its consequences have been emphasized in some detail by Q. Shafi and Ch. Wetterich, Univ. of Freiburg Preprint (Oct., 1977), and by V. Elias, J. C. Pati and Abdus Salam (Ref. 82).

87. J. C. Pati, S. Rajpoot and Abdus Salam, ICTP/76/11 (Phys. Rev. D to be published), ibid ICTP/76/15 (Physics Letters, To be published). H. S. Mani, J. C. Pati, S. Rajpoot and Abdus Salam, Trieste Preprint IC/77/88 (Physics Letters To be published).

88. R. Cahn and F. Gilman (SLAC-PUB 1977). The analogous calculations for the left-right symmetric model allowing for its generality (see Ref. 87), has been recently carried out by Mr. A. Janah at Univ. of Md. (To be published).

89. J. Leveille, Imperial College Preprint (Dec. 1977).

90. This is proved in its full generality for the $SU(2)_L \times SU(2)_R \times U(1)_{L+R}$-model by J. C. Pati, S. Rajpoot and A. Salam (Ref. 87); the equivalence for the special case of vanishing atomic parity violation was shown by H. Fritzsch and P. Minkowski (Ref. 85). Recently, the result has been further generalized to encompass bigger group structures by H. Georgi and S. Weinberg, Harvard Preprint (HUTP-77/A052).

91. K. Bardacki, M. B. Halpern, Phys. Rev. D6, 696 (1972). B. De Witt, Nucl. Phys. B51, 237 (1973).

92. R. N. Mohapatra and J. C. Pati, Physics Letters, 63B, 204 (1976).

93. Recently attempts have been made to give mass to color-gluons (with fractionally charged quarks), e.g. by A. De Rujula and R. Jaffe, Harvard Preprint (1977). Such a possibility, it seems, cannot however be entertained within a unified theory of quarks and leptons for reasons mentioned in the text (See Ref. 92 for details). A number of interesting results obtained by these authors can be carried over, however, to the case of unconfined integer-charge quarks, developed in Ref. 3.

94. J. C. Pati, Abdus Salam and S. Sakakibara, Phys. Rev. Lett. 36, 1229 (1976); J. C. Pati, S. Sakakibara and Abdus Salam, "The Missing Quark Mystery" Trieste Preprint IC/75/93 (unpublished). These two papers should be referred to for details on quark and proton-decay considerations.

95. The $q \rightarrow \ell\ell\bar{\ell}$ decay modes proceed through tree diagrams, which are damped by two heavy mass propagators, where as the loop-diagram Fig. 5(a) is damped by only one heavy mass-propagator m_X^{-2}. The decay mode- such as $q \rightarrow e^- \pi^+$ involving emission of a charged lepton, proceed either via tree or via loop-diagrams with pion-emission from inside the loop. (Compare with Fig. 5(c)). Given that pion is a $q\bar{q}$-composite, such loop-diagrams would also be damped (compared to the loop-diagram shown in Fig. 5(c)), if the emission of composite pion from the quark-line is governed by a form-factor $\propto (1/(k^2-m^2))$. See Ref. 94 for details of these arguments.

96. The ratio of the amplitudes of W_{14} versus $W_L^{\pm}$-loops (see Fig. 5(c) and 5(a)) is given by $[c_4^2 \ell n(m_X/m_{W_{14}})][c_1 c_4 \ell n(m_X/m_{W_L^+})]^{-1}$, assuming that W_L^+ mixes with X (rather than W_R^+ of the GIM $SU(2)_R$-group). [Noting that VEV of B and C-multiplets are interchangeable due to the $L \leftrightarrow R$ Symmetry in the theory (See Eq. (73)), W_R^+ can mix with X rather than W_L^+. Decays of charged red quarks into charged leptons via the two step process (see section 4.6) would exhibit V + A or V - A coupling depending upon whether it is W_L^+ or W_R^+, which mixes with X (see Ref. 94 for details)].

97. Hadrons of 2-3 GeV mass are not expected to have pair production cross-section at ISR-Fermilab energies much bigger than $\approx 10^{-30}$ cm^2, as judged from the empirical observation of J/ψ-production (e.g.).

98. The best empirical "lower limit" on proton-lifetime at present is $\approx 2 \times 10^{30}$ years (F. Reines and M. Crouch, Phys. Rev. Lett. 32, 493 (1974) and see references therein). The above search is sensitive to energetic μ^+ s, which may be expected to arise especially from an

assumed two-body decay mode of the proton ($p \to \mu^+ + \pi^\circ$). For a suggestion on decay-mode independent search based on geochemical methods, see S. P. Rosen, Phys. Rev. Lett. 34, 774 (1975). Recent decay-mode independent searches, yielding proton lifetime limits $\gtrsim 2.2 \times 10^{26}$yr years, have been made by R. I. Steinberg and J. C. Evans, Jr., Invited Paper, Neutrino-Conference, Elbrus, USSR (1977) and Science 197, 989 (1977). For an excellent up-to-date review, see M. Goldhaber, invited talk at Ben Lee Memorial Conference (Oct., 1977), To appear in the Proceedings.

99. See a recent estimate by A. J. Buras, J. Ellis, M. K. Gaillard and D. V. Nanopoulos, Preprint CERN TH 2403 (1977).

100. For an up-to-date review on the tests of unconfined color, see a recent paper by J. C. Pati and Abdus Salam," Design of Future Experiments: I - To Distinguish Between the Alternatives of Physical and Hidden Color; II - To Test if the Neutral Gauge Boson lies in the vicinity of PETRA-PEP-Region" Trieste Preprint IC/77/65. See also Ref. 46.

101. J. C. Pati, J. Sucher and C. H. Woo, Phys. Rev. D15, 147 (1977).

102. S. Sakakibara, B. Kayser, J. C. Pati and G. Zorn (unpublished).

103. This takes into account the photon-gluon-cancellation-effect in lepto-production of color (Ref. 45).

104. See the third paper in Ref. 45 for a derivation of $R_{Color} = (1/8)(m_U^2/\mu^2)^2$, with m_U denoting the gluon-propagator-mass and μ the effective mass of the gluon-partons ($\mu \leq m_N$).

105. Recent PLUTO-data, reported at Photon-Lepton Symposium, held at Hamburg, W. Germany (August, 1977).

106. T. Appelquist and J. Carrazone, Phys. Rev. D11, 2856 (1975).

107. Ref. 58 for other cases (e.g. $M' \sim m_W \sim 100$ GeV).

108. For recent attempts in this direction see F. Wilczek and A. Zee (Preprint, Princeton University, 1977), H. Fritzsch, CERN preprint TH2358 (1977) and S. Weinberg, Harvard Preprint (1977).

109. S. Coleman and E. Weinberg, Phys. Rev. D7, 1888 (1973); E. Gildener, Phys. Rev. D14, 1667 (1976). For a suggestion in the recent context, see H. S. Mani, J. C. Pati and Abdus Salam, Trieste Preprint IC/77/80 and Ref. 58.

Properties	$[SU(4)]^4$	SU(5)	SO(10)	E_7 ($\supset SU(6) \times SU(3)'_{col}$)
Minimal Fermionic Content	$2(4,1,\bar{4},1)+2(1,4,1,\bar{4})$=12 quarks + 4 leptons + 12 mirror quarks + 4 mirror leptons	$10+10+\bar{5}+\bar{5}$=12 quarks + 2 charged and 2 two component neutral leptons	16=12 quarks + 4 leptons	56 two-component fermions = 10 leptons + 18 quarks $= (20,1)_L+(6,3)_L+(6^*,3^*)_L$ under $SU(6) \times SU(3)'_{col}$
No. of gauge particles	60 = 15 + 15 + 15 + 15	24	45	$133=(35,1)+(1,8)+(15,3)+(15^*,3^*)$ under $SU(6) \times SU(3)'_{col}$
Possible Low Energy Symmetries	$G_W \times \begin{cases} SU(3)'_{L+R} \\ \text{or} \\ SU(3)'_L \times SU(3)'_R \end{cases}$ G_W = (i) $G_L = SU(2)_L \times U(1)$ or (ii) $\bar{G}_{LR}=SU(2)_L \times SU(2)_R \times U(1)_{L+R}$ or (iii) $G_{LR} = SU(2)_L \times SU(2)_R \times U(1)_L \times U(1)_R$	$SU(2)_L \times U(1) \times SU(3)'_{L+R}$	$(G_L$ or $\bar{G}_{LR}) \times SU(3)'_{L+R}$	$SU(2)_L \times U(1) \times SU(3)'_{L+R}$ and other possibilities (Ref.55)
$\sin^2\theta_W$ (Neutrino-data (expt ≈ 0.24 – 0.30)	$G_L \times SU(3)'_{L+R} \to \approx 0.20$ $\bar{G}_{LR} \times SU(3)'_{L+R} \to \approx 0.27$ $G_L \times SU(3)'_L \times SU(3)'_R \to \approx 0.30$ (Ref. 58)	≈ 0.20 (Ref. 53, 57)	$G_L \times SU(3)'_{L+R} \approx$ ≈ 0.20 $\bar{G}_{LR} \times SU(3)'_{L+R}$ ≈ 0.27 (Ref. 57)	$G_L \times SU(3)'_{L+R} \approx 0.67$ (Ref.55)
Naturalness of Quantum Numbers (Parity, Strangeness)	Yes	Yes	Yes	?
Intrinsic Parity Conservation	Yes	No	Yes	Yes
Intrinsic Baryon, Lepton & Fermion No. Conservation	Yes	No	No	No

Properties	$[SU(4)]^4$	SU(5)	SO(10)	E_7
Near Null Results of Atomic Parity Expt (if upheld)	OK Tests at PETRA & PEP	No	OK	OK But distinct from $[SU(4)]^4$ & SO(10). Tests at PETRA & PEP.
Number of Light Neutral Weak Bosons with mass of the lightest one	$G_L \to$ one ($m_Z \sim 83$ GeV) $\bar{G}_{LR} \to$ two ($m_{N_1} \sim 58$–70 GeV) $G_{LR} \to$ three ($m_{Z_1} \sim 35$–70 GeV)	$G_L \to$ one	$G_L \to$ one $\bar{G}_{LR} \to$ two	$G_L \to$ one
Quark-charges	Fractional or Integral	Fractional	Fractional	Fractional
Proton-Decay	Through SSB, In 6th order of gauge interactions, if quarks are integer-charged. $\tau_{proton} \sim 10^{29}$–$10^{32}$ years-Ref. (94)	Through Basic gauge interactions in 2nd order (need ultraheavy mass to account for proton-stability) $\tau_{proton} \sim 10^{38}$ years (Ref. 99)	Same as SU(5)	Same asSU(5)
Unifying Mass-Sacle (M)	$G \downarrow$ $G_W \times SU(3)'_L \times SU(3)'_R \to M \simeq 10^5$ GeV $G_W \times SU(3)'_{L+R} \to M \simeq 10^{15}$ GeV Ref. (58)	$M \sim 10^{15}$ GeV Ref. 53, 57	$M \sim 10^{15}$ GeV	$M \gtrsim 10^{15}$GeV

PART B: BASIC LEFT-RIGHT SYMMETRY IN NATURE: ITS IMPLICATIONS FOR ATOMIC PARITY, NEUTRINO AND HIGH-ENERGY e^-e^+-EXPERIMENTS†

Jogesh C. Pati*

Department of Physics and Astronomy
University of Maryland
College Park, Maryland

TABLE OF CONTENTS

---

† Part B supplements part A; the two parts may, however, be read independently of each other.

* Supported in part by the National Science Foundation under Grant No. GP43662X.

I. INTRODUCTION

1.1 Neutral-Current Interaction as a Probe to the Nature of Unification

It has recently been emphasized by several authors[1] that on the one hand all neutral-current interactions involving neutrinos appear to be remarkably consistent with the simple "left-handed" single parameter $SU(2)_L \times U(1)$-theory.[2] On the other hand, the null or near null results of atomic parity experiments,[3] when confronted with the present theoretical calculations,[4] appear to be grossly inconsistent with the predictions of the simple theory.

There would appear to be a possible dilemma in that one needs an alternative to the left-handed $SU(2)_L \times U(1)$-theory, which on the one hand preserves all the successes of this theory for neutrino-experiments, and yet on the other hand explains the null (or near null) results of atomic parity experiments.

This apparent dilemma finds a simple resolution within the left-right symmetric theory[5] in which (V-A) and (V+A)-currents occur in a symmetrical fashion coupled to distinct sets of gauge particles. The corresponding gauge-structure is of the form

$$L_{gauge} = g[W_L(V-A) + W_R(V+A)]. \tag{1}$$

Since available neutrinos are only left-handed (anti-neutrinos only right-handed), such a $(V-A) \leftrightarrow (V+A)$-symmetric gauge structure would appear to the available neutrino-beams as though it possessed only the left-handed (V-A)-piece; thus in its simplest version incorporating weak and electromagnetic interactions (see below); such a gauge-structure would still appear to left-handed neutrinos[6] as though it possessed just the left-handed $SU(2)_L \times U(1)$-structure. But electrons in atoms possessing all four components respond to (V-A) as well as (V+A)-interactions with equal affinity. Thus atomic interactions in such a theory would be parity conserving to order $G_F\alpha$ in the limit that neutral W_L and W_R develop symmetrical masses through spontaneous symmetry breaking. Departures from parity conservation may in general arise within such a theory in the neutral (as well as the charged sector) only to the extent however of the spontaneously induced mass-asymmetry[7] between W_L and W_R.

1.2 The Gauge-Structure

The simplest example of the (V-A)$\leftrightarrow$(V+A) symmetric gauge structure as mentioned above incorporating weak and electromagnetic interactions is provided by the local symmetry

$$\mathcal{G}_{LR} = SU(2)_L \times SU(2)_R \times U(1)_{L+R} \tag{2}$$

proposed some time ago[5] with the primary motivation that Nature must be intrinsically symmetric between left versus right. The components $SU(2)_L$ and $SU(2)_R$ generate parallel (V-A) and (V+A)-currents, respectively, while $U(1)_{L+R}$ generates an abelian vector current. The abelian factor is to be viewed as part of a non abelian symmetry. Example of such a symmetry is $SU(4)'_{L+R}$, which gauge the three quark-colors as well as lepton number, the corresponding symmetry describing weak, electromagnetic as well as "strong" interactions of quarks and leptons is given by the "basic" group-structure:[5]

$$G = SU(2)_L \times SU(2)_R \times SU(4)'_{L+R} \tag{3}$$

The left-right symmetric theory ($SU(2)_L \times SU(2)_R \times U(1)_{L+R}$ as well as <u>all</u> its quark-lepton unifying extensions, e.g. the one based on $SU(2)_L \times SU(2)_R \times SU(4)_{L+R}$ and others (to be mentioned later) have <u>three distinguishing features</u>:

(1) There exist right-matter parallel to every left-matter. In particular, corresponding to the left-handed neutrinos (ν_L^e, ν_L^μ), there must exist the right-handed <u>neutrinos</u>[8] (ν_R^e, ν_R^μ) dintinct from the right-handed antineutrinos ($\bar{\nu}_R^e$ and $\bar{\nu}_R^\mu$)

(2) For every left-handed (V-A)-current coupled to a set of gauge particles $\vec{W}_L$, there exists within the theory a <u>parallel</u> right-handed (V+A)-current coupled to a distinct set of gauge particles $\vec{W}_R$ with equal strength ($g_L^{(o)} = g_R^{(o)}$). Within the $SU(2)_L \times SU(2)_R$-substructure, there must exist two sets of charged particles $W_L^\pm$ and $W_R^\pm$ and two neutral ones W_L^3 and W_R^3.

(3) Observed distinction between left versus right in the charged current sector, i.e. the dominance of (V-A) over (V+A) in β-decay, μ-decay and μ capture, is brought about by the spontaneously induced mass-splitting between the charged W_R^+ and W_L^+,

$$m_{W_R^+} > m_{W_L^+}$$

Likewise left-right-asymmetry in the neutral current-sector (if any) must arise by the spontaneously induced mass-asymmetry between the neutral W's (W_L^3 and W_R^3). Such asymmetries in the charged as well as neutral sector must then disappear at asymptotic energies[5,9] ($>m_{W_R^+}$).

1.3 The Dichotomy

As alluded to earlier, the results of recent atomic parity experiments, subject to the present theoretical calculations, appear to show that the strength of atomic parity violation in neutral current interactions may perhaps be one to two orders of magnitudes smaller than G_F (if not smaller still), in contrast to charged current interactions, where the magnitude is known to be of order G_F. Such a dichotomy between charged and neutral current interactions is not permissible within the simple "left-handed" $SU(2)_L \times U(1)$-theory, but it can arise within the left-right symmetric theory $SU(2)_L \times SU(2)_R \times U(1)_{L+R}$ and <u>all</u> of its quark-lepton unifying extensions[5], if the spontaneously induced mass-asymmetry between the charged gauge particles $(W_L^{\pm}, W_R^{\pm})$ is large, while at the same time the mass-asymmetry between the neutral members is small or "zero". Such a difference between the mass-asymmetries of the charged and neutral sectors may come about simply as follows:[10]

Introduce Higgs-Kibble scalar-fields transforming as <u>vectors as well as spinors</u> under $SU(2)_{L,R}$:

$$\begin{aligned} E_R &= (1, \underset{\sim}{3}, Y=0);\; E_L = (\underset{\sim}{3}, 1, Y=0) \rightarrow \text{vectors} \\ B &= (1, \underset{\sim}{2}, Y=+1);\; C = (\underset{\sim}{2}, 1, Y=+1) \rightarrow \text{spinors} \end{aligned} \tag{4}$$

Y denotes the U(1)-quantum number for the group $\mathcal{G}_{LR}$. The vector-fields $E_{L,R}$ may contribute through their vacuum expectation values to the masses of only the charged $W^{\pm}_{L,R}$, but <u>not</u> to the masses of the neutral ones. The spinor fields (B,C) on the other hand contribute (through their VEV) to the neutral as well as the charged W-masses. Thus with

$$\langle E_R \rangle \gg \langle E_L \rangle, \quad \text{but} \quad \langle B \rangle \sim \langle C \rangle \tag{5}$$

one would obtain a large mass-asymmetry between the charged W's, even though that between the neutral ones $(W^3_{L,R})$ may be small or "zero".[11] Correspondingly, parity violation in the charged sector would be large $(\mathcal{O}(g_L^2/8m^2_{W_L^+}) \equiv \mathcal{O}(G_F/\sqrt{2}))$, while that in the neutral sector could be vanishingly small. In the extreme limit[12] $\langle B \rangle = \langle C \rangle$ and with $g_R = g_L$, neutral current interactions would acquire the effective parity conserving form

$$\mathcal{L}_{NC} = VV + AA \tag{6}$$

This is barring <u>finite</u> $\mathcal{O}(\alpha)$ radiative corrections[13] to $(g_L - g_R)/g_L$, which arise from $m_{W_R^+} \neq m_{W_L^+}$. Such corrections would induce in this case atomic parity violation in order $G_F^{(N)}\alpha$, where $G_F^{(N)}/\sqrt{2} \equiv g_L^2/8m^2_{N_1}$ and m_{N_1} is the mass of the lightest neutral gauge particle. At the present level of theory and experiments, this extreme limit $\langle B \rangle = \langle C \rangle$ as a zeroth order solution (rather than $\langle B \rangle \sim \langle C \rangle$) is of course not warranted. I return to this point in more detail later.

1.4 <u>Left-Right Symmetric Versus Left-Right Conjugated Vector-Like Theories</u>

It needs to be stressed that even in the extreme limit of parity conservation for neutral current interaction, the left-right symmetric theory possesses the VV as well as the AA-piece (see Eq. 5). Such a theory would always distinguish between left-handed neutrinos (ν_L) and right-handed antineutrinos ($\bar{\nu}_R$). This may be seen by writing down the parity conserving (VV + AA)-neutral current interaction of neutrinos as follows (Write $\nu \equiv \nu_L + \nu_R$ with $\nu_R \neq \bar{\nu}_R$):

$$L_\nu^{NC} = \alpha(\bar{\nu}\gamma_\mu\nu)V_\mu + \beta(\bar{\nu}\gamma_\mu\gamma_5\nu)A_\mu \qquad (7)$$

$$= [\bar{\nu}\gamma_\mu(\frac{1+\gamma_5}{2})\nu]\ \underline{(\alpha V_\mu + \beta A_\mu)} + [\bar{\nu}\gamma_\mu(\frac{1-\gamma_5}{2})\nu]\ (\alpha V_\mu - \beta A_\mu) \qquad (7.1)$$

$$\xrightarrow[\text{(For } \nu_L \text{ \& } \bar{\nu}_R)]{} [\bar{\nu}\gamma_\mu(\frac{1+\gamma_5}{2})\nu]\ \{\frac{\alpha-\beta}{2}\ (V+A)_\mu + \frac{\alpha+\beta}{2}\ (V-A)_\mu\} \qquad (7.2)$$

V_μ and A_μ denote relevant vector and axial vector hadronic currents. We have dropped the second term in (7.1), since it contributes only to the scattering of right-handed neutrinos (ν_R) and left-handed antineutrinos ($\bar{\nu}_L$). Noting that the square bracket in (7.2) acts like a (V-A) current for neutrinos and (V+A) for antineutrinos and that (V-A)(V-A) as well as (V+A)(V+A) interaction lead to constant $(d\sigma/dy)$, while the interference terms (V-A)(V+A) and (V+A)(V-A) lead to $(1-y)^2$-distribution, we obtain

$$(d\sigma/dy) \propto (\alpha-\beta)^2\ (1-y)^2 + (\alpha+\beta)^2 \qquad \text{(For } \nu_L)$$

$$\propto (\alpha-\beta)^2 + (\alpha+\beta)^2\ (1-y)^2 \qquad \text{(For } \bar{\nu}_R)$$

Thus due to the presence of VV as well as AA-pieces ($\alpha\neq 0, \beta\neq 0$), the left-right symmetric theory (<u>even in the parity conserving[14] limit</u>) would distinguish between left-handed ν_L and right-handed $\bar{\nu}_R$, in particular it predicts

$$\sigma(\nu_L p)_{NC} \neq \sigma(\bar{\nu}_R p)_{NC} \qquad (8)$$

as observed experimentally.[15]

This is to be contrasted from the so-called vector-like theories[16] based on the group structure $SU(2)_{L+R} \times U(1)_{L+R}$. In these theories (V-A) and (V+A) currents co-exist, but they couple to the same set of gauge-particles ($W^\pm, W^3$). Hence, we call them "restricted"[17] "<u>left-right conjugated</u>" theories. These theories naturally predict atomic parity violation to vanish in accordance with the data. But with only vectorial quark-current (i.e. having no AA-effective interaction; i.e. $\beta=0$), they also predict that $\sigma(\nu_L p)_{NC} = \sigma(\bar{\nu}_R p)_{NC}$ in clear contradiction with the data.

We thus see that the near null result of atomic parity experiments together with the observed ν-$\bar{\nu}$-distinction appear to suggest[18] that quite possibly (V-A) and (V+A)-currents occur in Nature in a symmetrical fashion coupled to distinct sets of gauge particles. The basic lagrangian is neither left-handed, nor "left-right-conjugated" in the sense mentioned above, but it is left-right symmetric.

1.5 Left-Right Symmetry and CP

A further distinct advantage of the left-right symmetric theory[5] $SU(2)_L \times SU(2)_R \times U(1)_{L+R}$ as well as all its quark-lepton-unifying extensions is that it allows one to generate a desirable milliweak theory of CP violation[19] through phase angles in the Cabibbo-mass matrix with only four quark-flavors. The attractive properties of this theory are listed below.

First one finds that within such a theory, one can link the smallness of CP violation to the observed smallness of (V+A) compared to (V-A)-interactions, one obtains:

$$|\eta_{+-}| \simeq |(m_{W_L^+}/m_{W_R^+})^2(\sin 2\theta_R/\sin 2\theta_L)\sin(\delta_R-\delta_L)|$$

where $\theta_{L,R}$ are the Cabibbo-angles and $\delta_{L,R}$ the phase angles defined in the space of (left, right) fermions via their mass-matrix.[19] Thus even if $|\delta_R-\delta_L| \sim \pi/2$ and $\theta_R \simeq \theta_L$, CP-violation is naturally suppressed because $m_{W_R^\pm} >> m_{W_L^\pm}$. Second, the effective nonleptonic $|\Delta S|=1$ CP conserving (P^+) and CP-violating (P^-)-interactions, arising via the gauge interactions of the model, are found to satisfy the so-called isoconjugate relation $[I_3,P^-] = 1/2\ i(\tan\xi)P^+$, where I_3 is the third component of isospin and $\tan\xi$ is related to the phase angles. Such a relation automatically yields (regardless of the choice of ξ)

$$\eta_{+-} = \eta_{oo}\ ,\quad \phi_{+-} = \tan^{-1}(2\Delta m/\Gamma_s)$$

where $\eta_{ij} = Amp(K_L \to \pi^i\pi^j)/Amp(K_S \to \pi^i\pi^j)$ and ϕ_{+-} is the phase of η_{+-}. Both these relations, which incidentally coincide with the predictions of the superweak theory of CP-violation,[20] are in full accord with the data.

Distinctions of the isoconjugate milliweak model of CP-violation lie in that on the one hand, it predicts milliweak CP-violation in $\Delta S = 1$ -processes such as $\Lambda \to N + \pi$-decays, like most milliweak models. However, its $\Delta S = 0$ gauge interactions are CP conserving. Hence in such a model, electric dipole moment of the neutrino arises (a) through Yukawa interactions of left over Higgs fields, and (b) through fourth order gauge interactions (i.e. in order $G_F^2\varepsilon$, where $\varepsilon \sim 10^{-3}$)). The latter make a contribution to the e.d.m. of the neutron in the range of $\simeq 10^{-29}$ ecm. But the former (given that[21] Higgs mesons are relatively heavy ($m_\sigma > 30$ GeV), but that they are not excessively heavy ($m_\sigma \lesssim 1000$ GeV), or else the Higgs system would become strongly interacting) lead to a dipole moment

$$d_n \simeq (10^{-24} - 10^{-27})\ \text{ecm}\quad .$$

This is on the one hand lower than the prediction of a number of milliweak models, which typically predict $d_n \geq 10^{-23}$ ecm, on the other hand it is orders of magnitude higher than the prediction of genuine superweak models; the latter predict $d_n \leq 10^{-29}$ ecm. The present experimental value[21] $d_n = (0.4 \pm 1.1) \times 10^{-24}$ ecm is compatible with the predictions of either model (isoconjugage milliweak[19] as well as superweak[20]). Improvement in this measurement in planning should be able to distinguish between these two classes of theories.

The point worth noting at this stage is that the isoconjugate milliweak CP-violation with the desirable features as noted above is special to the left-right symmetric theory $SU(2)_L \times SU(2)_R \times U(1)_{L+R}$ and its unifying extensions. Such a theory of CP-violation can not be generated within the $SU(2) \times U(1)$-group structure regardless of the choice of quark-flavors and the chirality of different couplings.

In view of this it is worthwhile to spell out the implications of the left-right symmetric theory for the various ongoing experiments as well as experiments in planning. In particular, I present in this talk the consequences of this theory for:

(I) Neutrino-Experiments

(II) Atomic-Parity Experiments, and

(III) e^-e^+-asymmetry Experiments at PETRA and PEP energies.

We find[6] that given the present results of the atomic parity experiments, the left-right symmetric theory typically implies a mass for the lightest weak neutral gauge boson (N_1) to be ≈ 58 to 70 GeV, <u>which is considerable lighter</u> than the mass (≈ 83 GeV) of Z° for the $SU(2)_L \times U(1)$-theory. Such light mass reflects itself in rather large forward-backward-asymmetry for $e^-e^+ \to \mu^-\mu^+$ compared to the value expected within the $SU(2) \times U(1)$-theory even at PETRA and PEP energies. This offers the possibility of a clear experimental distinction between the left-right symmetric versus left-handed $SU(2)_L \times U(1)$-theory in the near future devoid of the present uncertainty in atomic theoretical calculations.

In Chapter II, I review briefly the main features of the left-right symmetric theory; in Chapter III, I present its experimental consequences mainly for neutrino-scattering, for atomic parity experiments and for high energy electron-positron colliding beam experiments, with brief remarks on the consequences of the theory for deep inelastic ep-scattering and dilepton-production by hadrons. Chapter IV contains concluding remarks.

II. LEFT-RIGHT SYMMETRY (WITHIN THE BASIC MODEL) AND ITS SPONTANEOUS BREAKING

2.1 The Basic Model

Recall briefly the following salient features of the basic model.[5] The model introduces a minimal[22] set of sixteen four-component basic fermions $F_{L,R}$ possessing four flavors (p,n,λ,c) and four colors (red, yellow, blue and lilac); the fourth color is identified with lepton-number $L \equiv L_e + L_\mu$:

$$F_{L,R} = \begin{bmatrix} p_r & p_y & p_b & p_\ell = \nu_e \\ n_r & n_y & n_b & n_\ell = e^- \\ \lambda_r & \lambda_y & \lambda_b & \lambda_\ell \quad \mu^- \\ c_r & c_y & c_b & c_\ell \quad \nu_\mu \end{bmatrix}_{L,R} \tag{9}$$

$$\text{red} \quad \text{yellow} \quad \text{blue} \quad \text{lilac}$$

To implement left-right discrete symmetry as well as quark-lepton unification, the model assumes a minimal local symmetry

$$G = SU(2)_L \times SU(2)_R \times SU(4)'_{L+R} \tag{10}$$

The components $SU(2)_L$ and $SU(2)_R$ (subject to the discrete symmetry[5] $L \leftrightarrow R$) operate on the flavor doublets $\{(p,n)_L + (c,\lambda)_L\}$ and $\{(p,n)_R + (c,\lambda)_R\}$ respectively for each of the four colors (r,y,b,ℓ). Thus they generate (V-A) and parallel (V+A) flavor gauge-currents of quarks as well as leptons, coupled to distince sets of gauge particles $(W_L^\pm, W_L^3)$ and $(W_R^\pm, W_R^3)$ respectively with equal strength $(g_L^{(o)} = g_R^{(o)})$. Note the necessity for the existence of right-handed matter balancing <u>every</u> left-handed matter within this gauge structure. In particular, right-handed neutrinos (ν_R^e, ν_R^μ) must coexist with the left-handed ones (ν_L^e, ν_L^μ). The model is left-right symmetric in the sense defined before.

Simultaneously the symmetry G treats quarks and leptons as members of <u>one</u> multiplet by treating lepton-number as the fourth color. The corresponding SU(4)'-four color-symmetry of quarks and leptons operating on the quartet of indices (r,y,b,ℓ), for each of the four flavors, generates left-right symmetric color-vector currents coupled to a fifteen-plet of gauge particles ($V(\underset{\sim}{8})$, X, $\overline{X}$ and S°).

In short, judged in the light of its predecessor (i.e., the left-right asymmetric gauge model[23] $SU(2)_L \times U(1) \times SU(3)'_{L+R}$), the so-called basic model introduces two major new hypotheses: (i) A complete symmetry between left versus right and (ii) A unification of the basic fermions (quarks and leptons).

In this talk, I shall use the quark-lepton unifying group-structure G to illustrate how the left-right symmetric <u>substructure</u> $G = SU(2)_L \times SU(2)_R \times U(1)_{L+R} \times SU(3)'_{L+R}$

could emerge as a possible low energy step in the spontaneous breakdown of a larger unifying symmetry G containing $\bar{G}_{LR}$. The general experimental consequences of the left-right symmetric substructure $\bar{G}_{LR}$ to be presented here would hold for any embedding of $\bar{G}_{LR}$ within a unifying symmetry G, as long as G descends hierarchically through $\bar{G}_{LR}$. The choice of the unifying symmetry G and the nature of its break-down would play a crucial role however in that such a choice would fix the ratios of the effective coupling constants associated with the substructure $\bar{G}_{LR}$, as mentioned later.

2.2 The Gauge-Bosons

The gauge-fields of the theory $W_{L,R}$ and V generated by $SU(2)_{L,R}$ and $SU(4)'$-symmetries respectively are conveniently represented by the matrices:

$$W_{L,R} = \begin{bmatrix} \frac{\vec{\tau}}{2}\cdot\vec{W} & 0 \\ 0 & \tau_1\frac{(\vec{\tau}\cdot\vec{W})}{2}\tau_1 \end{bmatrix}_{L,R} \qquad V = \frac{1}{\sqrt{2}}\begin{bmatrix} V_{11} & V_\rho & V_{K^*} & \bar{X}_1 \\ V_\rho^\dagger & V_{22} & V'_{K^*} & \bar{X}_2 \\ V_{K^*}^\dagger & V'^\dagger_{K^*} & V_{33} & \bar{X}_3 \\ X_1 & X_2 & X_3 & \sqrt{3/4}\, S^\circ \end{bmatrix} \tag{11}$$

where $V_{11} = (1/\sqrt{2})(V_3 + V_8/\sqrt{3} - S^\circ/\sqrt{6})$, $V_{22} = (-V_3 + V_8\sqrt{3} - S^\circ/\sqrt{6})$, $V_{33} = (1/\sqrt{2})(-2V_8/\sqrt{3} - S^\circ/\sqrt{6})$ and $S^\circ \equiv V_{15}$. The color-octet of gauge-particles (gluons) $V(\underset{\sim}{8})$ given by ($V_\rho, V_\rho^\dagger, V_{K^*}, V_{K^*}^\dagger, V'_{K^*}, V'^\dagger_{K^*}, V_3$ and V_8) appearing in the top left 3 x 3 block of V are coupled to (red, yellow and blue) colors only (i.e. only to quarks). They are electrically neutral if quarks are fractionally charged; while they would possess four charged members ($V_\rho^\pm, V_{K^*}^\pm$) and four neutral ones ($V'_{K^*} = V^\circ_{K^*}$, $V'^\dagger_{K^*} = \bar{V}^\circ_{K^*}$, V_3 and V_8) if quarks are integer-charged. Depending upon the nature of spontaneous symmetry breaking, the gauge model of the type mentioned above could lead to <u>either quark-charge</u> - integral or fractional, as discussed later.

The triplet of X-particles (X_1, X_2 and X_3) couple quarks of three colors (red, yellow and blue) of any given flavor to the lepton of the same flavor, i.e. they couple to $(\bar{q}\ell)$ - currents. They would be fractionally charged ($Q_X = 2/3$) if quarks are fractionally charged and integer-charged ($Q_{X_1} = 0$, $Q_{X_2} = Q_{X_3} = +1$) if quarks are integer charged.

Defining $\nabla_\mu F = \partial_\mu F + igW_\mu F - ifFV_\mu$, the Fermi-lagrangian of the model is given by

$$L_{Fermi} = - T\gamma[\bar{F}_L(\gamma_\mu\nabla_\mu)_L F_L + (L \leftrightarrow R)] \tag{12}$$

where $F_{L,R} = (1/2)(1 \pm \gamma_5)F$. Thus

$$L_{int} = g \left[\sum_{\alpha=r,y,b,\ell} (\bar{p}_\alpha \bar{n}_\alpha \bar{\lambda}_\alpha \bar{c}_\alpha)_L \, (W_L)_\mu \, \gamma_\mu \begin{pmatrix} p_\alpha \\ n_\alpha \\ \lambda_\alpha \\ c_\alpha \end{pmatrix} + L \rightarrow R \right]$$

$$+ f \sum_{i=p,n,\lambda,c} (\bar{F}^i_r \; \bar{F}^i_y \; \bar{F}^i_b \; \bar{F}^i_\ell) \, V_\mu \gamma_\mu \begin{pmatrix} F^i_r \\ F^i_y \\ F^i_b \\ F^i_\ell \end{pmatrix} \tag{13}$$

We have assumed discrete $L \leftrightarrow R$-symmetry[24] in setting $g_L = g_R = g$ in Eq. (12). Note the manifest completely left$\leftrightarrow$right symmetric parity conserving nature of the gauge interactions. (The general prescription for preservation of the discrete left$\leftrightarrow$ right-symmetry as a "natural" symmetry is given later). Assuming that the basic lagrangian is generated by a unifying symmetry G, described in terms of a single basic coupling constant, the two effective constants g and f associated with the descendent group-structure G would be related to each other through gauge-mass dependent finite renormalisation effects. This is discussed in detail in Part A.

Eq. (12) presupposes that the canonical fields $(n,\lambda)_{L,R}$ would in general be related to the corresponding physical fields (defined by the spontaneously generated Fermi mass-matrix) by Cabibbo-rotations through angles $\theta_{L,R}$. (Such rotations would in general involve complex phases leading to normal Cabibbo-GIM-current as well as CP violation[19]).

2.3 The Two Mass-Restrictions

The charged flavor gauge particles ($W_R^\pm$ and $W_L^\pm$) generate V+A and V-A β-decay interactions respectively. The effective strength of (V+A) relative to (V-A)-interaction within the four flavor-model is given by the ratio $(m_{W_L^+}/m_{W_R^+})^2(\cos\theta_R/\cos\theta_L)$. From the experimental observation that the longitudinal polarisation of electrons emitted in β-decays is (-v/c) within about 1%, we may deduce

$$m_{W_R^+} \gtrsim 3 \left(\frac{\cos\theta R}{\cos\theta L}\right)^{1/2} m_{W_L^+} \tag{14}$$

Thus for[25] $\cos\theta_R \simeq \cos\theta_L$, the charged W_R^+ must be at least about three times heavier than the charged W_L^+.

The second restriction concerns the mass of the exotic X-particles, which transform quarks to leptons and vice versa. These induce (within the basic model) $K_L \rightarrow \bar{\mu}e$-decays. From the known upper limit on the rate of such decays, we may deduce that

$$m_X \gtrsim (10^4 - 10^5) \text{ GeV} \tag{15}$$

In other words within the basic model quark-lepton unification must be lost through a mass-scale exceeding $\approx 10^4$ GeV. Spontaneous breaking of the symmetry G must be such as to satisfy these two mass-restrictions (13) and (14).

2.4 "Natural" Left-Right Symmetry

Before discussing the scheme of spontaneous symmetry breaking it is necessary to spell out the full requirements of discrete left-right symmetry, as they are relevant to the discussion of symmetry breaking. First and foremost, the discrete symmetry requires that the gauge-structure be left-right symmetric[5] (e.g. of the form $SU(2)_L \times SU(2)_R \times U(1)_{L+R}$ with left and right handed fermions transforming in a parallel manner under $SU(2)_L$ and $SU(2)_R$. In turn such a gauge-structure (subject to the discrete symmetry demands that: (a) fermions as well as Higgs-Kibble scalar-mesons be introduced into the theory only as left-right symmetric pairs; and (b) the bare gauge coupling constants $g_L^{(o)}$ and $g_R^{(o)}$ associated with the left and right handed gauge interactions (e.g. $SU(2)_L$ and $SU(2)_R$) be equal.

The choice $g_L^{(o)} = g_R^{(o)}$ is permissible consistent with renormalisability provided the discrete left-right-symmetry is "natural" in the sense that radiative corrections involving the full lagrangian induce at most finite calculable corrections to the symmetry. Only in this case, one would obtain $(g_L^{ren} - g_R^{ren})/g_L^{ren}$ to be at most $\mathcal{O}(\alpha)$. Thus to impose $g_L^{ren} = g_R^{ren}\ (1 + \mathcal{O}(\alpha)\text{finite})$, regardless of the values of the renormalised parameters, it is imperative that the L $\leftrightarrow$ R-discrete symmetry be natural in the sense defined above.

The necessary and sufficient condition[13] for preserving L $\leftrightarrow$ R-discrete symmetry is this: The symmetry must be preserved not only in the basic gauge lagrangian, but also in the Higgs-scalar field potential. This in turn implies that the bare self couplings of the scalar fields must respect the L $\leftrightarrow$ R-symmetry, as also the gauge-interactions. Putting it differently, naturalness of L $\leftrightarrow$ R-symmetry is preserved as long as the symmetry is broken in the basic lagrangian in no other way except possibly in a "soft" manner through scalar-mass terms.

The result stated above may be seen easily by using a general theorem due to K. Symanzik.[26] The theorem states that when a symmetry of the lagrangian is broken by a certain term, vertices with dimensions higher than those of the symmetry breaking term suffer only finite renormalisations due to the symmetry breaking term. Here the symmetry in question is L $\leftrightarrow$ R-discrete symmetry; the symmetry-breaking terms, if any, are scalar-mass terms of dimension two. All vertices (involving gauge-interactions or scalar self couplings) have dimensions higher than two and thus suffer only finite renormalisations due to the symmetry breaking terms.

In summary, insisting on naturalness, the discrete L $\leftrightarrow$ R-symmetry should be preserved throughout in the basic lagrangian except possible in the mass-terms of the scalar fields. Two remarks are in order:

(i) If one did permit L $\leftrightarrow$ R-asymmetric scalar-mass terms it would of course be straightforward to induce L $\leftrightarrow$ R asymmetric vacuum-expectation values ($\langle\phi_{L,R}\rangle = \sqrt{-\mu^2_{L,R}/\ _{L,R}}$) leading to $m_{W_R^+} \gg m_{W_L^+}$. Apart from leading to such gauge-mass inequality, the scalar mass-asymmetries do not directly make any noticeable contribution to observed parity violation at low-energies[13] assuming that the left-over members of $\phi_{R,L}$ are relatively heavy ($\gtrsim$ 100 GeV). At very high energies all masses and their differences can be neglected in any case. Such mass-asymmetries, if present, would therefore be directly relevant to observable L $\leftrightarrow$ R-asymmetry only at medium high energies ~ $\mu_{R,L} \gtrsim$ 100 GeV.

(ii) On the other hand, if one did not introduce any mass-asymmetry whatsoever, i.e. even if the full theory was completely left-right symmetric in accordance with one's aesthetic demands,[5] left-right-asymmetry can well arise[27] entirely spontaneously due to asymmetric ground state (vacuum). This would be analogous to the spontaneous breaking of the gauge-symmetry and would be in line with the point of view that all asymmetries observed in Nature are low-energy phenomena deriving their origin through boundary conditions (properties of the ground state - the vacuum) rather than through the basic equations of motion.

It is good to bear in mind, however, that the two alternatives (with or without an explicit asymmetry in the scalar mass-terms) appear to coincide with each other for all practical purposes at low energies. In both cases, the left-right symmetry is "natural"; parity violation arises in either case almost entirely through the gauge-mass inequality ($m_{W_R} \neq m_{W_L}$) rather than through $g_L \neq g_R$.

In what follows, I outline the scheme of spontaneous symmetry breaking proposed in the second paper of Ref. 5, which (retrospectively) turns out to be in full accord with the requirements[13,27] of natural left-right-symmetry as outlined above.

2.5 Spontaneous Symmetry Breaking

Introduce the following sets of Higgs-Kibble multiplets together with the pattern of vacuum-expectation values as shown below to induce a rather general pattern of spontaneous breaking of the symmetry-structure $G = SU(2)_L \times SU(2)_R \times SU(4)'_{L+R}$. (See Sec. IV together with Footnotes 19 and 21 of the second paper of Ref. 5 for more details on potential and its minimisation leading to the pattern of vacuum-expectation values shown below):

$$A = (2+2,2+2,1) = \left|\begin{array}{ccc|c} a_1 & a_1 & a_1 & \\ \hline & & & a_4 \end{array}\right| \qquad \begin{array}{c}\underline{\text{Mass To}}\\ (W^{\pm},W^3)_{L,R}\end{array}$$

$$B = (1,2+2,\bar{4}) = \left|\begin{array}{cccc} o & & & \\ & o & & \\ & & o & \\ & & & b_4 \end{array}\right| \dashrightarrow \begin{pmatrix} W_R^{\pm}, W_R^3 \\ X, S^o \end{pmatrix}$$

$$C = (2+2,1,\bar{4}) = \left|\begin{array}{cccc} c_1 & & & \\ & c_1 & & \\ & & c_1 & \\ & & & c_4 \end{array}\right| \dashrightarrow \begin{array}{c} W_L^{\pm}, W_L^3 \\ V(\underset{\sim}{8}), X, S^o \end{array} \qquad (16)$$

Vector-Representations:

$$D = (1,1,\underset{\sim}{15}) = \left|\begin{array}{cccc} d & & & \\ & d & & \\ & & d & \\ & & & -3d \end{array}\right| \qquad \begin{array}{c}\underline{\text{Mass To}}\\ X's\end{array}$$

$$E_R = (1,\underset{\sim}{3},1) = \begin{pmatrix} 0 \\ \varepsilon_R \\ 0 \end{pmatrix} \qquad W_R^{\pm}$$

$$E_L = (\underset{\sim}{3},1,1) = \begin{pmatrix} 0 \\ \varepsilon_L \\ 0 \end{pmatrix} \simeq 0 \qquad W_L^{\pm} \qquad (17)$$

The representations of the scalar fields are characterised by the corresponding dimensions. Note that the choice of these scalar multiplets is made in accordance with the full requirements of discrete left-right-symmetry; i.e. for every "left" multiplet there is a corresponding "right" multiplet. (By "left" and "right" scalar multiplets, I mean those scalar multiplets, which couple to W_L's and W_R's respectively.) For example B pairs with D, E_R with E_L; while A and D are self left-right conjugates. The above pattern of vacuum expectation-values emerges as an allowed minimum of the potential subject to the requirements of "natural" left-right symmetry as outlined in the previous section.

Few general remarks are now in order:

(1) Integer Versus Fractional Quark-Charges: It should be emphasised that the unification-hypothesis based on the symmetry-structure G or $\bar{G}$ (or any extension thereof

retaining the form $G_{flavor} \times G_{color}$) permit quark-charges to be <u>either</u> integral <u>or</u> fractional. Which of these two solutions is actually realised depends simply upon whether $c_1 \neq 0$, or $c_1 = 0$, as exhibited schematically below.

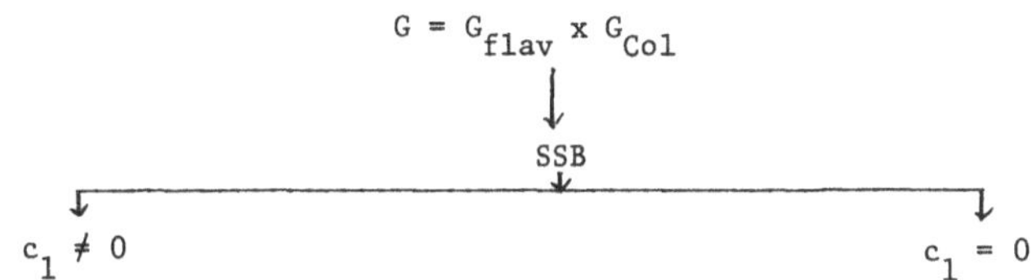

$c_1 \neq 0$:

1) only photon remains massless

2) Octet of Color gluons acquire mass (~1 to few GeV). Color-gluons mix weakly with flavor-gauge bosons. (Such mixing influences lepto-production of color.[28])

3) $Q = Q_{flav} + Q_{Col}$

$$Q_F = \begin{bmatrix} 0 & +1 & +1 & 0 \\ -1 & 0 & 0 & -1 \\ -1 & 0 & 0 & -1 \\ 0 & +1 & +1 & 0 \end{bmatrix}$$

4) A good global SU(3) color-symmetry is preserved[29].

5) Quarks decay rapidly into leptons; proton becomes unstable but sufficiently longlived ($\tau_p \sim 10^{30}$ years). No need[30] to assume confinement of quarks or gluons.

$c_1 = 0$:

1) Photon as well as the octet of color gluons remain massless.

2) $Q = Q_{flav} + Q^{(0)}_{Col.}$ Singlet

$$Q_F = \begin{bmatrix} 2/3 & 2/3 & 2/3 & 0 \\ -1/3 & -1/3 & -1/3 & -1 \\ -1/3 & -1/3 & -1/3 & -1 \\ 2/3 & 2/3 & 2/3 & 0 \end{bmatrix}$$

3) Need to assume confinement of at least massless gluons (unless they are seen).

These two patterns of quark-charges and the consequences thereof are allowed alternatives within the same theory[5]. Thus whether the parameter is finite or zero makes a world of difference in general. However, considering that c_1 (even when it is nonvanishing) is small (~ 1 GeV) compared to other relevant mass-scales in the problem (~ 100 GeV), it is immaterial insofar as we would be interested in studying neutral current phenomena below threshold for color-production, as to whether c_1 is chosen to be zero or finite but small.

So, in what follows, for simplicity of writing we will set $c_1 = 0$, bearing in mind however that <u>all our considerations regarding neutral and charged current interactions below threshold for color-production are equally valid for either quark-charges (integral or fractional)</u>.

(2) Gauge-Mass Matrices; Smallness of Atomic Parity Violation: Setting $c_1 = 0$, the gauge-mass-matrices are given by[5]:

$$m^2_{X,\bar{X}} = (f^2_{X/2})\ (b^2 + c^2 + d^2) \tag{18.1}$$

$$m^2_{W^{\pm}_L,\, W^{\pm}_R} = \frac{g^2}{4} \begin{array}{c} \begin{array}{cc} \qquad W^{+}_L & \qquad\qquad W^{+}_R \end{array} \\ \begin{bmatrix} a^2 + c^2 + \varepsilon^2_L & 2(a^2_1 + a_1 a_4) \\ 2(a^2_1 + a_1 a_4) & a^2 + b^2 + \varepsilon^2_R \end{bmatrix} \end{array} \tag{18.2}$$

$$m^2_{\text{neutral}} = \frac{1}{4} \begin{array}{c} \begin{array}{ccc} \quad W^3_L & \qquad\quad W^3_R & \qquad\quad S^{\circ} \end{array} \\ \begin{bmatrix} g^2(a^2 + c^2) & -g^2 a^2 & \sqrt{3/2}\ f_{15}\, g c^2 \\ -g^2 a^2 & g^2(a^2 + b^2) & \sqrt{3/2}\ f_{15}\, g b^2 \\ \sqrt{3/2}\ f_{15}\, g c^2 & \sqrt{3/2}\ f_{15}\, g b^2 & 3/2\ f^2_{15}(b^2 + c^2) \end{bmatrix} \end{array} \tag{18.3}$$

where $a^2 \equiv 3a^2_1 + a^2_4$, $b^2 \equiv b^2_4$ and $c^2 \equiv c^2_4$. We have inserted f_{15} and f_X (rather than f) in the entries corresponding to the particles $S_o \equiv V_{15}$ and X respectively to denote the fact that the effective renormalised coupling constants associated with these gauge particles are in general expected to be quite different from that associated with the octet of color gluons due to differing finite renormalisation effects. (This is elaborated in the last chapter of part A).

Note the important feature that the vector multiplet $D = (1, 1, \underset{\sim}{15})$ provides mass only to the exotic X-particles coupled to $\bar{q}\gamma_{\mu}\ell$-currents, but not to $V(\underset{\sim}{8})$ gluons and S°. Thus D breaks $SU(4)'_{L+R}$ naturally to $SU(3)'_{L+R} \times U(1)_{L+R}$, where $U(1)_{L+R}$ denotes the symmetry of the 15th generator of SU(4)'. Likewise the vector multiplets $E_R = (1, \underset{\sim}{3}, 1)$ and $E_L = (\underset{\sim}{3}, 1, 1)$ provide masses only to the charged $W^{\pm}_R$ and $W^{\pm}_L$ but not to the neutral ones.

Thus the interesting feature of the model is that it is possible to destroy the left $\leftrightarrow$ right symmetry in the charged current-sector altogether by choosing $\langle E_R\rangle \gg \langle E_L\rangle, \langle A\rangle, \langle B\rangle, \langle C\rangle$ so that $m_{W^{+}_R} \gg m_{W^{+}_L}$ and yet preserve it approximately or "exactly" in the neutral-current sector by choosing $\langle B\rangle \simeq \langle C\rangle$ or $\langle B\rangle = \langle C\rangle$, which ensures that the neutral gauge mass-matrix is still approximately or exactly left-right-symmetric (i.e. symmetric under the interchange $W^3_L \leftrightarrow W^3_R$). In other words as long as $\langle B\rangle = \langle C\rangle$, the neutral current sector of the theory and therefore atomic interactions would remain parity conserving. Thus departures from $L \leftrightarrow R$ symmetry in the neutral current sector, and therefore atomic parity violation, in the left-right symmetric theory[5] are simply proportional to $(b^2 - c^2)/(b^2 + c^2)$ (barring finite $O(\alpha)$-corrections to $(g_L - g_R)/g_L$). Atomic parity violation would be small ($<<O(GF)$) if $(b^2-c^2)/(b^2+c^2)<<1$, while it will vanish[11] to $O(G_F\alpha)$ if $b^2=c^2$.

The question whether the solution of atomic parity conservation (i.e. $\langle B\rangle-\langle C\rangle = 0$) is "natural," i.e. whether it is radiatively stable <u>despite the mass-symmetry in the charged sector ($m_{W_R^\pm} >> m_{W_L^\pm}$)</u> is fundamentally important. A full discussion of this question is beyond the scope of these lectures. I refer you to the recent work of Mani, Pati and Salam[31] and Mohapatra, Paige and Sidhu[32] for a full discussion on this point. Briefly, the results of these authors are the following:

The previously considered scheme of spontaneous symmetry breaking suggesting[11] Q_W(Atomic) = 0 (i.e. $\langle B\rangle = \langle C\rangle \neq 0$ as a zeroth-order-solution) is <u>not</u> "natural". A natural solution guaranteeing that the renormalised value of the relevant asymmetry-parameter $x \equiv (\langle B\rangle^2-\langle C\rangle^2)/\langle C\rangle^2$ be order α can be obtained, if the left-right symmetry breaking in the <u>neutral-sector</u> had a dynamically radiative origin[31] in the sense of Coleman and Weinberg[33], while that in the charged-sector is triggered by the Higgs-mechanism; or alternatively if suitable scalar-multiplets (transforming like composites of BC) are introduced[32] into the theory. In what follows, we will explore the phenomenological consequences of the model by allowing for both $b \sim c$ (small Q_W) as well as $b = c$ ($Q_W = 0$ barring finite $O(\alpha)$-correction).

(3) Hierarchy: We now exhibit two alternative solutions for the low energy weak and electromagnetic symmetry, both consistent with the mass restrictions on $m_{W_R^+}$ and m_X as mentioned before. Both solutions may emerge from the <u>same</u> pattern of SSB (exhibited in eq. (18)) depending upon the relative magnitudes of the parameters associated with this pattern:

<u>Case I</u>: $\quad (G \xrightarrow{\langle B\rangle} G_0 = SU(2)_L \times U(1) \times SU(3)'_{L+R} \xrightarrow{\langle A\rangle,\langle C\rangle} G_{obs})$.

This case arises if we choose B to be very large - in fact the major contributor to the X-mass. This corresponds to $\langle B\rangle \gtrsim 10^4$ GeV $>> \langle C\rangle$, $\langle E_{R,L}\rangle$ and $\langle A\rangle$. In this case quark-lepton-unification as well as left-right symmetry (in the neutral <u>and</u> in the charged scetors) are lost in one stroke through $\langle B\rangle$ being very large. The descendent "left-handed" symmetry $G_0 = SU(2)_L \times U(1) \times SU(3)'_{L+R}$ breaks subsequently through $\langle A\rangle$ and $\langle C\rangle \lesssim 100$ GeV to lead to the "observed" symmetry $G_{obs} = U(1)_{\text{flavor E.M.}} \times SU(3)'_{L+R}$; there arises only <u>one</u> relatively light weak neutral gauge particle - the familiar Z^0. Note that in this case there is no real necessity for the introcudtion of the vectorial scalar fields D and $E_{R,L}$, since no dichotomy between the charged (non-diagonal) versus neutral (diagonal)-current sectors is being called for.

<u>Case II</u>:
$$(G \xrightarrow{\langle D\rangle} \bar{G} = SU(2)_L \times SU(2)_R \times U(1)_{L+R} \times SU(3)'_{L+R}$$
$$\downarrow \langle E_R\rangle$$
$$\bar{\bar{G}} = SU(2)_L \times U(1)_R \times U(1)_{L+R} \times SU(3)'_{L+R} \xrightarrow[\langle E_L\rangle \approx 0]{\langle B\rangle,\langle C\rangle,\langle A\rangle\neq 0} G_{obs}).$$

This second case arises if the scalar field D = (1,1,15) is the major contributor to the X-mass and if its vacuum expectation value far exceeds those of <A>,<B>,<C>, and $\langle E_{R,L}\rangle$. The descendent left-right symmetric group $\bar{G}$ breaks subsequently through the vacuum expectation value of the "right-handed" vectorial scalar field E_R being relatively large. This makes the charged $W_R^{\pm}$ heavy compared to $W_L^{\pm}$ as well as compared to the neutral gauge-fields ($W_{L,R}^3$ and S^o). Subsequently, the descendent symmetry $\bar{\bar{G}}$ breaks through[34] $\langle B\rangle \approx \langle C\rangle \gg \langle A\rangle$ down to the observed symmetry G_{obs}; there arise in this case two relatively light[5,6] "weak" neutral gauge particles N_1 and N_2; each having a mass $\sim m_{W_L^+}$ (for details see later).

The two cases are exhibited schematically in Fig. 1. It needs to be stressed that other alternative patterns of spontaneous break-down of the unifying symmetry G are, in general, permissible. For example $G = [SU(4)]^4$ may descend[35] to

$$G_{chiral} = SU(2)_L \times SU(2)_R \times SU(3)'_L \times SU(3)'_R \times U(1)_{L+R}$$

or into

$$G'_{chiral} = SU(2)_L \times SU(2)_R \times SU(3)'_L \times SU(3)'_R \times U(1)_L \times U(1)_R$$

(where $U(1)_{L,R}$ denote the 15th generators of the four-color symmetries $SU(4)'_{L,R}$). The phenomenology of neutral current-interactions discussed here would remain the same[36] as long as G_{chiral} or G'_{chiral} descend to G_{obs} passing through either $G_o = SU(2)_L \times U(1) \times SU(3)'_{L+R}$ or $\bar{G} = SU(2)_L \times U(1)_R \times U(1)_{L+R} \times SU(3)'_{L+R}$. These different alternatives would of course differ from each other as regards the ratios of low-energy effective coupling constants (see remarks later as well as the last two chapters of part A).

III. EXPERIMENTAL CONSEQUENCES OF THE LEFT-RIGHT SYMMETRIC THEORY

3.1 Neutral Eigenstates

Setting $c_1 = 0$, the three eigenstates of the neutral mass-matrix including the massless photon A_μ are given by:[6]

$$A_\mu = \frac{W_L^3 + W_R^3}{\sqrt{2}} \cos\alpha - S^o \sin\alpha \qquad \text{(Frac. ch. quarks } c_1 = 0)$$

$$N_1 = Z_A \cos\beta + Z_V \sin\beta$$

$$N_2 = -Z_A \sin\beta + Z_V \cos\beta \tag{19}$$

where,

$$Z_A = (W_L^3 - W_R^3)/\sqrt{2} \tag{20}$$

$$Z_V \equiv (W_L^3 + W_R^3)/\sqrt{2} \sin\alpha + S^o \cos\alpha \tag{21}$$

$$\cos\alpha = \sqrt{2}\,(e/g); \; \sin\alpha = \sqrt{2/3}\,(e/f_{15}) \tag{22}$$

Thus,

$$1/e^2 = 2/g^2 + (2/3)\,(1/f_{15}^2) \tag{23}$$

$$\tan 2\beta = \frac{(3r+1)^{1/2}(\eta-\zeta)}{(3r/2)(\eta+\zeta)-2(1-\zeta)} \tag{24}$$

$$\begin{pmatrix} m_{N_1}^2 \\ m_{N_2}^2 \end{pmatrix} = \frac{m_{W_L^+}^2}{4} \Big[(3r+2)(\eta+\zeta)+4(1-\zeta) \mp \{3r(\eta+\zeta)-4(1-\zeta)\}\sec 2\beta \Big] \tag{25}$$

$$m_{W_L^+}^2 = [(3r+1)/3r](e^2/\sqrt{8})G_F^{-1} \simeq \Big[(3r+1)/3r\Big]\,(53 \text{ GeV})^2 \tag{26}$$

$$\simeq (g^2/4)(c^2+a^2) \tag{27}$$

where,

$$r \equiv (f_{15}^2/g^2); \; \zeta \equiv c^2/(c^2+a^2); \; \eta \equiv b^2/(c^2+a^2) \tag{28}$$

$$0 \lesssim \zeta \lesssim 1; \quad 0 \lesssim \eta < \infty \tag{29}$$

The fields Z_A and Z_V - given by (20) and (21) - couple to purely axial and purely vector currents respectively. The eigenstates N_1 and N_2 are mixtures of the parity eigenstates Z_A and Z_V, the mixing being determined by $\tan 2\beta$. Note for $b=c$, $\tan 2\beta \to 0$, $N_1 \to Z_A$, $N_2 \to Z_V$. Thus, as expected, in this limit ($b=c$), neutral current-interactions of the theory would conserve parity (to $\mathcal{O}(G_F\alpha)$); the effective interaction would have the parity conserving form VV + AA. In general, the theory would permit parity violation in the neutral-current sector to the extent that $\tan 2\beta \neq 0$.

[In writing Eqs. (19) and (23), we have set $c_1 = 0$ for convenience. This corresponds to quarks being fractionally charged. For integer-charged quarks ($c_1 \neq 0$), the composition of the photon would be given by: $(1/e)A_\mu = (W_L^3 + W_R^3)/g - \sqrt{2/3}\, S^o/f_{15} + (V_3 + V_8/\sqrt{3})/f_s$. The normalisation condition is: $(1/e^2) = 2/g^2 + (2/3)\, f_{15}^2 + (4/3)\, f_s^2$; where f_s denotes the effective strong color-gluon-coupling constant. In this case, there would be small admixtures ($\lesssim (m_V/m_{W_L^+}) \sim 1\%$) of the gluonic state $(\sqrt{3}\, V_3 + V_8)/2$ in N_1 and N_2. Such admixture would have no effect on neutral-current-processes below threshold for color-production. Thus as mentioned before, the analysis presented here holds essentially to the same extent for either quark-charge.]

The neutral current interactions of the theory depend in general upon three independent parameters: (i) the coupling constant-ratio $r = f_{15}^2/g^2$; (ii) $\eta \equiv b^2/(c^2+a^2)$ and (iii) $\zeta \equiv c^2/(c^2+a^2)$. The coupling constant ratio r relates to the familiar weak angle $\sin^2\theta_W$ of the SU(2) x U(1)-theory (see later). As we will see, <u>in practice</u>, the theory would depend upon only two parameters r and η. For neutrino-physics a single parameter (r) will suffice as in SU(2) x U(1)-theory.

3.2 Interaction; The Equivalence Theorem

The neutral current interaction induced by N_1 and N_2 is given by[6]

$$L_{Nc} = (J_\mu^A \cos\beta + J_\mu^V \sin\beta)N_{1\mu} + (-J_\mu^A \sin\beta + J_\mu^V \cos\beta)N_{2\mu} \tag{30}$$

$J_\mu^{A,V}$ are the source currents of $Z_\mu^{A,V}$. In the four-flavor four-color-model, their fermionic contents are:

$$\begin{aligned} J_\mu^A = (g/2\sqrt{2})\, [(\bar{\nu}_e\gamma_\mu\gamma_5\nu_e - \bar{e}\gamma_\mu\gamma_5 e) + (e \to \mu) \\ + (\bar{p}\gamma_\mu\gamma_5 p + p \to c) - (\bar{n}\gamma_\mu\gamma_5 n + n \to \lambda)] \end{aligned} \tag{31}$$

$$\begin{aligned} J_\mu^V = g/(2\sqrt{2}\ \sqrt{3r+1})\ [(3r+1)\bar{\nu}_e\gamma_\mu\nu_e + (3r-1)\bar{e}\gamma_\mu e + (e \to \mu) \\ + (1-r)\,(\bar{p}\gamma_\mu p + p \to c) - (1+r)\,(\bar{n}\gamma_\mu n + n \to \lambda)] \end{aligned} \tag{32}$$

Note that neutrinos are in general four component fields with arbitrary mass[8] in the theory (alternatively one may substitute $\nu \equiv \nu_L + \nu_R$).

(a) Amplitudes for Neutrino-Processes:

Define the relevant amplitudes for processes involving <u>left-handed neutrinos</u> and <u>right-handed antineutrinos</u> by the effective lagrangian as follows:

$$L^{eff}_{(\nu_L,\bar{\nu}_R)} = \frac{G_F}{\sqrt{2}} (\bar{\nu}_\mu \gamma_\mu (1+\gamma_5)\nu_\mu)[\bar{p}\gamma_\mu (g_V^p + g_A^p\gamma_5)p + \bar{n}\gamma_\mu (g_V^n + g_A^n\gamma_5)n + \bar{e}\gamma_\mu (M_V^e + M^e{}_A\gamma_5)e + \ldots] + \frac{G_F}{\sqrt{2}} \bar{\nu}_e\gamma_\mu (1+\gamma_5)\nu_e[\bar{e}\gamma_\mu (E_V^e + E_A^e\gamma_5)e + \ldots] \tag{33}$$

Here p and n denote quark-fields. The effective amplitudes $g^{p,n}_{V,A}$ etc. may be obtained straightforwardly from Eqs. (24), (25), (30), (31) and (32). They are:[6]

$$\begin{aligned} g_V^p &= (r-1)G_V \quad ; & g_A^p &= -G_A \\ g_V^n &= (r+1)G_V \quad ; & g_A^n &= G_A \\ M_V^e &= -(3r-1)G_V \quad ; & M_A^e &= G_A \\ E_V^e &= -1-(3r-1)G_V; & E_A^e &= -1+G_A \end{aligned} \tag{34}$$

<u>All</u> neutrino-amplitudes are determined in terms of just <u>two</u> master-functions G_V and G_A; they are:

$$G_V = \frac{\eta + 2(1-\zeta)}{(3r+1)(\zeta+\eta-\zeta^2)} \tag{35}$$

$$G_A = \frac{\eta}{(\zeta+\eta-\zeta^2)} \tag{36}$$

(b) <u>Amplitudes for Atomic Parity Violation:</u>

Define $L_{Atomic} = (G_F/\sqrt{2})\ (\bar{e}\gamma_\mu\gamma_5 e)\ (C_V^p\bar{p}\gamma_\mu p + C_V^n\bar{n}\gamma_\mu n)$ (37)

Using Eqs. (24) - (32), we obtain:

$$C_V^{(p\atop n)} = -\frac{(r\mp 1)m^2_{W_L^+}\ (1/m^2_{N_1} - 1/m^2_{N_2})\sin 2\beta}{2(3r+1)^{1/2}} \tag{38}$$

The effective charge Q_W characterising atomic parity violation for an atom (A,Z) is given by:

$$Q_W = -2\{(2C_V^p + C_V^n)Z + (2C_V^n + C_V^p)\ (A-Z)\}$$

$$= \frac{2Z - (3r+1)A}{(3r+1)} \frac{(\eta-\zeta)}{\zeta+\eta-\zeta^2} \tag{39}$$

(c) Amplitudes for e^-e^+-Processes:

Define the four-fermion amplitude for $e^-e^+ \to \mu^-\mu^+$ induced by one photon plus N_1 and N_2-exchanges by:

$$A(e^-e^+ \to \mu^-\mu^+) = (\frac{e^2}{s} + G^{\mu}_{VV})\ (\bar{e}\gamma_\mu e)(\bar{\mu}\gamma_\mu \mu) + G^{\mu}_{VA}(\bar{e}\gamma_\mu e)(\bar{\mu}\gamma_\mu\gamma_5\mu)$$
$$+ G^{\mu}_{AV}(\bar{e}\gamma_\mu\gamma_5 e)(\bar{\mu}\gamma_\mu \mu) + G^{\mu}_{AA}(\bar{e}\gamma_\mu\gamma_5 e)(\bar{\mu}\gamma_\mu\gamma_5\mu) \qquad (40)$$

Using (24)-(32), we obtain:

$$G^{\mu}_{VV} = (G_F/\sqrt{2})(3r-1)^2/(3r+1)\ [D_1(s)\ \sin^2\beta + D_2(s)\cos^2\beta]$$
$$G^{\mu}_{VA} = G^{\mu}_{AV} = (G_F/2\sqrt{2})\ \frac{(3r-1)}{\sqrt{3r+1}}\ \sin^2\beta\ (D_2(s) - D_1(s))$$
$$G^{\mu}_{AA} = (G_F/\sqrt{2})\ (D_1(s)\ \mathrm{Cos}^2\beta + D_2(s)\ \mathrm{Sin}^2\beta) \qquad (41)$$

where s = Total Center of Mass (energy)^2, $D_{1,2}(s) = m^2_{W_L^+}/(s-m^2_{N_{1,2}})$. For s in the vicinity of $m^2_{N_1}$ of $m^2_{N_2}$, the N_1 or N_2 widths (~1 GeV) needs to be inserted into $D_{1,2}$.

Likewise one may write the amplitudes for $e^-e^+ \to p\bar{p}(=c\bar{c})$ and $e^-e^+ \to n\bar{n}(=\lambda\bar{\lambda})$ due to N_1 and N_2-exchanges (without photon) as:

$$G^{\binom{p}{n}}_{VV} = \begin{bmatrix} (1-r) \\ -(1+r) \end{bmatrix} G^{\mu}_{VV}; \qquad G^{\binom{p}{n}}_{AV} = \begin{bmatrix} (1-r) \\ -(1+r) \end{bmatrix} (G^{\mu}_{AV})/(3r-1)$$

$$G^{p}_{VA} = -G^{n}_{VA} = -G^{\mu}_{VA}; \qquad G^{p}_{AA} = -G^{n}_{AA} = -\ G^{\mu}_{AA} \qquad (42)$$

The pure vector amplitudes should be supplemented by photon-contributions (for example $-(2/3)e^2/s$ for p and $+(1/3)(e^2/s)$ for n-quarks in the fractionally charged quark-model. For the integer-charge quarks photon as well as color-gluons will contribute above threshold for color-production.[28] This case needs to be worked out separately).

Results:

Equipped with all the relevant amplitudes, we are now ready to present the results. I first draw attention to two general features of the model.

(I) The $SU(2)_L \times U(1)$-Limit ($b^2 >> c^2 + a^2$; or $\eta >> 1$):

The vacuum-expectation value b of the "right-handed" spinorial scalar field $B = (1,2+2,\bar{4})$ reduces the $SU(2)_R \times U(1)_{L+R}$ - component (of the left-right symmetric theory $G = SU(2)_L \times SU(2)_R \times SU(4)'_{L+R}$ or $\bar{G} = SU(2)_L \times SU(2)_R \times U(1)_{L+R} \times SU(3)'_{L+R}$) into just U(1), which is the diagonal sum of the neutral generators within $SU(2)_R$ and the $U(1)_{L+R}$. Thus when b tends to be very large $\to \infty$ ($\therefore \eta \to \infty$), the left-right symmetric theory G or $\bar{G}$ reduces to its left-handed limit the familiar $SU(2)_L \times U(1) \times SU(3)'_{L+R}$. In this limit ($\eta >> 1$ with $r \neq 0$), we obtain (using Eqs. (19)-(25)):

$$A = W_L^3 \sin\theta_W + B^\circ \cos\theta_W \tag{43.1}$$

$$N_1 = W_L^3 \cos\theta_W - B^\circ \sin\theta_W \tag{43.2}$$

$$N_2 = (\tfrac{3}{2} r + 1)^{-1/2} [\sqrt{3/2r}\, V_{15} + W_R^3] \quad (V_{15} \equiv S^\circ) \tag{43.3}$$

where,

$$B^\circ \equiv (f_{15}^2 + \tfrac{2}{3} g^2)^{-1/2} [f_{15} W_{3R} - \sqrt{2/3}\, g\, V_{15}] \tag{44}$$

$$\sin^2\theta_W = 3r/(6r + 2) \tag{45}$$

$$m_{N_1}^2 \simeq m_{W_L^+}^2 \; (6r + 2)/(3r + 2) = m_{W_L^+}^2/\cos^2\theta_W \tag{46}$$

$$m_{N_2}^2 = (m_{W_L^+}^2/2)(3r + 2)\eta \to \infty \tag{47}$$

Note that the composition of B° (Eq. (44)) is the same as that of the U(1)-gauge field in the $SU(2)_L \times U(1)$-theory. Thus N_1 (in the limit $\eta \to \infty$) reduces to the Z_o of the $SU(2)_L \times U(1)$-theory with the identification $\sin^2\theta_W = 3r/(6r+2)$; N_2 in the same limit becomes infinitely heavy. Note also the familiar relationship (Eq.(46)) between the masses of N_1 (i.e. Z_o) and W_L^+.

The master functions G_V, G_A determining all neutrino-amplitudes (see Eq. (34)-(36)) and the atomic parity violation-parameter Q_W (Eq. (39)) reduce in the limit $\eta \to \infty$ to:

$$G_V \xrightarrow[0\le\zeta\le1]{\eta \to \infty} 1/(3r + 1); \qquad G_A \xrightarrow[0\le\zeta\le1]{\eta \to \infty} 1 \tag{48}$$

$$Q_W \xrightarrow[0\le\zeta\le1]{\eta \to \infty} [2Z-(3r+1)A]/(3r+1) \tag{49}$$

Note that G_V, G_A and Q_W are <u>independent</u> of the parameter $\zeta \equiv c^2/(c^2+a^2)$ in the $\eta \to \infty$ limit. <u>All amplitudes are determined in this case in terms of the single parameter $r = (f_{15}^2/g^2)$ or equivalently $\sin^2\theta_W' = 3r/(6r+2)$.</u> This is the familiar <u>result of $SU(2)_L \times U(1)$.</u>

(II) <u>The Chiral Symmetric Limit ($a^2 \to 0$, i.e. $\zeta = \frac{c^2}{c^2+a^2} \to 1$);</u>

In the limit $a^2 \to 0$ (i.e. $\zeta \to 1$), W_L-W_R-mixing mass as well as all fermion-masses vanish. Appropriately, this limit will be referred to as the global (γ_5-invariant) chiral symmetric limit. It is important to note one (remarkable) general property of this limit. I state this as a theorem. The theorem in its general form, stated below, was noted by Pati, Rajpoot and Salam (Ref. 6). For the special case of Q_W=0, it was noted by Fritzsch and Minkowski (Ref. 11). Its extension to groups bigger than $\bar{G}_{LR}$ is due to Georgi and Weinberg (Ref. 11).

<u>The Equivalence Theorem</u>: In the limit of exact global chiral symmetry ($a^2 \to 0$, i.e. $\zeta \to 1$), even though the masses of the two neutral eigenstates N_1 and N_2, the mixing angle β, as well as the atomic parity violation-parameter Q_W vary with b^2 (i.e. with $\eta = b^2/(c^2 + a^2)$; <u>the amplitudes of all left-handed neutrino (or right-handed antineutrino)-induced processes at Zero momentum-transfers are independent of the vacuum-expectation-values of "right-handed" scalar-fields such as B</u>. In the present case, they are independent of b^2 and therefore of $\eta = b^2/(c^2 + a^2)$. They depend just on the single parameter $r = f_{15}^2/g^2$. As they are independent of η, they remain unaltered even if we pass to the limit $\eta \to \infty$, i.e. even when the left-right-symmetric theory approaches its $SU(2)_L \times U(1)$ limit. (See discussion under I).

As a consequence, in the limit of chiral symmetry ($a^2 \to 0$), the predictions of the left-right symmetric theory with two light neutral gauge particles (N_1 and N_2) amazingly coincide with those of the left-handed $SU(2)_L \times U(1)$-theory with only one light gauge particle Z° for all low-momentum transfer processes induced by left-handed neutrinos and right-handed antineutrinos. The two theories even in the chiral limit would still differ from each other in general in respect of (i) atomic parity experiments, (ii) rates and asymmetry parameters in $e^-e^+ \to \mu^-\mu^+$ and hadrons and (iii) rates and asymmetry-parameters in $pp \to \mu^-\mu^+ + X$, i.e. for processes in which all four components of fermions get a chance to play their roles. In addition, they will differ from each other in respect of the masses of the neutral gauge particles.

The qualitative reason for this result is the fact that left-handed neutrinos (in the chiral, symmetric world) are simply blind to right-handed (V+A)-gauge interactions, as explained in the introduction. This may be seen explicitly by taking the chiral symmetric limit ($a^2 \to 0$, i.e. $\zeta \to 1$) of the neutrino-amplitudes, i.e. of the master functions G_V and G_A (Eqs. (35) and (36)) for the left-right symmetric theory. We obtain

$$G_V\ (\zeta = 1) = 1/(3r+1) \tag{50}$$

$$G_A\ (\zeta = 1) = 1 \tag{51}$$

The neutrality of neutrinos also turns out to be a relevant factor. Thus in the chiral symmetric limit G_V and G_A of the left-right symmetric theory become <u>independent</u> of η and <u>coincide</u> with the G_V and G_A of the left-handed $SU(2)_L \times U(1)$-theory (Eq. (48). By contrast in the same limit $\zeta = 1$, we obtain

$$Q_W(\zeta=1) = \frac{\{2Z-(3r+1)A\}}{(3r+1)} \left(\frac{\eta-1}{\eta}\right) \tag{52}$$

Thus the atomic parity violation-parameter Q_W for the left-right symmetric theory depends upon η (even in the limit of chiral symmetry $\zeta = 1$); it differs from the $SU(2)_L \times U(1)$-prediction for Q_W except of course in the limit $\eta \to \infty$, when the two theories become identical to each other in all respects.

With these two general properties I and II, we are ready to present specific experimental predictions:

3.3 <u>Results on Neutrino-Scattering</u>

As noted above, predictions of the left-right symmetric theory for left-handed neutrino-scattering differs from those of the left-handed $SU(2)_L \times U(1)$-theory only to the extent that chiral symmetry is not exact in nature ($a^2 \neq 0$). Since the smallness of fermion-masses (i.e., m_q and $m_{lepton} << m_{W_L^+}$) depends on the parameter a^2 as well as on the relevant Yukawa Coupling Constants, it is not possible to put apriori a stringent upper limit on the chiral symmetry breaking parameter $a^2/(c^2+a^2) = 1 - \zeta$ on the basis of fermion-masses only.

Examining the neutrino-data as well as the atomic parity violation-parameter Q_W, however, one may draw the following important conclusion[6] for the chiral symmetry-breaking-parameter:

<u>If</u> Q_W is indeed small (say at least a factor of 3 lower in magnitude compared to the $SU(2)_L \times U(1)$-value), and <u>if</u> the neutrino-parameters retain their present values (especially R_ν, $R_{\bar\nu}$ and $R = \sigma^{\nu N}_{Nc}/\sigma^{\nu}_{Nc}$); <u>both these features can be well accounted for within the left-right symmetric theory provided the chiral symmetry-breaking parameter (1-ζ) is small compared to unity</u>. To be precise, one needs[6]:

$$1 - \zeta = a^2/(c^2+a^2) < .2$$

For such small values of $(1-\zeta)$, predictions of the theory for neutrino-parameters remain essentially the same as those for $(1-\zeta) \to 0$.

For simplicity, the predictions of the theory pertaining to neutrino-experiments are presented in Table I for the case of exact chiral symmetry ($a^2 \to 0$, i.e. $1 - \zeta = 0$) with two different values of the coupling constant ratio $r \equiv f_{15}^2/g^2 = 1/3$ and $4/9$. (Correspondingly $\sin^2\theta_W = (3r/6r+2) = .25$ and $.29$). The predictions are based on the use of the quark-model with the neglect of sea-partons.

We find that the overall agreement of the left-right symmetric theory (as also of the $SU(2)_L \times U(1)$-theory) with neutrino-data is rather impressive for r lying in the range (see Fig. 2):

$$0.3 \lesssim r = f_{15}^2/g^2 < 0.66 \tag{53}$$

Correspondingly,

$$0.23 \lesssim \sin^2\theta_W = \frac{3r}{6r+2} \lesssim 0.33 \tag{54}$$

The lower value of $\sin^2\theta_W \simeq 1/4$ (i.e. $r \simeq 1/3$) would be preferred if the data settles near the recent results of the CDHS-measurements[37] $R_\nu = 0.30 \pm .015$.

It is remarkable that the range of the coupling constant-ratio r exhibited above, deduced phenomenologically from the neutrino-data, lies well within the rather limited range of the same, deduced on <u>theoretical grounds</u>. The latter are based on the hypothesis that the left-right symmetric substructure $\bar{G} = SU(2)_L \times SU(2)_R \times U(1)_{L+R} \times SU(3)'_{L+R}$ descends through hierarchical breaking of a unified symmetry G (e.g. $= [SU(4)]^4$ or SO(10)) described by a single coupling constant; the coupling constants of the low-energy subgroup being related to each other through finite renormalization effects. For two interesting and typical cases of the pattern of spontaneous symmetry breaking, these considerations yield:[38]

Hierarchy	$r = f_{15}^2/g^2$	$\sin^2\theta_W$
$G \xrightarrow{M} \bar{G} = SU(2)_L \times SU(2)_R \times U(1)_{L+R} \times SU(3)'_{L+R}$	(0.33~0.40)	(0.25~0.28)
$G(\text{e.g.}[SU(4)]^4) \xrightarrow{M} G'_{\text{Chiral}} = SU(2)_L \times SU(2)_R \times U(1)_L \times U(1)_R \times SU(3)'_L \times SU(3)'_R$	(0.66~0.70)	(0.33~0.36)

The above values of r are obtained for the ratio of the effective strong versus electromagnetic coupling (α_s/α) varying between ∞ down to ≈ 15. The chirality L,R refer to the coupling of the normal fermions; the chiral coupling of the mirror fermions of the $[SU(4)]^4$-theory being opposite to that of the basic fermions $F_{L,R}$.[22]

The observed value of r appears to be in the range 1/3 to 2/3 (i.e. $\sin^2\theta_W$ between 1/4 to 1/3) just as derived above theoretically under the hypothesis that the unifying symmetry G descends into the left-right symmetric low-energy substructures $\bar{G}$ or G'_{chiral}. (By contrast, if we postulate that G descends spontaneously to the left-handed symmetry $SU(2)_L \times U(1) \times SU(3)'_{L+R}$, in the manner considered in prior work[39], we obtain $(\sin^2\theta_W)$ren = (1/6 to 0.20) for (α_s/α) varying between ∞ down to ≈ 15.

As specific examples, the SU(5)-theory[40] can descend only through the left-handed subgroup $SU(2)_L \times U(1) \times SU(3)'_{L+R}$, hence it predicts lower values of $\sin^2\theta_W$ (varying between 1/6 and ≈ 0.20). By contrast the $[SU(4)]^4$-theory can descend through the left-right symmetric subgroups $\bar{G}$ or G'_{chiral} and correspondingly predicts higher values of $\sin^2\theta_W$ as noted above.

An accurate determination of $\sin^2\theta_W$ can help distinguish between different uniconstant unifying symmetries as also between alternative patterns of spontaneous breaking of the same symmetry G.

I now turn to the question of atomic parity violation in the left-right symmetric theory.

3.4 Atomic Parity Violation Parameter Q_W and the Masses of the Two Neutral Gauge Particles N_1 and N_2

The variation of Q_W for atomic bismuth as a function of $x \equiv (\eta-\zeta)/\zeta = (b^2-c^2)/c^2$ is exhibited in Fig. 3), while the variations of the masses of N_1 and N_2 as a function of $[Q_W/Q_o]$ are exhibited in Fig. 4; Q_o denotes the $SU(2)_L \times U(1)$-prediction for Q_W. Both these figures are drawn for a typical value of $r = f_{15}^2/g^2 = 0.44$. (This is equivalent to setting $\sin^2\theta_W \simeq 0.29$).

As we see, Q_W passes through zero[41] at $x = 0$ (i.e. b=c) approaching its $SU(2)_L \times U(1)$-value in the limit $x \to \infty$ (i.e. $b^2 \to \infty$ (with $\zeta = 1$). While for negative x $(b^2 < c^2)$, Q_W rises rapidly with $|x|$ acquiring unbounded positive values, (see Fig. 3). This is opposite in sign compared to Q_o, the $SU(2)_L \times U(1)$ value for Q_W. Thus $[Q_W/Q_o]$ acquires unbounded negative values as x decreases below zero (i.e. as b^2 becomes smaller than c^2). Such negative signs for (Q_W/Q_o) has the potential interest in that it implies a particularly light neutral gauge particle N_1 $(m_{N_1} < m_{W_L^+})$, see Fig. 4.

From the recent experimental values of the optical rotation parameters for atomic Bismuth, together with the existing theoretical calculations, one infers[3] that (Q_W/Q_o) lies in the range:

$$-\frac{1}{2} \lesssim (Q_W/Q_o) \lesssim + 1/3 \tag{55}$$

Either sign for (Q_W/Q_o) appears to be consistent with the data at present. For illustration, we list the predicted masses of N_1 and N_2 for a range of values of (Q_W/Q_o) with $r = 0.44$ (i.e. $\sin^2\theta_W \approx 0.29$):

(Q_W/Q_o)	m_{N_1} (GeV)	m_{N_2} (GeV)	$m_{W_L^+}$ (GeV)
+1	83	∞	70
0.33	76	120	70
0	70	107	70
-0.5	58	100	70
-1	50	99.5	70

In view of the particularly light N_1 ($m_{N_1} \leq m_{W_L^+}$), suggested by the atomic parity data, it is to be expected that some of the consequences of the left-right symmetric theory with such a light neutral gauge particle would be much more dramatic than those of the $SU(2)_L \times U(1)$-theory with a Z° of mass $\approx$ 83 GeV ($> m_{W_L^+}$). High Energy electron-positron colliding beam experiments provide a clear demonstration of such differences. I now turn to a discussion of such differences.

3.5 Asymmetries in High Energy e^-e^+-Collisions at PETRA and PEP

Assuming time reversal invariance, the differential cross section for $e^-e^+ \rightarrow \mu^+\mu^-$ involving in general a polarized incident beam (but no final state polarization measured) is given by five invariant quantities[42] $(F_1, F_2, \ldots, F_5)$, which in turn are given by the effective amplitudes $G^\mu_{VV}, G^\mu_{VA}, G^\mu_{AV}, G^\mu_{AA}$ defined before (Eqs. (40)-(41)). The F_i's are related[43] to the G's by:

$$\begin{aligned}
F_1 &= (1 + (s/e^2)G^\mu_{VV})^2 + (s^2/e^4)](G^\mu_{AA})^2 + 2\,(G^\mu_{VA})^2] \\
F_2 &= (1 + (s/e^2)G^\mu_{VV})^2 - ((s/e^2)G^\mu_{AA})^2 \\
F_3 &= 2(s/e^2)G^\mu_{AA} + 2(s/e^2)^2\,[G^\mu_{VA}G^\mu_{AV} + G^\mu_{VV}G^\mu_{AA}] \\
F_4 &= F_5 = -2(s/e^2)G^\mu_{VA}\,[1 + (s/e^2)(G^\mu_{VA} + G^\mu_{AA})]
\end{aligned} \tag{56}$$

The relevant asymmetry parameters are:

(A) Longitudinal Asymmetry in $e^-e^+ \to \mu^-\mu^+$ with longitudinally polarized incident beam: suppressing kinematic polarization factor;

$$L_p^{\mu\mu}(\theta) = [2\cos\theta)F_5 + (1+\cos^2\theta)F_4]/[(1+\cos^2\theta)F_1 + 2\cos\theta F_3]$$

This is a parity violating parameter. A measure of this ratio is given by (F_4/F_1).

(B) The mean longitudinal polarization of μ^+ with unpolarized incident beam: $\bar{P}_L(\mu) = -F_4/F_1$.

(C) The forward/backward-asymmetry with unpolarized incident beam (no final state polarization measured): $A^{\mu\mu} = (N_f - N_b)/(N_f + N_b) = (3/4)(F_3/F_1)$. Here N_f and N_b are the number of μ^+'s in the forward and backward hemispheres relevant to incident e^+. This is a parity conserving parameter.

(D) Longitudinal asymmetry for $e^+e^- \to \pi^+\pi^-$: suppressing kinematic polarization factor; $\rho_L^{\pi\pi} = -2\sigma/(1+\sigma^2)$, where $\sigma = (G_{AV}^p - G_{AV}^n)/(e^2/s + G_{VV}^p - G_{VV}^n)$.

The asymmetry-parameters for two centre of mass energies $\sqrt{s} = 28$ GeV and 38 GeV covering the PETRA and PEP energy regions for a range of values of the atomic parity violation parameter Q_W are given in Table II (with r = 0.55, i.e. $\sin^2\theta_W \simeq 0.32$ and $\zeta = 1$). These predictions are compared with those of $SU(2)_L \times U(1)$.

TABLE II (Ref. 43)

e^-e^+ ASYMMETRIES AS A FUNCTION OF Q_W †

		$SU(2)_L \times SU(2)_R \times U(1)$ THEORY					SU(2) x U(1) LIMIT
Q_W INPUT		245	147	88	0	-49	-146
m_{N_1} (GeV)		47	53	58	67	73	81
m_{N_2} (GeV)		95	98	100	110	124	∞
$L_p^{\mu\mu}$	$\sqrt{s}$=28	5%	3%	1.8%	0	-1%	-2.5%
	$\sqrt{s}$=38	-8%	3.1%	2.9%	0	-1.8%	-4.9%
$A^{\mu\mu}$	$\sqrt{s}$=28	-42%	-31%	-25%	-17%	-14%	-8%
	$\sqrt{s}$=38	-65%	-71%	-58%	-38%	-30%	-16.5%
$\rho_L^{\pi\pi}$	$\sqrt{s}$=28	-22%	-11%	-6%	0	+2.9%	7.9%
	$\sqrt{s}$=38	-86%	-37%	-18%	0	+6.7%	16%

† $\sqrt{s}$ is given in GeV (r = 0.55, ζ= 1). m_{N_1} and m_{N_2} are predictions of the theory, as also the asymmetry-parameters for a given Q_W. $L_p^{\mu\mu}$ denotes F_4/F_1.

DISCUSSION

(A) Note that as long as Q_W is either small[3] in magnitude (compared to its SU(2) x U(1)-value), or opposite in sign,[3] the left-right symmetric theory[5] would inevitably predict that the parity conserving forward-backward asymmetry parameter $|A^{\mu\mu}|$ must be strongly enhanced by a factor ($\simeq$ 2 to 5) compared to its SU(2) x U(1)-value even at PETRA and PEP energies.

(B) Second, in general there is a correlation between the signs of the parity violating asymmetry-parameters for $e^-e^+ \rightarrow \mu^-\mu^+$ and $\pi^-\pi^+$ on the one hand and the sign of Q_W on the other.

Thus e^-e^+ asymmetry parameters (as also atomic parity violation-parameter Q_W) unlike neutrino data, provide clear distinctions between the left-right symmetric theory with a light neutral mass ($\lesssim m_{W_L^+}$) on the one hand and the SU(2) x U(1)-theory on the other. Should experiments show relatively large forward-backward asymmetry (see Table II) or parity violating asymmetry parameters with signs opposite to those of SU(2) x U(1), they would provide clear signatures in favour of the left-right symmetric theory together with the characteristic prediction of a light N_1.

It should be stressed that atomic parity experiments, though indicative of a left-right symmetric theory, may not be regarded as definitive largely due to theoretical uncertainty associated with the relevant atomic calculations. The $e^-e^+ \rightarrow \mu^-\mu^+$-asymmetry calculations are totally devoid of any such dynamical uncertainty; thus such measurements at PETRA and PEP would be crucial to a clear distinction between the left-right-symmetric theory with the solution of two light neutral gauge particles on the one hand and the left-handed $SU(2)_L$ x U(1)-theory on the other.

Extended Left-Right Symmetry: Asymmetry Measurements at SPEAR and DORIS Energies:

It should be recalled (see remarks before and also part A) that unifying symmetries such as $[SU(4)]^4$ permit an extended low energy weak/electromagnetic symmetry of the form[36]

$$G_{LR} = SU(2)_L \times SU(2)_R \times U(1)_L \times U(1)_R$$

which is left-right symmetric like $\bar{G}_{LR} = SU(2)_L \times SU(2)_R \times U(1)_{L+R}$. G_{LR} in fact contains $\bar{G}_{LR}$ as a subgroup. Following the results of Ref. 6, it may be argued[36] that G_{LR} (like $\bar{G}_{LR}$) retains all the successful features of the "left-handed" $SU(2)_L$ x U(1)-symmetry for neutrino-experiments; simultaneously (like $\bar{G}_{LR}$) it can accommodate the null or near null results of atomic parity experiments. In these two respects, the symmetry G_{LR} resembles its subgroup $\bar{G}_{LR}$. The major difference between the two symmetries is this: $\bar{G}_{LR}$ permits in general two relatively light neutral weak bosons (N_1 and N_2); the mass of the lightest one (N_1) is constrained by present atomic parity experiments to lie above about 58 GeV (see Eq. (55) and the table following the same equation). By contrast the extended symmetry G_{LR} permits in general three relatively light neutral weak bosons (Z_1, Z_2 and Z_3); atomic parity experiments do not set any constraint on how light the lightest among them (Z_1) can be.

It turns out[36] that Z_1 (especially for $m_{Z_1} \lesssim 50$ GeV) is mostly like a massive axial photon. Such a gauge-particle contributes significantly to hyperfine splitting in hydrogen, (g-2) of muon, as well as to forward-backward $e^-e^+ \to \mu^-\mu^+$-asymmetry parameter even at existing SPEAR and DORIS energies. Considerations based on hyperfine splitting and (g-2) allow one to set the following lower limits on the mass of Z_1

$$m_{Z_1} \gtrsim 15 \text{ GeV (hfs) (Ref. 36)}$$
$$m_{Z_1} \gtrsim 35 \text{ GeV } ((g-2)_\mu) \text{ (Ref. 44)} \qquad (57)$$

This permits the possibility that Z_1 may even be discovered at PETRA and PEP. At any rate, even if it lies in the mass-range 40 to 60 GeV range, its virtual effects at PETRA and PEP would be dramatic. The presence of a light Z_1 ($m_{Z_1} \leq 60$ GeV) can be felt sensitively even at SPEAR and DORIS energies. This may be seen from table III, which gives the values of the forward-backward asymmetry parameter $A^{\mu\mu}$ as a function of m_{Z_1} at SPEAR-DORIS and PETRA-PEP energies (for $\sin^2\theta_W = 0.3$).

TABLE III

m_{Z_1} (GeV)	$A_{\mu\mu}$ (for $SU(2)_L \times SU(2)_R \times U(1)_L \times U(1)_R$) $\sqrt{s} = 7$	28	40
21	-0.18	.56	.73
35	-0.06	-.63	.41
45	-0.03	-.64	-.36
49	-0.03	-.52	-.64
64	-0.01	-.20	-.48

We see that accurate measurement[45] of $A^{\mu\mu}$ at SPEAR and DORIS energies can decide on the mass or set an improved lower limit on the mass of the lightest Z_1. In turn, this could help determine whether a real or a virtual Z_1 will be seen prominently at PETRA and PEP.

As a general remark, it is worth noting that the symmetry $G_{LR} = SU(2)_L \times SU(2)_R \times U(1)_L \times U(1)_R$, the left-right symmetric subgroup $\bar{G}_{LR} = SU(2)_L \times SU(2)_R \times U(1)_{L+R}$ and the standard left-handed symmetry $SU(2)_L \times U(1)$ are three alternative low energy manifestations of the same parent symmetry $[SU(4)]^4$. As to which of these manifestations may correspond to nature, depends largely on the mass of the lightest Z^o. Neither neutrino, nor atomic parity experiments can help resolve a choice between G_{LR} and $\bar{G}_{LR}$, but forward-backward asymmetry measurements at SPEAR and DORIS and at PETRA and PEP can help choose unambiguously between the three.

3.6 Asymmetry Parameters in ep, pp and $\bar{p}p$-Experiments

The left-right symmetric theory with a light neutral mass-solution ($m_{N_1} \lesssim m_{W_L^\pm}$) may be distinguished from the $SU(2)_L \times U(1)$-theory by other means, notably via measurements of parity violating effects in polarized deep inelastic eN-scattering on the one hand and via measurements of rates as well as asymmetry parameters of dilepton production in high energy pp and $\bar{p}p$-collisions on the other.

A calculation of the asymmetry-parameter for deep inelastic eN-scattering has recently been done by Cahn and Gilman[46] for $SU(2)_L \times U(1)$-theory and by Janah[47] for the $SU(2)_L \times SU(2)_R \times U(1)_{L+R}$-theory permitting the general possibility[6] of Q_W(Atomic) $\neq 0$.

For $\bar{p}p \to \mu^-\mu^+ + X$, a recent calculation by Mani, Rajpoot, Salam and myself[48] shows that the parity conserving forward-backward asymmetry turns out to be a particularly sensitive parameter (with dilepton-mass $\gtrsim$ 30 GeV), to help distinguish between the two theories. (This is similar to the situation encountered for $e^-e^+ \to \mu^-\mu^+$). An optimum center of mass-energy, where the dilepton-production-rate is descent (i.e. $(d\sigma/dm_{\mu\mu}) \gtrsim 10^{-37}$ cm^2/GeV for $m_{\mu\mu} \simeq$ 40-70 GeV) and the forward-backward asymmetry-parameter $A_{\mu\mu}$ is also generally large (though varying with $m_{\mu\mu}$) appears to be around 200 GeV. A plot of $A_{\mu\mu}$ varying with $m_{\mu\mu}$ is given in Fig. 3 for 3 different values of Q_W for the center of mass-energy $E_{Cm} = \sqrt{s}$ = 200 GeV.

It appears that (a) $e^-e^+ \to \mu^-\mu^+$-measurements at PETRA and PEP, (b) deep inelastic ep- and μp-scatterings at SLAC and Fermilab, (c) measurements of dilepton-forward backward asymmetry-parameter as well as the dilepton-rate, and last but not the least (d) accurate measurements of atomic parity violation not only in Bismuth, but also in other atoms such as Cesium and in particular in hydrogen should help us decide which of the two (if any) - the left-right symmetric theory with the light-neutral mass solution and the left-handed $SU(2)_L \times U(1)$-theory - may correspond to nature. They should also help fix the mass of the light neutral weak gauge particle, if not really produce it. Such experiments are eagerly awaited.

IV. CONCLUDING REMARKS

The advantages of the postulate of left-right symmetry[5] are many: (1) It helps us generate through spontaneous symmetry breaking a desirbale milliweak theory of CP-violation[19], the strength of which is directly related to the observed left-right-asymmetry (dominance of V-A over V+A) in Nature.

(2) It provides the scope for accommodating small atomic parity violation (should this be desired experimentally). It simultaneously encompasses the left-handed $SU(2)_L \times U(1)$-theory as a limiting possibility. In either case it provides distinctive testable predictions, for example for $e^-e^+ \to \mu^-\mu^+$-experiments.

(3) From a practical standpoint, it offers the possibility of a particularly light neutral gauge particle (N_1 could be as light as about 58 GeV consistent with present atomic parity experiments). This enhances the expected effects of the virtual neutral weak gauge particle in many experiments in planning, for example for PETRA and PEP. If the left-right symmetry has its origin within the extended group-structure G_{LR} (rather than $\bar{G}_{LR}$), the mass ot the lightest neutral weak boson can be even lighter than 58 GeV (see Eq. (57)).

(4) From a theoretical standpoing, the left-right symmetric gauge-structure renders a number of important parameters to be finite and calculable; notably the mass-difference between n and p-quarks, relevant to isospin-conservation[49] and the Cabibbo angle.[50]

(5) Last but not the least, it meets one's aesthetic requirement[5] that Nature is intrinsically parity conserving; the observed left-right asymmetry being a "low"-energy phenomenon to disappear at high energies.

Looking ahead, the left-right symmetric theory raises two important questions: (1) Where in the mass-scale should the charged right handed gauge-particle W_R^+ lie? (2) How and at what level should one expect to see right handed neutrinos? The answer to the second question is, of course, related to that of the first. Unfortunately neutral current-phenomena do not help put any experimental constraint on $m_{W_R^+}$. I end with two remarks pertaining to this question. Both signify extreme possibilities. On the one hand, the fact that the left-right symmetric theory naturally generates a milliweak theory of CP-violation[19] with

$$|\eta_{+-}| = |(m_{W_L^+}/m_{W_R^+})^2(\sin 2\theta_R/\sin 2\theta_L)\sin(\delta_R-\delta_L)| \approx 10^{-3}$$

suggests (assuming that the phase angle is not tiny and that $\theta_R \approx \theta_L$), that perhaps W_R^+ is a factor of 10 to 30 heavier than W_L^+. This would be in line with the attractive possibility[31] that spontaneous symmetry breaking is triggered at one place through the vacuum-expectation value of the "right-handed" vectorial scalar field E_R contributing to the mass of W_R^+ being non zero, which then feeds through finite order α-radiative corrections into the neutral and the charged W_L^+-sectors (in this picture $(m^2_{W_L^+}/m^2_{W_R^+})$ is naturally order α). This however makes the possibility of observing right handed neutrinos in β-decay or π^+-decay remote:

$$\mathrm{Rate}(\nu_R)/\mathrm{Rate}(\nu_L) = (m_{W_L^+}/m_{W_R^+})^4 \quad .$$

On the optimistic side, there is the possibility that CP-violation within the left-right symmetric theory may itself be a radiative phenomenon, i.e. it vanishes in the tree-approximation, but arises at the one-loop level. In this case, the phase $(\delta_R-\delta_L)$ might be naturally order α permitting the ratio $(m_{W_L^+}/m_{W_R^+})^2$ to be $\gtrsim (1/10)$. This might suggest that right-handed neutrino-emission (i.e. the rate of (V+A)-interaction) in β-decay and π^+-decay may actually be occurring at about 1% level. In this connection, improvement in the measurement of longitudinal polarization of β-decay electrons can not be overemphasized.

REFERENCES AND FOOTNOTES

1. Refer for example to the talks of C. Baltay and J. Steinberger at the European Physical Society meeting, Budapest, Hungary (1977), to appear in the Proceedings.
2. S. Weinberg, Phys. Rev. Letters 19, 1264 (1967); Abdus Salam in Elementary Particle Physics, ed. N. Svartholm (Almkvist and Wicksell, Stockholm, 1968), p. 367; S. L. Glasvhow, J. Iliopoulos and L. Maiani, Phys. Rev. D2, 1285 (1970). The $SU(2)_L \times U(1)$-gauge structure without the Higgs-mechanism for generation of gauge masses was proposed by S. L. Glashow, Nucl. Phys. 22, 579 (1961) and Abdus Salam and J. C. Ward, Phys. Letters 13, 168 (1964).
3. The recent results of L. L. Lewis et al., (Phys. Rev. Lett. 39, 795 (1977)) and P. E. G. Baird et al. (Phys. Rev. Lett. 39, 798 (1977)) for the optical rotation parameters in atomic Bismuth are: $R(876\ nm) = (-.7 \pm 3.2) \times 10^{-8}$ (Washington), $R(648\ nm) - (+2.7 \pm 4.7) \times 10^{-8}$ (Oxford). Given the previous atomic theoretical calculations based on relativistic central field approximation (see Ref. 4) together with the shielding effect (discussed by P. G. Sandars at the International Symposium on Lepton and Photon-interactions at high energies, held at Hamburg, W. Germany, August, 1977), the above numbers correspond to a basic atomic parity violation parameter Q_W for Bi lying between $-50 \leq Q_W \leq +75$, while the simple $SU(2)_L \times U(1)$-theory predicts $Q_W = -145$ (for $\sin^2\theta_W = 0.30$). Quite clearly, improvements in theoretical accuracy, or alternatively measurements in hydrogen are crucial to determine Q_W accurately.
4. I. B. Kriplovich, Soviet Physics - JETP Lett. 20, 315 (1974); E. M. Henley and L. Wilets, Phys. Rev. A19, 1911 (1976); M. Brimicombe, C. E. Loving and P. G. H. Sandars, J. Phys. B9, L237 (1976). Recently shielding effects have been considered (see Ref. 3).
5. The first suggestion of the left $\leftrightarrow$ right symmetric theory $SU(2)_L \times SU(2)_R \times U(1)_{L+R}$ comprising all matter (quarks as well as leptons) was made by J. C. Pati and Abdus Salam, Phys. Rev. Letters 31, 661 (1973); ibid. Phys. Rev. D10, 275 (1974). The motivations in these work were the realizations of basic left-right symmetry and quantization of electric charge. Motivation based on considerations of CP violation was provided by R. N. Mohapatra and J. C. Pati, Phys. Rev. D11, 566 (1975). The second paper (Phys. Rev. D10, 275 (1974), Ch. IV and VI and Footnote 21) proposes two alternative patterns of spontaneous symmetry breaking; both patterns are consistent with the hypothesis of "natural" left-right symmetry. One of the patterns permits only one, while the other permits two relatively light neutral weak gauge particles. It is this second alternative, which is relevant to present atomic parity experiments (Ref. 3), and is pursued recently by several authors (Ref. 11).
6. J. C. Pati, S. Rajpoot and Abdus Salam, Imperial College Preprint ICTP/76/11, Phys. Rev. D (to be published). The left-right symmetric theory

$SU(2)_L \times SU(2)_R \times U(1)_{L+R}$ is equivalent to the $SU(2)_L \times U(1)$-theory for left-handed neutrino-processes only to the extent that W_L-W_R mixing mass may be neglected compared to $m_{W_L^+}$. This is discussed later.

7. Barring possible small corrections due to Higgs-boson exchanges. Note if W_L and W_R have equal mass with no mixing between them, $(W_L \pm W_R)/\sqrt{2}$ are mass eigenstates. The gauge interaction given by Eq. (1) may be written in terms of these eigenstates as $(g/\sqrt{2}) \frac{(W_L+W_R)}{\sqrt{2}}(V) + \frac{(W_L-W_R)}{\sqrt{2}}(-A)$. This generates in second order parity conserving interaction $\propto$ (VV+AA).
8. We expect neutrinos to be in general four-component objects within the theory, unless ν_L and ν_R remain disjoint and therefore massless despite spontaneous symmetry breaking. Such four-component neutrinos may still have arbitrarily small mass. A natural understanding of the smallness of neutrino-mass or its masslessness is yet a challenge to the theory.
9. In a different context such a postulate was made by M. A. Bég and A. A. Zee, Phys. Rev. Lett. Vol. 30, 675 (1973).
10. J. C. Pati and Abdus Salam, Phys. Rev. D10, 275(1974)(Ch. IV, VI & Footnote 21).
11. The consequences of the zeroth-order solution $\langle B\rangle = \langle C\rangle \neq 0$ arising within the model proposed in Ref. 5 was first examined by H. Fritzsch and P. Minkowski, Nucl. Phys. B103, 61 (1976), and more recently by R. N. Mohapatra and D. P. Sidhu, Phys. Rev. Letters 38, 667 (1977). The more general case comprising $\langle B\rangle \sim \langle C\rangle$ and $\langle B\rangle = \langle C\rangle$ has been examined by J. C. Pati, S. Rajpoot and Abdus Salam, Imperial College, London, preprint ICTP/76/11 (to be published in Phys. Rev. D), and Physics Letters 71B, 387 (1977). For a somewhat different treatment of spontaneous symmetry breaking of the group structure proposed in Ref. 5 see, A. De Rujula, H. Georgi and S. L. Glashow, Harvard preprint, 1977. For an analogous phenomenological discussion containing the basic ingredients of the gauge-framework, see B. Kayser, Phys. Rev. D15, 3407 (1977). The equivalence theorem stating the equivalence of the left-right symmetric theory $SU(2)_L \times SU(2)_R \times U(1)_{L+R}$ with the left-handed theory $SU(2)_L \times U(1)$ for neutrino processes (see Ref. 6 and Chapter 3) has been further generalized to comprise extended group-structures by H. Georgi and S. Weinberg (Harvard preprint, HUTP-77/A052).
12. The choice $\langle B\rangle = \langle C\rangle \neq 0$ is in general not a "natural" solution of the theory in the technical sense; this is the case for example for Ref. 11. Mechanisms permitting $\langle B\rangle = \langle C\rangle \neq 0$ naturally despite radiative corrections are briefly mentioned later.
13. R. N. Mohapatra and J. C. Pati, Phys. Rev. D11, 566, 2558 (1975). For a calculation of such $O(\alpha)$ radiative corrections, see Q. Shafi and Ch. Wetterich, University of Freiburg preprint (THEP 77/3).
14. It needs to be stressed that contrary to common impression $(\sigma_{\nu_L p} \neq \sigma_{\bar{\nu}_R p})_{NC}$ does not imply parity non-conservation since $\nu_L \nrightarrow \bar{\nu}_R$ under parity. Such a

distinction between ν_L and $\bar{\nu}_R$ cross-sections eliminates only the class of parity-conserving theories, which are vector-like with no AA piece (see discussion below).

15. A. Benvenutti et al., Phys. Rev. Letters 37, 1039 (1976); J. Blietschau et al., Preprint CERN/EP/Phys. 76-55; B. C. Barish, Ca. Tech. preprint CALT-68-544. The clearest distinction is shown by measurements of $\sigma(\bar{\nu}_R p \rightarrow \bar{\nu}_R p)$ versus $\sigma(\nu_L p \rightarrow \nu_L p)$. See D. Cline et al., Phys. Rev. Letters 37, 252, 648 (1976) and W. Lee et al., Phys. Rev. Letters 37, 186 (1976).
16. M. A. B. Bég and A. Zee, Phys. Rev. Letters 30, 675 (1973). For a list of references on other vector-like models, see R. M. Barnett, Review talk, Brookhaven, APS Meeting, SLAC-PUB 1850.
17. We call them "restricted" vector-like theories to distinguish them from the mirror theory (J. C. Pati and A. Salam, Physics Letters 58B, 333 (1975)), in which (V-A) current of the basic fermions $(p,n,\lambda,c)_L$ and (V+A) current of the mirror fermions $(p',n',\lambda',c')_R$ couple to the same gauge particles W_A. Such a theory is vector-like in a broader sense; but within this theory there exist a parallel and distinct set of gauge particles W_B with their (V-A) and (V+A) coupling reversed compared to W_A; which have no counterparts within the restricted $SU(2)_{L+R} \times U(1)_{L+R}$-vector-like theories.
18. It is possible to construct SU(2) x U(1)-models satisfying atomic parity-data as well as neutrino-data. In particular, the E_7 model descending through SU(2)x U(1)-component (F. Gursey and P. Sikivie, Phys. Rev. Lett. 36, 775 (1976), ibid, Phys. Rev. D. 16, 816 (1977) and P. Ramond, Nucl. Phys. B 110, 214 (1976)) is interesting in this connection. However there is a possible question regarding "natural" suppression of strangeness changing neutral current-processes in such models (see S. L. Glashow and S. Weinberg, Harvard Preprint HUTP-75/A158).
19. R. N. Mohapatra and J. C. Pati, Phys. Rev. D11, 566 (1975).
20. L. Wolfenstein, Phys. Rev. Lett. 13, 562 (1964); Nucl. Phys. B77, 375 (1974). For a gauge theory version see R. N. Mohapatra, J. C. Pati and L. Wolfenstein, Phys. Rev. D 11, 3319 (1975).
21. The lower limit ($\approx$ 30 GeV) on the mass (m_σ) of the relevant left over Higgs-boson corresponds to the experimental constraint that the relation $\eta_{+-} = \eta_{oo}$ is known to hold to better 5% (see Ref. 19). In reference 19, m_σ was allowed to be as high as 10^4 GeV or higher. This led to a prediction for d_n varying between 10^{-24} to 10^{-29} e cm. The constraint imposed here that m_σ should not exceed about 1000 GeV (or else Higgs-fields would begin to interact strongly, see e.g. M. Veltman, Utrecht preprint, and B. W. Lee, C. Quigg and H. B. Thacker, Phys. Rev. Lett. 38, 883 (1977)) is important: it makes the dipole moment dn to necessarily exceed about 10^{-27} ecm for the isoconjugate model, bringing the same to an experimentally accessible range. The recent

experimental value $(.4 \pm 1.1) \times 10^{-24}$ ecm, obtained by W. B. Dress et al., Phys. Rev. D15, 9 (1977), is expected to be improved to the level of $\approx 10^{-26}$ ecm in the near future.

22. Extension to mirror-model (J. C. Pati and A. Salam, Physics Letters 58B, 333 (1975)) would need in addition a 16-fold mirror set of heavy quarks and heavy leptons. This is needed for the sake of complete unification. Such extensions do not however alter any of our discussions on neutral current-phenomena.

23. J. C. Pati and Abdus Salam, Phys. Rev. D8, 1240 (1973). C. Itoh, T. Minamikawa, K. Miura and T. Watanabee, Preprint (1973), unpublished.

24. In general, if the basic group G or its subgroup $\bar{G}$ descend from a higher unifying group G (e.g. $[SU(4)]^4$), the coupling constants associated with $SU(2)_L$ and $SU(2)_R$ may differ from each other through finite renormalization effects, which may in principle introduce large left-right asymmetry at the low energy level, even though the group structure G is left-right symmetric.

25. Extended models with more than four flavors (e.g. the mirror model, Ref. 22) may in general permit large skewness angles (see Ref. 22) leading to $\cos\theta_R \ll \cos\theta_L$. In such models one could obtain $m_{W_R^\pm} \sim m_{W_L^\pm}$. However, within unified theories with no abelian factor, such loss of manifest left-right symmetry in physical currents (in the sense defined recently by M. A. B. Bég, R. Budny, R. N. Mohapatra and A. Sirlin, Phys. Rev. Lett. 38, 1252 (1977)) is not permissible (see remark by J. Pati, 1976 Scottish University Lecture Notes, Page 109, Ed. I. M. Barbour and A. T. Davis).

26. K. Symanzik, in Fundamental Interactions at High Energies, Ed. A. Perlmutter et al. (Gordon and Breach, 1970).

27. This has been demonstrated for the case of a pair of left-right symmetric Higgs doublets B and C by G. Senjanovic and R. N. Mohapatra, Phys. Rev. D 12, 502 (1975). The result holds more generally in the presence of additional left-right symmetric Higgs-pairs with mutual couplings between them (see e.g. Ref. 31 and 32).

28. G. Rajasekharan and P. Roy, Pramana 6, 303 (1975), J. C. Pati and A. Salam, Phys. Rev. Lett. 36, 11 (1976), V. Elias, J. C. Pati, A. Salam and J. Strathdee, Pramana 4, 303 (1977).

29. R. N. Mohapatra, J. C. Pati and A. Salam, Phys. Rev. D13, 1733 (1976).

30. For a discussion of the consistency and the experimental consequences of the unconfined unstable integer-charge quark-hypothesis, see J. C. Pati and A. Salam, Comments on Nuclear and Particle Physics (1976) and in particular the recent Trieste Preprint, "Design of Experiments to Test..." IC/77/65.

31. H. S. Mani, J. C. Pati and A. Salam, "Naturalness of Atomic Parity Conservation Within Left-Right Symmetric Unified Theories" - Trieste Preprint IC/77/80, Phys. Rev. D (to be published).

32. R. N. Mohapatra, F. E. Paige and D. P. Sidhu, BNL-preprint (1977).

33. S. Coleman and E. Weinberg, Phys. Rev. D7, 1888 (1973).

34. In this case, one needs <A> << <C> in order that the $W_L^{\pm} - W_R^{\pm}$ mixing be small, as seems to be required by the neutrino-data together with the present atomic parity data.

35. Here the subscripts L and R refer to the gauge pattern of the basic fermions F. For the mirror fermions (Ref. 22) L and R are interchanged.

36. J. C. Pati, Proc. Second Orbis Scientae, Coral Gables, Florida, Jan, 1975 (P253-256), ed. by A. Perlmutter and S. Widmayer; J. C. Pati, S. Rajpoot and A. Salam (Ref. 6, Footnote 15). The experimental consequences of this group structure have recently been emphasized by Q. Shafi and Ch. Wetterich, Univ. of Freiburg Preprint (1977) and by V. Elias, J. C. Pati and Abdus Salam, Univ. of Md. Tech. Rep. No. 78-043 (1977), Physics Letters to be published.

37. CDHS result, M. Holder et al., CERN Preprint (1977).

38. V. Elias, Phys. Rev. D 14, 1896 (1976); Md. Tech. Rep. No. 77-253, Phys. Rev. D (To be published), Md. Tech. Rep. No. 78-040 (1977), V. Elias, J. Pati and A. Salam, U. of Md. Tech. Rep. 78-041 (1977).

39. H. Georgi, H. R. Quinn and S. Weinberg, Phys. Rev. Letters 33, 451 (1974).

40. H. Georgi and S. L. Glashow, Phys. Rev. Letters, 32, 438 (1974).

41. Barring finite order α-differences between $(g_L - g_R)/g_L$.

42. See for example, R. Budny, Physics Letters, 55B, 227 (1975); A. McDonald, Nuclear Physics B75, 343 (1974) and E. Lendavi and G. Pocsik, Physics Letters, 56B, 462 (1975). We follow McDonald's notations.

43. J. C. Pati, S. Rajpoot and Abdus Salam, Physics Lett. 71B, 387 (1977), ICTP/76/15, Physics Letters (To be published).

44. J. Leveille - Imperial College Preprint (1978).

45. Unpublished results of such a preliminary measurement carried out at SPEAR exist, which indicate that the asymmetry is less than 3% in magnitude at a mean center of mass energy ≈ 6.8 GeV (Private communications: B. Richter).

46. R. Cahn and F. Gilman (SLAC-Pub., 1977).

47. A. Janah, U. of Md. Preprint (In preparation).

48. H. S. Mani, J. C. Pati, S. Rajpoot and A. Salam, Trieste Preprint IC/77/88, Physics Letters (1977).

49. S. Weinberg, Phys. Rev. Letters 29, 1698 (1972). In this work, the symmetric gauge-structure is introduced only for quarks for the sake of natural isospin conservation; but leptons are assumed to be singlets of $SU(2)_R$.

50. F. Wilczek and A. Zee, Preprint (1977, Princeton Univ.), S. Weinberg, Preprint (1977); H. Fritzsch, CERN Preprint No. 2358 (1977).

TABLE I[†]

Predictions of Neutral-Current Neutrino Parameters in the Chiral Limit ($a^2 = 0$, or $\zeta = 1$)

PROCESS	PARAMETERS	PREDICTIONS $r \equiv f_{15}^2/g^2$ 0.33	0.44	EXPERIMENT GGM	HPWF	CITF	CDHS
$\frac{(\nu_\mu N)_{NC}}{(\nu_\mu N)_{CC}}$	R_ν	0.30	0.27	0.25 ±0.04	0.29 ±0.04	0.24 ±0.02	0.29 ±.01
$\frac{(\bar{\nu}_\mu N)_{NC}}{(\bar{\nu}_\mu N)_{CC}}$	$R_{\bar{\nu}}$	0.39	0.40	0.39 ±0.06	0.39 ±0.10	0.34 ±0.09	0.34 ±.03
$\left(\frac{\bar{\nu}N}{\nu N}\right)_{NC}$	R	0.44	0.48	0.59 ±0.14	≤ 0.61 ±0.25		
$\bar{\nu}_e e \to \bar{\nu}_e e$	$\frac{C_{\bar{\nu}_e e}}{0.57}$	1.00	1.15	0.87 ± 0.25 1.70 ± 0.44		1.5 ~ 3 MeV 3 ~ 4 MeV	Reines *et al*.
$\bar{\nu}_\mu e \to \bar{\nu}_\mu e$	$C_{\bar{\nu}_\mu e}$	0.14	0.17	$0.11^{+0.21}_{-0.09}$ GGM 0.54 ± 0.17 Aachen-Padova			
$\nu_\mu e \to \nu_\mu e$	$C_{\nu_\mu e}$	0.14	0.13	0.24 ± 0.12 Aachen-Padova			
$\nu_\mu P \to \nu_\mu P$	R^{el}_ν	0.13	0.12		0.17 ± 0.05 (HPWF) 0.23 ± 0.09 (CIR)		
$\bar{\nu}_\mu P \to \bar{\nu}_\mu P$	$R^{el}_{\bar{\nu}}$	0.18	0.19		0.2 ±0.10 (HPWF)		
$\left(\frac{\bar{\nu}_\mu P}{\nu_\mu P}\right)$ Elastic	R^{el}	0.46	0.53		0.4 ±0.2 (HPWF)		

[†]The predictions of elastic νP and $\bar{\nu}P$-scattering parameters are for a choice of axial mass $m_A = 1.08$ GeV. The values of $r = .33$ and $.44$ correspond to $\sin^2\theta_W = 3r/(6r+2) = 0.25$ and 0.28 respectively.

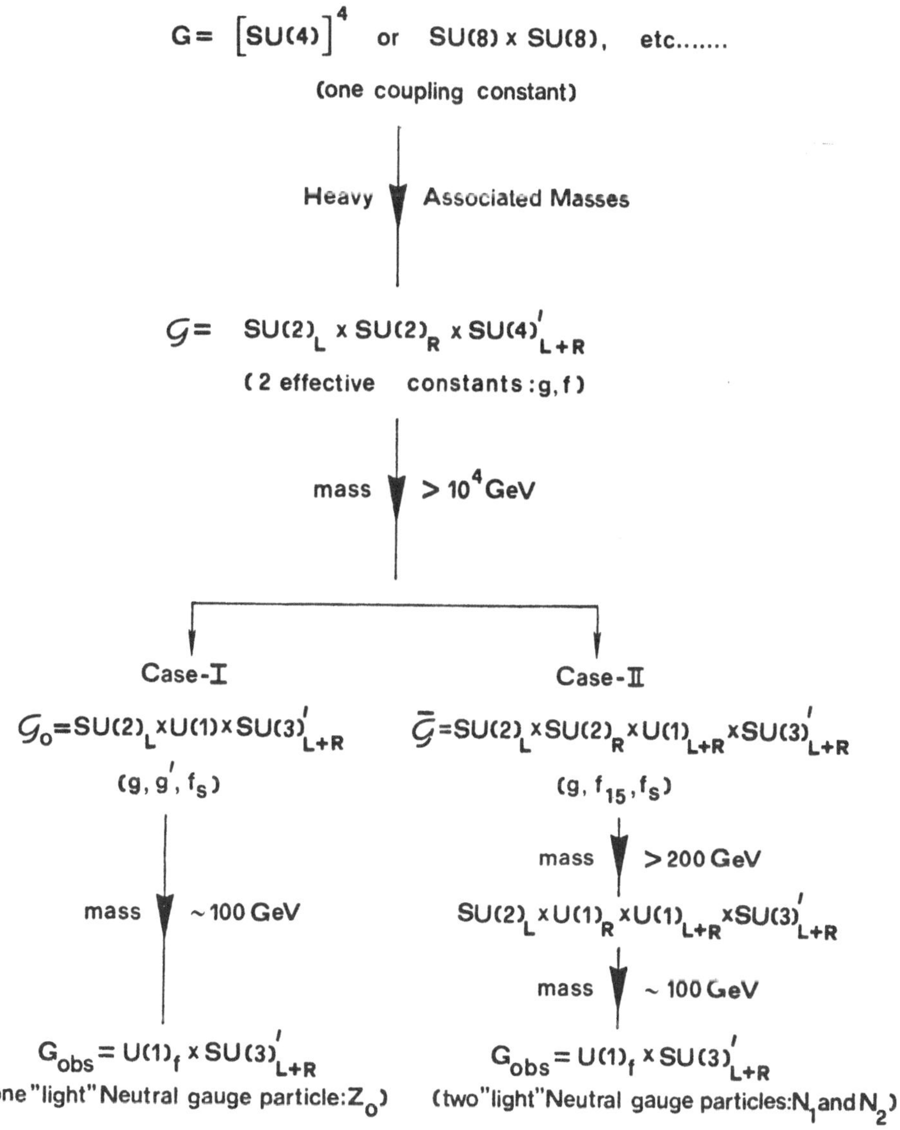

G= [SU(4)]^4 or SU(8) x SU(8), etc.......
(one coupling constant)
Heavy Associated Masses
𝒢= SU(2)_L x SU(2)_R x SU(4)'_L+R
(2 effective constants: g, f)
mass > 10^4 GeV
Case-I
𝒢_0=SU(2)_L x U(1) x SU(3)'_L+R
(g, g', f_s)
mass ~100 GeV
G_obs = U(1)_f x SU(3)'_L+R
(one "light" Neutral gauge particle: Z_0)
Case-II
𝒢̄=SU(2)_L x SU(2)_R x U(1)_L+R x SU(3)'_L+R
(g, f_15, f_s)
mass > 200 GeV
SU(2)_L x U(1)_R x U(1)_L+R x SU(3)'_L+R
mass ~100 GeV
G_obs = U(1)_f x SU(3)'_L+R
(two "light" Neutral gauge particles: N_1 and N_2)

Figure 1

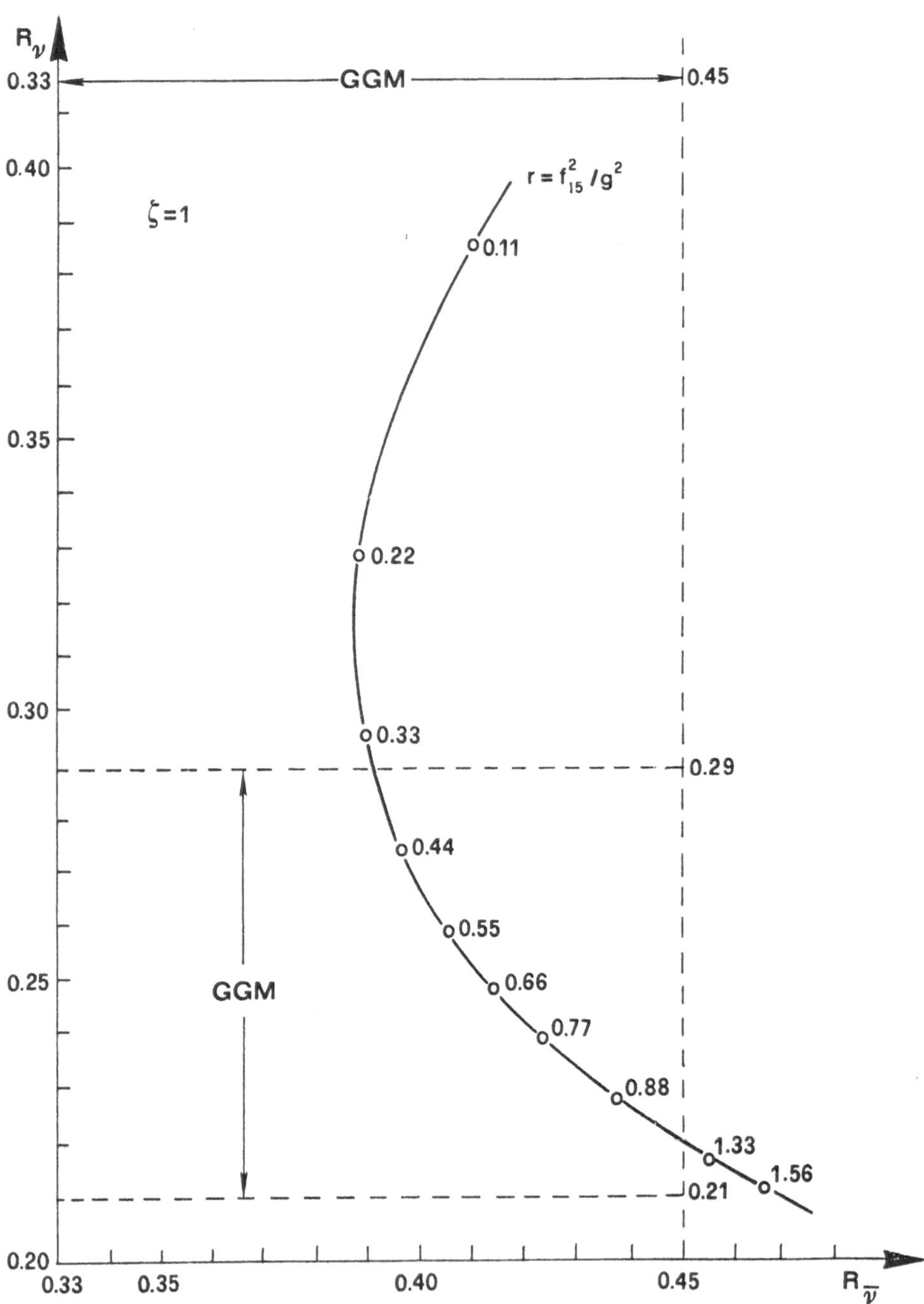

Figure 2
($\sin^2\theta_W = 3r/(6r+2)$, see text)

Figure 3

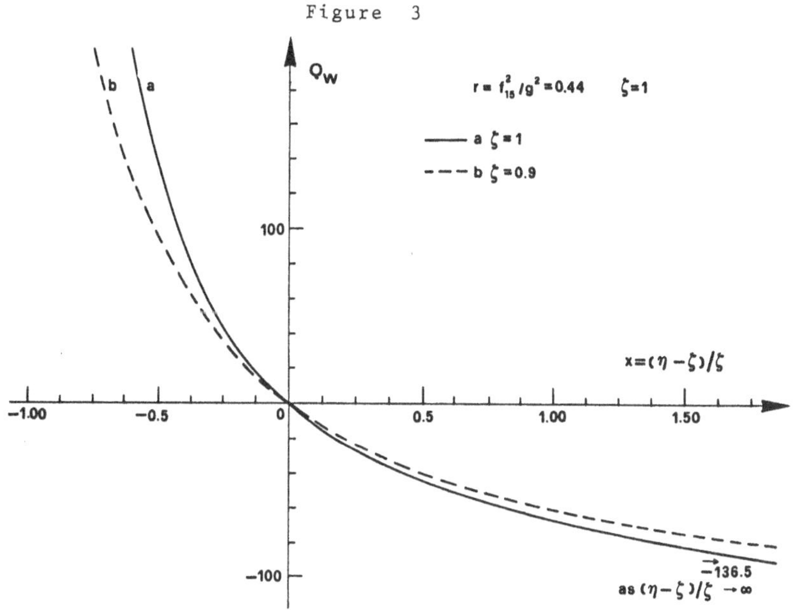

A plot of the atomic parity violation parameter Q_W vs. the neutral current-parity violating parameter $x \equiv (b^2 - c^2)/c^2$.

Figure 4

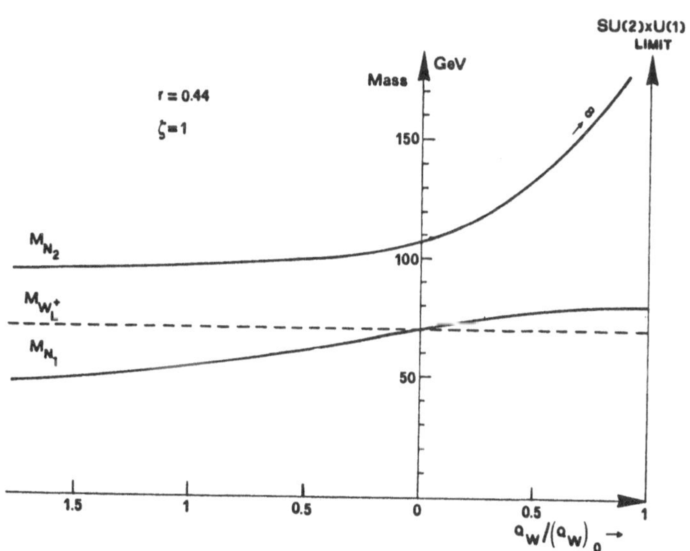

A plot of the masses of the two neutral weak gauge particles N_1 and N_2 vs. (Q_W/Q_0). Q_W and Q_0 denote the atomic parity violation parameters for SU(2) x SU(2) x U(1) and SU(2) x U(1)-theories respectively.

Figure 5

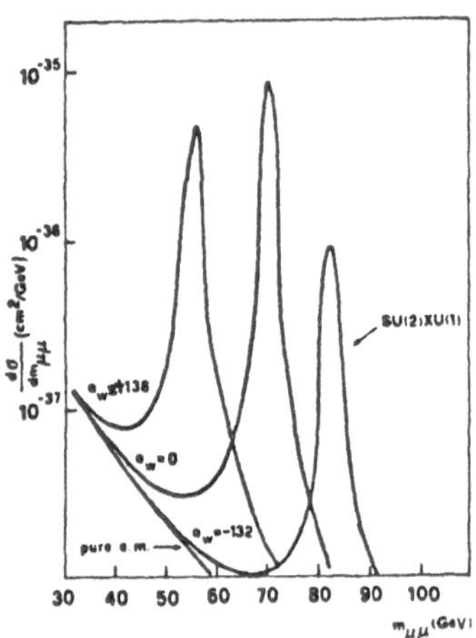

A plot of $d\sigma/dm_{\mu\mu}$ vs. $m_{\mu\mu}$ for $p\bar{p}\to\mu\bar{\mu}$ x ($\sqrt{s}$ = 200 GeV)

Figure 6

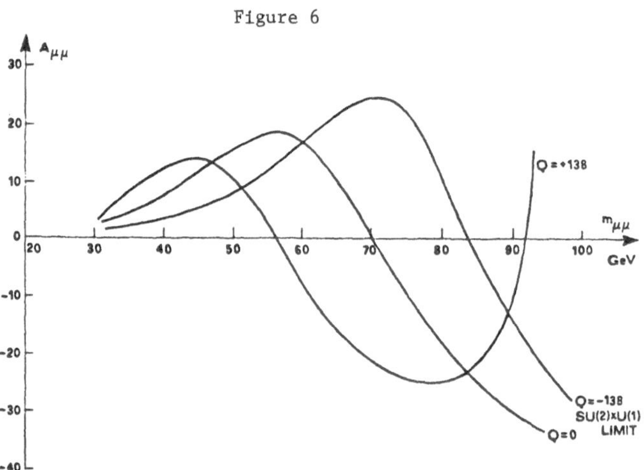

A plot of $A^{\mu\mu}$ vs. $m_{\mu\mu}$ for $P\bar{P}\to\mu^+\mu^-X$ ($\sqrt{s}$ = 200 GeV)

Alvaro De Rújula
The Physics Laboratories

Harvard University
Cambridge - MA 02138
U. S. A.

ABSTRACT

I lectured on two subjects: the origin of the Cabibbo angle and other weak mixing angles and quark liberation. My lectures were based, rather verbatim, on recently published literature, and I did not think it would be useful to rewrite the material here, due to its lack of originality in content and/or presentation.

On the subject of the weak mixing angles I followed the article "Theory of Flavour Mixing" by S.L. Glashow, Howard Georgi and myself, Ann. Physics 109, 258 (1977). One of the results of this investigation is the well-satisfied relation $tn^2\theta_c \simeq m_\pi^2/2m_k^2$. This relation has been considered in the past by R. Gatto, G. Sartori and M. Tonin, Phys. Letters 28B, 128 (1968); N. Cabibbo and L. Maiani, ibid. 131, and R. Jackiw and H.J.Schnitzer - Phys.Rev. D5, 2008 (1972). In a context similar to ours, it has also been discussed by S. Weinberg, HUTP-77/A057, F. Wilczek and A. Zee, Princeton preprint, and H. Fritzsch, CERN preprint.

The subject of quark liberation was discussed along the lines of the article "Unconfined Quarks and Gluons" by L.C. Giles, R.L. Jaffe and myself, MIT preprint CPT 635 (1977).

QUARKS, COLOUR AND HADRON SPECTROSCOPY

R. H. Dalitz

Department of Theoretical Physics
University of Oxford
1 Keble Road, Oxford OX1 3NP
England

TABLE OF CONTENTS

1. INTRODUCTION - QUARKS

We base our discussion of hadronic states on the Gell Mann-Zweig spin-1/2 quarks q with fractional charge values and baryon number B=+1/3, supplemented now and in the future by new quarks as called for by the experimental data. At present, we have the following list:

$$
\begin{array}{llllll}
q = & (u & d) & s & c & b? \\
Q = & (+2/3 & -1/3) & -1/3 & +2/3 & -1/3?
\end{array}
$$

The pair (u,d) are an isodoublet, whereas the further quarks listed are isosinglet. The quarks (u,d,s) are those involved in the well-known unitary symmetry SU(3), while the further quarks are SU(3)-singlet. The latter may be involved in higher symmetries; for example, there may be an SU(4) symmetry for the quarks (u,d,s,c), although this is not yet established. For any hadron state, the baryon number B is given by $\{N(q)-N(\bar{q})\}/3$, where $N(q)$ and $N(\bar{q})$ denote the number of quarks q and antiquarks $\bar{q}$, respectively, in this state. Similarly, its "strangeness" S is given by $\{N(\bar{s})-N(s)\}$, its "charm" C by $\{N(c)-N(\bar{c})\}$, and so on. We suppose that the forces between quarks, which bind them to make hadrons, are due to the exchange of neutral bosons, called "gluons", which are isosinglet and have B=S=C=0.

We identify the "partons" of Feynman's models for deep-inelastic interaction processes with these quarks q (their antiparticles $\bar{q}$ being "antipartons"): the gluons are "neutral partons" which carry energy and momentum but which do not otherwise take part in lepton-hadron deep-inelastic scattering. For the intermediate energy range, for incident energies 5-100 GeV, lepton-nucleon deep-inelastic scattering has been observed to have the property of "scaling", the scaling variable being

$$x = (\underline{k}^2-k_0^2)/2Mk_0, \tag{1.1}$$

where k_0 and $\underline{k}$ are the energy and momentum transferred from the lepton to the target nucleon (mass denoted by M). In this situation, the scattering data can be expressed phenomenologically in terms of scaling structure functions $F_2(x)$ and $F_3(x)$ defined for the appropriate scattering process in terms of the following expression for the measured cross-section,

$$\frac{d^2\sigma}{dxdy} = \frac{g^4ME}{2\pi(M_W^2+k_\mu^2)^2}\{\ (1-\tfrac{1}{2}y)^2F_2(x) + \lambda xy(1-\tfrac{1}{2}y)F_3(x)\ \}, \tag{1.2}$$

where $y=k_0/E$, E being the incident lepton energy, and $k_\mu^2=(\underline{k}^2-k_0^2)$. $F_3(x)$ arises from the interference between parity-conserving and parity-reversing amplitudes; hence $\lambda=0$ holds for electron or muon scattering due to the electromagnetic interactions (which conserve parity), whereas $\lambda= \pm 1$ holds for the weak processes $\nu\to\ell^+$ and $\bar{\nu}\to\ell^-$, respectively. With expression (1.2), we obtain the electromagnetic case by taking $g^2=e^2$ and $M_W=0$; for the weak processes of present interest, we may neglect the k_μ^2 term in the denominator and replace g^2/M_W^2 by the weak interaction constant $G_F/\surd 2$.

For the various kinds of partons, u,d,s,c, etc, in a proton, there will be distribution functions u(x), d(x), s(x), etc, respectively; the proton will also contain some antipartons, $\bar{u},\bar{d},\bar{s},\bar{c}$, etc, and their distribution functions may be denoted by $\bar{u}(x)$, $\bar{d}(x)$, $\bar{s}(x)$, etc, respectively. The structure functions occurring in the expressions (1.2), with which the experimental data are to be compared, are then expressible in terms of these distributions. For example, for electron-proton scattering, the structure function takes the form

$$F_2^{eP} = x\ \{\ \tfrac{4}{9}(u(x)+\bar{u}(x)) + \tfrac{1}{9}(d(x)+\bar{d}(x)) + \tfrac{1}{9}(s(x)+\bar{s}(x)) + \dots\ \} \tag{1.3}$$

from which we may obtain the expression F_2^{eN} by charge symmetry, making the interchange $u(x)\leftrightarrow d(x)$ and leaving s(x), c(x), etc, unchanged, with the result

$$F_2^{eN} = x\ \{\ \tfrac{4}{9}(d(x)+\bar{d}(x))\ +\ \tfrac{1}{9}(u(x)+\bar{u}(x))\ +\ \tfrac{1}{9}(s(x)+\bar{s}(x))\ +\ \ldots\ \} \qquad (1.4)$$

From these expressions, and those for the structure functions F_2 and F_3 for the inelastic processes $\nu \to \ell$ and $\bar{\nu} \to \ell^+$, it is in principle possible to deduce empirically the forms of all these structure functions; in practice, at least, with the present data, this requires that some simplyifying assumptions be made, and analyses have been presented by various authors[1,2], which we need not go into here.

The above remarks assume that partons are point-like, as indicated by the scaling property. More recently, clear evidence has been found for scaling violations, in the interaction data obtained with both muon and neutrino beams of very high energy ($\gtrsim$ 100 GeV). In principle, such effects might correspond to a form factor $G(k_\mu^2 R^2)$ for the parton, where R is an effective radius, but it appears that this is not the case but that the observed scaling violations are more probably associated with "asymptotic freedom" effects (see Sec.2 below) due to the non-Abelian gauge character of the underlying hadronic interactions[3]. Their existence does not invalidate the parton-sum rule arguments we shall now use, based on the use of data for energies much below this regime of scaling violations.

The firmest conclusions which we can draw from the parton-model analysis of the deep-inelastic data are those expressed as sum rules, involving integration over all x. We mention three of these sum rules:

(i) The use of the expressions $F_3^{\nu P} = 2(d-\bar{u})$ and $F_3^{\nu N} = 2(u-\bar{d})$ leads us to the result[4]

$$\tfrac{1}{2}\int_0^1 (F_3^{\nu P}(x)\ +\ F_3^{\nu N}(x))\,dx = \int_0^1 (u+d+s-\bar{u}-\bar{d}-\bar{s})\,dx\ =\ N(q)-N(\bar{q}), \qquad (1.5)$$

where $N(q)$ and $N(\bar{q})$ denote the number of quarks and antiquarks in the proton. Note that $\int_0^1 (s-\bar{s})dx=0$, since the proton has zero strangeness. The quantity on the left-hand side of (1.5) is exactly that which would be measured for the difference between the interactions of ν and $\bar{\nu}$ beams with complex nuclei having N=Z as target material; the measurements reported from the CERN experiments lead to the value[5]

$$N(q)\ -\ N(\bar{q})\ =\ 3.2\pm 0.35 \qquad (1.6)$$

This is consistent with the view that the nucleons consist of three valence quarks, together with a sea (presumably SU(3) singlet) of $q\bar{q}$ pairs.

(ii) The coefficients occurring in eqs. (1.3) and (1.4) are the squares of the corresponding quark charges, whereas the expressions $F_2^{\nu P} = 2x(d+\bar{u})$ and $F_2^{\nu N} = 2x(u+\bar{d})$ have unit coefficients. Hence the ratio[5]

$$\int_0^1 (F_2^{eP}\ +\ F_2^{eN})\ dx/\int_0^1 (F_2^{\nu P}+F_2^{\nu N})\,dx\ =\ 0.29\pm 0.03 \qquad (1.7)$$

provides a measure of the quark charge values. With the result (1.6) and charge +1 for the proton, we expect charge +2/3 for the u quark (as assumed in writing down eq.(1.3), of course); this requires the value 5/18 = 0.277 for the ratio (1.7), which is compatible with the empirical ratio given on the right side of (1.7).

(iii) The mass (or momentum) carried by all the partons is given by

$$\int_0^1 xdx(u+\bar{u}+d+\bar{d}+s+\bar{s}) = \frac{3}{(1-\delta/3)}\int_0^1 (F_2^{eP}+F_2^{eN})\ dx = 0.55\pm0.05, \tag{1.8}$$

where δ gives the fraction of strange quarks in the proton. The uncertainty in δ is included in the uncertainty quoted in (1.8); for an SU(3)-singlet sea, the data would suggest $\delta \approx 0.06$, but its value could well be lower than this. The result (1.8) indicates that about 45% of the energy-momentum carried by a nucleon must reside in neutral partons which are inactive for both electromagnetic and weak interactions, and it is generally believed that these are the neutral gluons which transmit the strong hadronic forces between the quarks, as we shall discuss again below.

As indicated following eq.(1.6), the nucleons are believed to be systems involving three valence quarks, and are typical of the full $1/2^+$ baryon octet $B=(N,\Lambda,\Sigma,\Xi)$. They lie quite close in mass to the $3/2^+$ baryonic decuplet states $D=(\Delta,\Sigma^*,\Xi^*,\Omega)$ and these two SU(3)-multiplets are believed to be closely related, comprising all the low-lying baryonic states corresponding to the lowest configuration $(1s)^3$ in a shell-model representation [6] for the three valence quark system. In so far as the spin-spin forces between quarks may be neglected, these B and D unitary multiplet states would comprise the full set of (2x8+4x10)=56 spin x unitary spin states belonging to an SU(6)-multiplet $\underline{56}$. As is clearly the case for the D component of this $\underline{56}$- representation, since its substate Δ^{++} with $m_J= +3/2$ consists of three u-quarks with parallel spins, the permutation symmetry of the SU(6) factor in the $\underline{56}$-representation is even. This observation constituted a serious problem for the quark model, for the full wavefunction for these three quark systems in the $(1s)^3$ configuration,

$$\psi(1,2,3) = \underset{\text{space}}{\phi(1,2,3)} \times \Big[\underset{\text{spin}}{\chi(1,2,3)} \times Q_{SU(3)}(1,2,3)\Big]_{\underline{56}} \tag{1.9}$$

then clearly had even permutation symmetry. Since the quark-partons are necessarily spin-½ objects, as is known both from the baryon and decuplet spin values and from the satisfying of the Callan-Gross relationship[7] by the data on deep-inelastic electron-nucleon scattering, the Spin-Statistics Theorem of Pauli[8] excludes Bose statistics for quarks. In any case, this is excluded on empirical grounds since protons are observed to obey Fermi statistics. It was therefore necessary to conclude that there exists some further degree of freedom in the specification of the state of a quark. With this, the complete wavefunction $\psi(1,2,3)$ for the three-quark system representing a baryon or decuplet resonance would have an additional factor $\mathcal{C}(1,2,3)$ multiplying those given in (1.9), which could be antisymmetric and so restore the antisymmetry required for $\psi(1,2,3)$ by the Pauli principle.

This difficulty was first appreciated by Greenberg[6], who proposed that quarks might obey parastatistics of order 3, with the additional implicit assumption that only qqq states whose space x SU(6) wavefunction (1.9) had even permutation symmetry could occur in the low-mass regime at present under exploration. Thus, we were led to the "symmetric quark model" for baryons. Han and Nambu[9], with a different purpose in mind (the development of a quark theory with integral charge values), proposed that there might exist three distinct quark triplets and showed that the forces between them could well be such as to push all quark states other than those corresponding to the symmetric quark model up to very high mass values. It was clear that the formation of an antisymmetric factor $\mathcal{C}(1,2,3)$ required that the new degree of freedom should have at least three distinct states and/or act in some three-dimensional space. Further, if the dimensionality were greater than 3, it would be possible to construct more than one antisymmetric factor $\mathcal{C}(1,2,3)$, in which case the proton would have statistical weight ω greater than 1, contrary to fact. The group symmetry corresponding to this new quantum number could be either discrete or continuous, in principle, but it has been found that its identification with the SU(3)' group of unitary transformations of modulus unitary in a "colour space" with axes labelled red (r), white (w) and blue (b) proves to be the most fruitful hypothesis.

The dynamical use of the colour variable has been discussed in a simple way by Lipkin[11]. This involves the introduction of a colour octet of vector gluon fields $(g_\alpha)_\mu$, where $\alpha=1,\ldots 8$ labels the octet space associated with $SU(3)_C$, where C stands for colour, and $\mu=0,1,.3$ is the space-time label. These gluons are singlet in ordinary SU(3). Their coupling with the quark fields has the form

$$\mathcal{L}_{int} = \sum_{\alpha,\mu} f(\bar{q}\gamma_\mu\lambda_\alpha^c q)(g_\alpha)_\mu, \tag{1.10}$$

with coupling amplitude f. One-gluon exchange, in the non-relativistic limit, then leads to a potential interaction of the general form

$$V_{ij} = (f^2/4\pi)(\Sigma_\alpha\lambda_\alpha^c(i)\lambda_\alpha^c(j))v_{ij}(r) \tag{1.11}$$

between two quarks (or a quark and an antiquark, or two antiquarks). For a given system of N(q) quarks and $N(\bar{q})$ antiquarks, this interaction has the property that it gives greatest attraction for states which are colour singlet. The simplest of these colour singlet states are the structures $(\bar{q}q)$, the meson states, and (qqq), the baryon states. These two states have triality zero, where triality is defined for a system as $[N(q)-N(\bar{q})]_{mod3}$, and it is a property of the $SU(3)_C$ group that all colour singlet states have zero triality. Since only colour singlet states can lie low in mass, it follows then that all low mass states must have zero triality, as we know empirically to be the case, so that the introduction of the $SU(3)_C$ symmetry and the colour octet gluons provides an explanation of this fact. Further, the interaction between an isolated quark or antiquark and any colour-singlet system of quarks and antiquarks is zero, in first approximation, since it involves averaging the factor $\lambda_\alpha^c(j)$ in (1.11) over all the quarks and antiquarks in the colour-singlet system, whereas we necessarily have $\langle\lambda^c_\alpha(j)\rangle_{C=0} \equiv 0$ for a colour-singlet state. Since the interactions V_{ij} are additive, it follows that the interaction between two colour-singlet systems is also zero, in first approximation, and thus relatively weak, when viewed on the scale of the qq and $q\bar{q}$ forces themselves. A colour singlet system of quarks and antiquarks will therefore separate out into as many colour-singlet systems as it can, these systems being only weakly bound, if bound at all. The deuteron provides an illustration. It contains 6 quarks, but for the major part of the time, the system has the structure (P+N), consisting of two colour-singlet 3-quark systems orbiting around each other. Unfortunately, although this qualitative picture is very suggestive, it carries conviction quantitatively only for a situation where the quarks are very heavy and the binding is very high, whereas it appears that, especially for the non-strange mesons and baryons, the hadronic states we have to deal with mostly consist of light quarks moving under the influence of interactions which are relatively weak at short distances, as we shall see below.

2. COLOUR GAUGE THEORY SPECULATIONS

The present view is that the colour symmetry is exact. In part, this is the simplest view; in part, it is a consequence of the empirical absence of any non-singlet colour states, although we have already attributed this to a specific property of the qq and $\bar{q}q$ forces, which causes all such states to lie very high in mass. An exact $SU(3)_C$ symmetry requires the existence of a colour-octet of gauge fields g_α, which are space-vectors and SU(3)-singlets, and which couple with the exactly-conserved colour-current $(j^c_\alpha)_\mu$. These gauge fields then represent the gluons themselves. They are neutral in charge. Being colour octet, these gluon fields contribute to the colour current with which they interact. Their interaction energy is therefore explicitly non-linear, and such gauge theories are termed "non-Abelian". The gauge theories which have been long known to us are those of electromagnetism and of General Relativity (gravity). The former is an example of an Abelian gauge theory; the gauge particles (photons) do not carry charge and so do not contribute to the exactly-conserved electromagnetic current. The latter is a non-Abelian gauge theory; the gauge particles (gravitons) do carry energy and momentum and do contribute to the conserved stress tensor to which the graviton is coupled.

There are two speculations current concerning the non-Abelian gauge theory of colour:

(i) Asymptotic Freedom holds. By this we mean that the coupling amplitude f specified in eq.(1.10) depends on the momentum transfer k^2_μ carried by the gluon in such a way that $f \to 0$ as k^2_μ increases indefinitely[12]. This property would imply, for example, that quark-quark (and quark-antiquark) interactions are relatively weak in (a) high-energy and high-momentum-transfer collisions, or in (b) qq (or $\bar{q}$q) interactions at very short distances. The first of these is an attractive possibility when we attempt to account for the successes of the parton-model picture, a situation where parton-parton interactions within each hadron are neglected during the high-k_μ^2 interaction between an incident and a target parton. The second allows us the possibility of discussing the fine and hyperfine structure of mesonic and baryonic states which arises from short distance interactions, in terms of the parameter $\alpha_s = f^2(k_\mu^2)/4\pi$ by using the procedures already developed for quantum electrodynamics, as we shall discuss in Secs. 4 and 5.

The property of Asymptotic Freedom does not hold for all classes of field theory. For example, it is known not to hold for Abelian gauge theories. Asymptotic Freedom can hold for non-Abelian gauge theories, and it is plausible on the basis of perturbation theory arguments that it might hold for all such gauge theories, although this has not been rigorously demonstrated yet.

(ii) Quark-Confinement holds. Free quarks have not yet been observed with any certainty. There has been a report from Stanford[13] that fractional charge values consistent with $\pm e/3$ have been measured in an experiment using superconducting niobium balls of mass about 10^{-4} mg, but this needs further experimental research for its corroboration. Morpurgo[14] has made studies with carbon pellets, and more recently with iron balls, but has found no balls with fractional change. Perhaps there may still exist some primordial quarks from the time of the Big Bang; certainly, the cosmic ray studies which have been made make it difficult to understand how such quarks could be produced by the present cosmic radiation at the rates which would be required by the Stanford evidence. Clearly, there will have to be much more work carried out in the near future, to follow up these experiments of La Rue et al[13] and of Morpurgo[14]. For the present, we shall accept the current view that free quarks cannot exist. All present models require quarks within the well-known hadrons to be light, with masses $\approx$ 300-500 MeV, so that, if it is possible to create q$\bar{q}$ pairs, it is difficult to understand why they are not observed to be produced in current accelerator experiments. Of course, an isolated quark has non-singlet colour, and we have already argued above that such a state may be limited to very high mass values. In other words, our use of colour actually requires isolated quarks to be very heavy, whereas there is a variety of evidence (e.g. nucleon magnetic moments, hadron spectroscopy, etc) which indicates that quarks within hadrons are light (masses 0-350 MeV for (u,d) quarks and about 500 MeV for the s quark). It is possible to construct theories of quarks and hadrons which have such a property but they have no other advantages.

The present view is that the hadronic internal forces may simply have properties which do not allow an isolated quark to be extracted from a hadron. This is the hypothesis of "quark confinement". To remove a quark from a hadron would require essentially infinite energy, but if a large energy U were available and used for this purpose, it would simply go into the production of bound $\bar{q}q$ states, i.e. mesons, since there is a very large number of states of the form (hadron + mesons) which have energy lying below the energy U. No mechanism to generate quark confinement is yet known for the three dimensional space in which we live.

Our present situation is then that we have a specific theory, known as Quantum Chromodynamics (QCD)[15] which is a gauge theory based on colour as an exact symmetry, including quark triplets with colour, the interaction Langrangian being of the form

$$\mathcal{L}_{int} = \sum_{\alpha,\mu} f(g_\alpha)_\mu \left\{(\bar{q}\gamma_\mu\lambda_\alpha^c q) + j_{\mu\alpha}^C(\text{gluons})\right\}, \tag{2.1}$$

where $j_{\mu\alpha}^C$ denotes the contributions of the gluon fields g_α to the colour current and q represents the quark field in colour space, SU(3)-space, Dirac space and ordinary space, the free Lagrangian for the quarks having the form

$$\mathcal{L}_q = \bar{q}(\gamma.p + m + |S|\delta m)q, \tag{2.2}$$

where S denotes strangeness, so that the mass assigned for the s quark differs from that assigned to the (u,d) quarks. This last assumption, $m_s \neq m_u = m_d$, is the only violation of SU(3) in the full Lagrangian

$$\mathcal{L} = \mathcal{L}_q + \mathcal{L}_g + \mathcal{L}_{int}. \tag{2.3}$$

The terms (2.1) and (2.2) are readily generalized to include c-quarks, b-quarks and any further quarks which may be required by the experimental data, as long as these each enter in a qualitatively similar way to the s-quark, differing primarily in mass value. Initially, we shall speak only in terms of the quarks (u,d,s).

The questions concerning this QCD Lagrangian are as follows:

(i) does this Lagrangian $\mathcal{L}$ lead to a theory which predicts quark confinement? This is a difficult question, in view of the high degree of non-linearity in the theory, but what preliminary indications there are are not promising. We may note that, in (1+1) dimensions, this theory does indeed lead to quark confinement, but this cannot be taken to imply quark confinement for the physical case of (3+1) dimensions. For one thing, for the Abelian gauge theory of electromagnetism, we know that the Coulomb interaction has linear dependence on the charge-charge separation r in (1+1) dimensions, but 1/r dependence in the case of (3+1) dimensions.

(ii) does this theory $\mathcal{L}$ have the property of asymptotic freedom?

(iii) does this theory $\mathcal{L}$ lead to a spectrum of hadrons and to a hadron dynamics which are in accord with the empirical situation, for suitable choices of the quark masses $m_s, m_u = m_d = m$ and the coupling constant f? This is not impossible, although the simplicity of the basic theory would then be rather breath-taking. However, if it were the case, the next question would be to ask what is the origin of this mass difference (m_s-m) and why the underlying cause of this violation of SU(3)-symmetry should affect only this one term in $\mathcal{L}$.

At present, the tacit speculation is that the answer is "yes" to the first two questions and "probably" to the third question, but these are not yet proven, either way. Consequently, the phenomenological analyses which base themselves on these general (supposed) features of QCD have many detailed uncertainties and corresponding non-uniqueness in their fitting to the data.

3. THE LONG-RANGE (QUARK-CONFINING) FORCES

A harmonic potential between quarks,

$$v(r) = ar^2+b, \tag{3.1}$$

although without any theoretical basis, does correspond to confined quarks, of course. It leads to wavefunctions of a well-known form, convenient for explicit calculations. For the (qqq) system, its use leads directly to wavefunctions for which the c.m. motion can be factored out explicitly, leaving wavefunctions of a shell-model form for the wavefunction describing the internal motion, a very convenient situation. However, the relationship between the qq potentials and the total energy E is not that which would be given by the (non-relativistic) Schrodinger equation, since this would predict the excitation energy to increase linearly with N, the number of quanta, whereas the empirical observations show that it is E^2 which increases linearly with N, at least for the mesonic and baryonic leading trajectories:

(a) $\rho(733)-A_2(1310)-g(1680)-$

(b) $\Delta(1232)-\Delta(1925)-\Delta(2415)-\Delta(2850)-\Delta(3275)-$

With the confinement potential (3.1), this would require the use of a wave equation where the total energy appears only in the form E^2, such as the Blankenbecler-Sugar equation.

The most popular confining potential form used for the qq and $\bar{q}q$ systems today is the linear form

$$v(r) = ar+b \tag{3.2}$$

which has been widely used for the mesonic systems $\bar{c}c$. For the 3q baryonic systems, its use would require a more elaborate discussion for the space wavefunction, since the c.m. motion no longer factors out. For example, one could assume a space wavefunction of general form $F(r_{12},r_{23},r_{31})\times\phi(\theta,\phi,\psi)$, where r_{ij} is the distance between quarks i and j and (θ,ϕ,ψ) denote the Euler angles for the normal to the plane (123) and the azimuthal rotation of the (123) particle configuration around this normal. No calculation taking the potential form (3.2) seriously has yet been made for the 3q-system. As noted above, the theoretical basis for the form (3.2) is quite uncertain but its use for the discussion of mesonic states has worked reasonably well to date.

What happens when two quark systems come into contact? For definiteness, let us speak in terms of two protons, which is a 6q-system $\{(uud)+(uud)\}$. At large distances, the potential (1.11) averages to zero, so there is no long-range PP force, in first order. This is the case even if one considers the Van der Waal's forces between them, arising from the interaction between the electric dipoles excited in each by the presence of the other. The reason is that the low-lying 3q states reached by a dipole excitation are necessarily colour singlet, from the considerations given in Sec.2, so that the colour averaging again reduces the dipole excitation matrix-elements to zero. Dipole-dipole forces are possible from the excitation of both protons to non-singlet colour states, but these states lie so high in mass that such forces are strongly suppressed, not only for the PP system but for any system consisting of low-mass hadrons.

The PP system is a 6q system, and the quarks obey Fermi statistics, so that the quark wavefunction for the PP state must be totally antisymmetric with respect to interchange of any two quarks. This antisymmetry has no physical effect when the PP separation r is large, since the wavefunctions of two quarks in different protons then have no overlap. However, at short distances, there will be exchange effects resulting from this antisymmetry, which will contribute to the PP forces. When the two protons overlap strongly, the quarks of one proton find that many states are filled by quarks of the other proton, and some quarks must then be excited to higher orbitals; at the quark level, then, the overlapping PP configuration is energetically unfavourable, which is equivalent to the statement that a strong repulsion is predicted for the PP system at short distances. This situation is qualitatively similar to that well-known for the ^{4}He-^{4}He system in nuclear physics[17], where the 1s orbitals are completely filled by the nucleons of one α-particle, its configuration being $(1s)^4$. Another instructive example is provided by the N-^{4}He system. When $\ell_{N\alpha}=1$, the N-^{4}He interaction is very strongly attractive [18], leading to a marked $J^P=3/2^+$ resonance ($\Gamma\sim0.6$ MeV) at N-α c.m. energy 0.89 MeV and a broad $J^P=1/2^+$ resonance ($\Gamma\sim1.5$ MeV) at N-α c.m. energy about 6.4 MeV, which is not particularly surprising in view of the strongly attractive nature known for the nucleon-nucleon interaction. On the other hand, for $\ell_{N\alpha}=0$, the N-^{4}He interaction is found empirically to be strongly repulsive [18]. The explanation is that in the effective potential well provided by the other nucleons, there are only two $(1s)_N$ orbitals available for the neutrons and these are already filled by the neutrons in the target ^{4}He. The lowest state available for the incident $\ell_{N\alpha}=0$ neutron is the $(2s)_N$ orbital and this lies quite far above the $(1p)_N$ orbital responsible for the observed $\ell_{N\alpha}=1$ N-^{4}He resonances mentioned just above. The wavefunctions for the 1s and 2s orbitals are sketched on Fig.1(a). On Fig.1(b), we illustrate how the 2s orbital may appear as a 1s orbital for an attractive interaction with a hard core repulsion of radius a. Of course, for positive Nα energies T, the wavefunctions sketched on

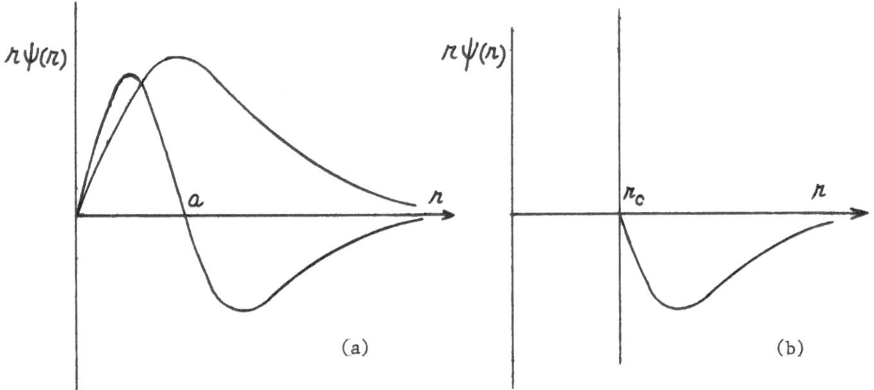

Fig. 1. 1s and 2s wavefunctions in an attractive potential: (a) $r\psi(r)$ is plotted vs r, and the 2s wavefunction has a node at r = a, and (b) the 2s wavefunction $r\psi(r)$ is plotted vs r for $r \geqslant a$, as if it were the ground state in a potential with an infinitely repulsive hard core with radius $r_c = a$.

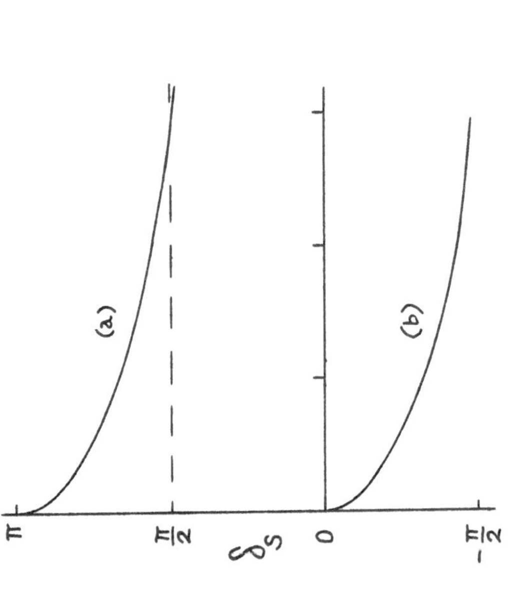

Fig. 2. The S-wave N-α phase shift as function of the c.m. energy, deduced from the elastic scattering and polarization data, by (a) Satchler et. al.[63], and (b) Arndt and Roper[64]. The former corresponds to an optical model which has one S-wave bound state; the latter was obtained from a model-independent phase-shift analysis.

Figs.1 oscillate with increasing r, while the form of the wavefunctions at short distances, and consequently the "effective hard core" radius a, vary relatively slowly with T. The negative phase-shift δ_0 for $\ell_{N\alpha}=0$, which is shown on Fig.2 and which is normally interpreted as due to repulsion, should be replaced by the positive phase-shift $(\delta_0+\pi)$ which starts from value π at T=0 and falls through $\pi/2$ with increasing T to reach an asymptotic value 0 as $T \to \infty$. This is the behaviour which is well known for cases where there exists one bound state for the scattering system, e.g. for ^3S NP scattering where the bound state in question is the deuteron, and which is expressed by the Levinson theorem:

$$\delta(T=0) - \delta(T=\infty) = (N_b - N_r)\pi, \tag{3.3}$$

where N_b and N_r denote the number of bound states and resonances (the latter being defined by the requirement that δ should then increase through $\pi/2$ with increasing T), respectively, for this system. For the physical N-^{4}He system, there is no bound state, of course. On the other hand, the effective N-^{4}He potential, as deduced from the $\ell_{N\alpha}=1$ scattering, for example, is quite strong and sufficient to lead to a deeply bound state for the neutron. As far as the effective N-^{4}He potential is concerned, the Levinson theorem (3.3) should clearly hold, with $N_b=1$ and $N_r=0$. One might question the validity of an "effective potential" for the interaction of systems with internal structure, but the fact is that this effective potential (including a spin-orbit term) is at least qualitatively valid for all $\ell_{N\alpha}\neq 0$. The resolution of the apparent contradiction is the recognition that this $\ell_{N\alpha}=0$ N-^{4}He bound state in the effective potential is "excluded" by Pauli-principle requirements which go beyond it and that the Levinson theorem (3.3) should be amended to read[19]

$$\delta(T=0)-\delta(T=\infty) = (N_{ex} + N_b - N_r)\ \pi, \tag{3.4}$$

where N_{ex} denotes the number of "excluded states".

To return to the PP system, we see that the key question is whether or not there are any "excluded" states for this system, considered at the quark level. As discussed by Ribeiro[20], the answer is that there is no excluded state for the PP system (nor for the $\Delta^{++}\Delta^{++}$ system, for that matter). The reason is the additional flexibility in the wavefunction permitted by the colour degree of freedom. For a given quark type, there are six 1s orbitals available, the number being 2x3, the factors arising from the spin labels (up, down) and the colour labels(red, white, blue), respectively. For the S-wave $\Delta^{++}\Delta^{++}$ system, we can form the antisymmetric wavefunction

$$\psi(\Delta^{++},\Delta^{++}) = \mathrm{Det}(\uparrow u_r(1),\uparrow u_b(2),\uparrow u_w(3),\downarrow u_r(4),\downarrow u_b(5),\downarrow u_w(6))\,\phi_{space}((1s_u)^6), \tag{3.4}$$

as required. This wavefunction is colour singlet, has spin S=0 and belongs to the SU(3) representation $\underline{10}\times\underline{10}$; it is the only $(1s)^6$ wavefunction which can so be formed for the system $(\Delta^{++}\Delta^{++})$. Since an antisymmetric wavefunction (3.4) can be formed for this very symmetric case, the Δ^{++} wavefunction being symmetric in spin

(S=3/2), unitary spin ($\underline{10}$) and space ($(1s)^3$) factors, it follows that an antisymmetric wavefunction can also be obtained for the $(1s)^6$ configuration representing the S-wave PP system with S=0, unitary spin $\underline{8}x\underline{8}$, and singlet colour.

(In passing, we may remark that the 1s orbital can accommodate 18 quarks before becoming saturated, the fully filled state having spin S=0 and being both SU(3) singlet and colour singlet.) Thus, since there is no excluded state, we have $N_{ex}=0$ in the Levinson relation (3.4). However, even if there is no completely excluded 6q state for the $\ell=0$ PP system, there is much antisymmetrization required in the 6q wavefunction which is not present in the (3q)x(3q) part of the wavefunction appropriate to large separation r and which both increases the kinetic energy and reduces the effectiveness of the attractive components of the qq potential. We note that any qq potential which saturates, or which averages to zero between a quark and a colour singlet system, must have a repulsive component to go with the strong attraction which forms the baryonic states. Only the $\bar{3}_c$ states are effective for qq in a colour singlet baryon, but in the baryon-baryon state the 6_c qq states are also effective. Since there is attraction in the $\bar{3}_c$ state, there must be repulsion in the 6_c state, to achieve saturation, and this is clearly the case for the potential (1.11), since $\lambda_c(1).\lambda_c(2)$ takes the values -2 for the $\bar{3}_c$ state and +1 for the 6_c state. Hence, even with no excluded 6q state, the effect of the Pauli principle for quarks is that the confining potential introduces much repulsion into the $\ell=0$ PP interaction at short distances, and perhaps enough to account for the hard core repulsion believed to exist in the PP interaction.

Ribeiro[20] has carried through the PP and NP scattering calculations for $\ell=0$, using the Resonating Group Method due to Wheeler[21] and assuming harmonic qq potentials of the form (1.11) with v(r) given by (3.1). Taking the $(1s)^3$ wavefunction for the nucleon to have the form

$$\phi(1,2,3) = N_\phi \, \mathrm{Exp}(-\alpha(r_{12}^2+r_{23}^2+r_{31}^2)/2), \tag{3.5}$$

where N_ϕ is the normalization constant, Ribeiro is led to an integro-differential equation of the general form

$$(-\frac{1}{M}\nabla^2 + V(R))\psi(\underline{R}) + \int W(\underline{R},\underline{R}')\psi(\underline{R}')d_3\underline{R}' = E\psi(\underline{R}) \tag{3.6}$$

for the wavefunction $\psi(\underline{R})$ of the nucleon-nucleon system, $\underline{R}$ being their separation in space. The kernel $W(\underline{R},\underline{R}')$ has the form $-4a\alpha(5R^2-3\underline{R}\cdot\underline{R}')\,K(\underline{R},\underline{R}')/3$, where

$$K(\underline{R},\underline{R}') = N_K \, \mathrm{Exp}\,\{-5(9\alpha(R^2+R'^2)-6\alpha\underline{R}\cdot\underline{R}')/16\} \tag{3.7}$$

and N_K is the normalization constant $27(3\alpha/\pi)^{3/2}/64$. The potential V(R) in (3.6) gives the non-exchange terms of the potential, and may include the long-range potential due to pion exchange, for example, whereas $K(\underline{R},\underline{R}')$ is due to quark-exchange resulting from the antisymmetrization. Although quark exchange between two protons may be considered as equivalent to some superposition of meson exchange processes, in a sense, it corresponds only to the exchange of an uncorrelated $q\bar{q}$ pair, which does not reflect the true meson mass spectrum known for the interacting $q\bar{q}$ system. Ribeiro's calculation includes no long-range potential in V(r) and gives pure repulsion for the S-wave nucleon-nucleon system. The strength a in the kernel $K(\underline{R},\underline{R}')$ was determined from properties of the 3q system, specifically to obtain the observed excitation energy for the L=2 state NF15(1638) relative to the L=0 state P = NP11(938). Taking the quark mass to be m_q=M/(2.79), to fit the proton magnetic moment, the parameter α in the wavefunction (3.5) is given by $\sqrt{(2m_q\alpha/3)}$; this value corresponds to an r.m.s. radius of $1/\sqrt{(3\alpha)} \approx 0.6$fm for the quark structure within a nucleon (the proton charge radius would then be given by $\sqrt{(R_q^2+(3\alpha)^{-1})}$, where R_q represents the charge distribution within the quark itself, due perhaps to the ordinary vector mesons ρ and ω). The L=0 phase shift δ_0 calculated by Ribeiro for the PP system is always negative, starting with δ_0=0 at threshold and reaching δ_0=0 at high energy, as illustrated in Fig.3. Up to c.m. energy $\sim$ 200 MeV, δ_0 corresponds closely to the effect of a hard core repulsion, with radius 0.35fm. This may be seen most clearly in the form of the radial wavefunction ψ_0(R), plotted in Fig.4. For $T_{c.m.} \lesssim$ 200 MeV, ψ_0(R) is strongly suppressed for $R \lesssim 0.35$fm and has the value zero for a separation $R \approx 0.35$fm; as R increases beyond this value, ψ_0(R) then increases rapidly, so that we have a situation comparable with that illustrated in Fig.1(b) for a hard core repulsion of radius r_c. There is, of course, no reason to expect or prefer a hard core representation for this repulsion. Indeed, the phase shift δ_0 plotted corresponds to a soft core repulsion, i.e. a repulsion with a smooth form and a finite magnitude.As the energy is increased, the effect of the repulsion decreases and the effective hard core radius r_c correspondingly shrinks. A similar situation is found to hold for the I=0 nucleon-nucleon system. At $T_{c.m.}$=10 MeV, we have again $r_c \approx 0.35$fm, but r_c now shrinks more rapidly with increasing energy, being only 0.22fm for $T_{c.m.}$=175 MeV. This rapid shrinking of r_c simply means that there is less repulsion due to quark antisymmetrization in the I=0 system than in the I=1 system, and this is to be expected, since the antisymmetrization constraints are more severe for the PP system than for the NP(I=0) system. We may also note here that, with Eq.(3.6), the excluded states may be defined as the eigenstates with eigenvalue unity, for the operator K; for the present case, Ribeiro has calculated the eigenvalues of K to be $1/3^n$, for integral n ≥ 1, so that there is no excluded state for the nucleon-nucleon system, in this model. To conclude, let us emphasize that Ribeiro's calculation does not pretend to give a realistic account of the nucleon-nucleon interaction, but its treatment of the short distance domain does suggest that this mechanism may well be the origin of the short-range repulsion in the nucleon-nucleon system.

4. BARYON SPECTROSCOPY WITH NON-CHARMED QUARKS

The SU(3) multiplets known for baryonic states are all $\underline{\alpha}$= $\underline{1}$, $\underline{8}$ and $\underline{10}$, or mix-

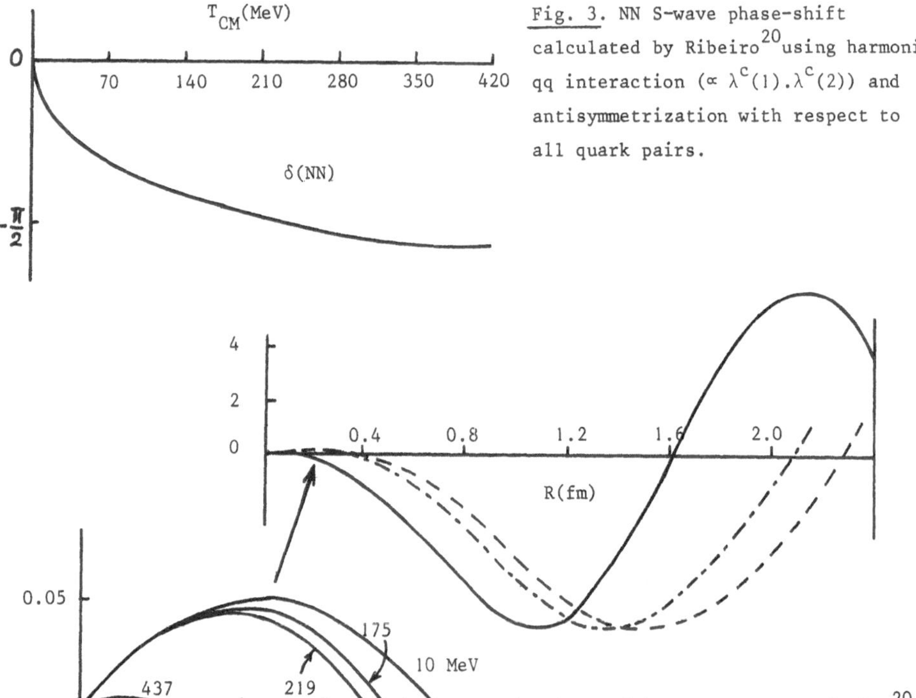

Fig. 3. NN S-wave phase-shift calculated by Ribeiro[20] using harmonic qq interaction ($\propto \lambda^c(1).\lambda^c(2)$) and antisymmetrization with respect to all quark pairs.

Fig. 4. $r\psi_s(r)$ calculated by Ribeiro[20] plotted vs. r for NN energies T_{CM} from 10 to 437 MeV. The small figure shows $r\psi(r)$ on an expanded scale, to show its smallness and the variation of the "effective hard core radius" with T_{CM}.

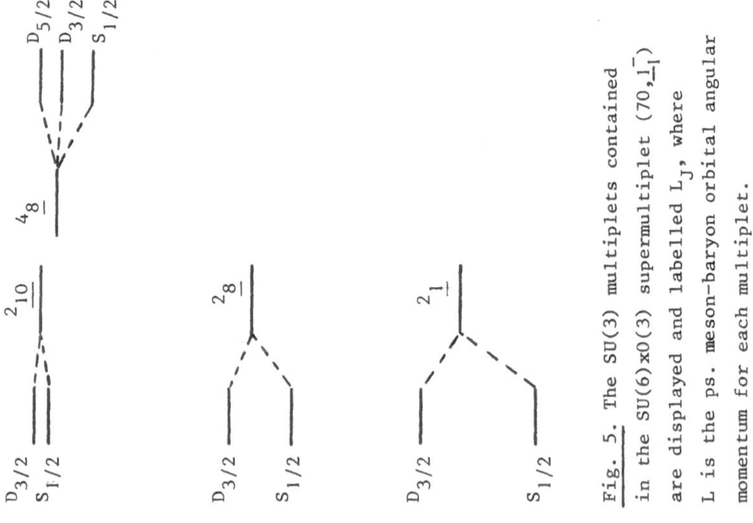

Fig. 5. The SU(3) multiplets contained in the SU(6)x0(3) supermultiplet $(70,\underline{1}^-)$ are displayed and labelled L_J, where L is the ps. meson-baryon orbital angular momentum for each multiplet.

tures of them, as expected in the 3q-model. They occur in groups of SU(3) multiplets with various $\underline{\alpha}$ and various J, but with the same parity P, and it has been possible to group these together to form SU(6)x0(3) supermultiplets. The parity P reflects the number of quanta of internal excitation n, being given by $P=(-1)^n$, and the total spin J is given by $\underline{J}=\underline{L}+\underline{S}$, where $\underline{L}$ corresponds to the 0(3) representation and $\underline{S}$ to the relevant element of the SU(6) representation considered.

The best known example is the ground state supermultiplet ($\underset{\sim}{56}$,L=0^+) consisting of the $1/2^+$ baryon octet and the $3/2^+$ decuplet, first proposed by Gursey and Radicati in 1964. Their mass formula was the following generalisation of the Gell Mann-Okubo mass formula:

$$M \;=\; M_0 + bY + c\,(I(I+1)-\tfrac{1}{4}Y^2) + dS(S+1), \tag{4.1}$$

where J=S, since L=0, and Y is the hypercharge (B+s)/2. In this formula the coefficient b can be deduced from $(N-\Xi)/2=-190$ MeV, the coefficient c from $(\Sigma-\Lambda)/2=$ 39 MeV, and the coefficient d from $(\Sigma^*-\Sigma)/3=64$ MeV, where the particle names have been used to stand for their mass values. The decuplet spacing is then predicted to be $-(b+c)=151$ MeV, which is in good accord with the observed value $(\Omega-\Delta)/3=147$ MeV. We note that the SU(3)-breaking parameters b and c are not small relative to the SU(6)-breaking parameter d.

The first-excited supermultiplet is ($\underset{\sim}{70}$,L=1^-). The value L=1 is expected from the shell-model excitation $1s\rightarrow 1p$; the L=1 space wavefunction has mixed permutation symmetry and so naturally occurs conjoined with the SU(6) $\underset{\sim}{70}$ representation, which also has mixed permutation symmetry, so that the full wavefunction (including the colour singlet factor) has the required antisymmetry. Since the $SU(6)\rightarrow SU(2)_\sigma xSU(3)$ reduction is here $\underset{\sim}{70}\rightarrow {}^2\underline{1}+{}^2\underline{8}+{}^2\underline{10}+{}^4\underline{8}$, this supermultiplet comprises the (SU(3),JP) multiplets shown on Fig.5, all of which are known empirically, at least from their N^*, Δ^* or Λ^* member. The gross separation of the multiplets due to SU(6)- and SU(3)-breaking is in accord with a spin-spin force of the same sign as is indicated for the ($\underset{\sim}{56}$,0^+) supermultiplet above and a quark mass difference (m_s-m) of about $(\Xi^*(1940)-N^*(1670))/2 \approx 135$ MeV, comparable with the value 147 MeV obtained from the L=0 decuplet (see above). The new feature of the spectrum is the evidence for spin-orbit coupling, whose effect is most clear in the mass difference between the $3/2^+$ and $1/2^+$ SU(3) singlet states, $\Lambda^*(1520)$ and $\Lambda^*(1405)$. The magnitude of the spin-orbit coupling differs from multiplet to multiplet in an unexplained way, unless one invokes spin-orbit couplings whose SU(3)-breaking terms dominate the SU(3)-invariant terms.

We now turn to consider the qq potential from the viewpoint of QCD, as envisaged by De Rujula et al.[23]. For large r, this potential is well approximated by the confinement potential V_{conf}, which may be taken to have some form such as (3.1) or (3.2). The region of small r is where asymptotic freedom should reign and where the coupling constant $\alpha_s=f^2(k_\mu^2)/4\pi$ is therefore small enough for the neglect of multi-gluon exchange terms in the qq potential; hence the potential is approximated here by the one-gluon-exchange potential V_g of the form (1.11) with $v_g(r)=\mathrm{Exp}(-m_g r)$, which may be taken as $v_g(r) \approx 1/r$, since $m_g r$ is small in the region where $v_g(r)$ is an adequate representation of V_{qq}. The form of the potential for intermediate r is not known and is therefore estimated by interpolating between these two limiting forms. The general form of the qq central potential V(r), given by

$$V(r) = V_{conf} + k\alpha_s/r \tag{4.2}$$

where $\alpha_s=f^2/4\pi$, and $k=\lambda^c(1)\cdot\lambda^c(2)=-2/3$ for the $\bar{3}$ qq system, which is that effective in a (colour-singlet) baryon. The expression (4.2) has made the tacit assumption that V_{conf} involves a vector coupling.

The reduction of the Dirac matrices in (1.11) leads to a spin-dependence in the qq potential:

$$V_{qq}=V(r)+\delta V(r)+\frac{3}{2m_1m_2}\left(\frac{1}{r}\frac{dV}{dr}\right)\{\underline{\ell}_{12}\cdot(\underline{S}_1+\underline{S}_2)\}+\frac{2}{3m_1m_2}\nabla^2 V\underline{S}_1\cdot\underline{S}_2+$$
$$\frac{1}{3m_1m_2}\left(\frac{d^2V}{dr^2}-\frac{1}{r}\frac{dV}{dr}\right)\left(\underline{S}_1\cdot\underline{S}_2-3\,\frac{\underline{S}_1\cdot\underline{r}\;\underline{S}_2\cdot\underline{r}}{r^2}\right), \tag{4.3}$$

where $\delta V(r)$ collects together some spin-independent correction terms, and the term $\underline{\ell}_{12}\cdot\underline{S}^+_{12}$ given is an approximation which will be sufficient for our present discussion. We note the following points:

(i) this interaction (4.3) is based on SU(3)-invariant couplings. The only SU(3)-breaking effects are those which result from the difference between the masses $m_u=m_d=m$ and m_s. These SU(3)-breaking terms generally occur with spin-dependent coefficients, so that they break SU(6)-symmetry at the same time. The pattern of symmetry breaking is therefore rather specific. Indeed, the only new parameter in V_{qq}, not already occurring in the SU(3) limit with the potential (4.2), is this mass difference (m_s-m). The fitting of the baryon mass data by V_{qq} therefore provides a serious test for it.

(ii) for $\ell_{qq}=0$, the only spin-dependence in V_{qq} is the hyperfine term, proportional to $\underline{S}_1\cdot\underline{S}_2$. Since $\nabla^2(1/r)=-4\pi\delta(\underline{r})$, the leading term at short distances is the Fermi contact term $(8\pi/3m_1m_2)\,\delta(\underline{r}_{12})\underline{S}_1\cdot\underline{S}_2$. The values of $\underline{S}_1\cdot\underline{S}_2$ are $-3/4$ for $S_{12}=0$ and $+1/4$ for $S_{12}=1$. Following De Rujula et al.[23], we consider the three states Σ^{*0}, Σ^0 and Λ of the $(\underset{\sim}{56},L=0^+)$ supermultiplet. Schematically, we can represent these states as follows:

		u d s	(3/8π) x hyperfine energy	
Σ^{*0}	I=1	(↑ ↑) ↑	$+\frac{1}{4}\left(\frac{1}{m^2}+\frac{2}{mm_s}\right)$	(4.4a)
Σ^0	I=1	(↑ ↑) ↓	$+\frac{1}{4m^2}+\frac{1}{mm_s}(\underline{S}_u+\underline{S}_d)\cdot\underline{S}_s = +\frac{1}{4m^2}-\frac{1}{mm_s}$	(4.4b)
Λ	I=0	(↑ ↓) ↑	$-\frac{3}{4m^2}$	(4.4c)

The energies given reflect the fact that the (ud) subsystem has spin 1 for Σ^{*0} and

Σ^0, but spin 0 for Λ; also that Σ^{*0} has spin 3/2, three parallel spins, while Σ^0 and Λ have spin $\frac{1}{2}$.

We note that the Σ^{*0} mass necessarily lies above those for Σ^0 and Λ, owing to the sign of the Fermi contact term. The mass difference between Σ^0 and Λ has the observed sign, as long as $m_s > m$. The relative magnitude of the mass differences $(\Sigma^*-\Sigma)$ and $(\Sigma-\Lambda)$ is correctly given for $m_s/m \approx 1.6$. With the value m chosen to fit the proton magnetic moment, we have the estimates

$$m = M_p/\mu_p = 0.34 \text{ GeV}, \qquad m_s = 0.54 \text{ GeV} \tag{4.5}$$

(iii) the $\ell_{qq}=0$ interaction is also effective in the submultiplets of the $(\underset{\sim}{70}, L=1^-)$ supermultiplet. Consider the N^* state from the $^4\underline{8}$ submultiplet and the Δ^* state from the $^2\underline{10}$ submultiplet, and neglect all spin-orbit and tensor interactions. Both states are (uud). Their space wavefunctions have mixed permutation symmetry, which means that they consist of two equally-probable terms $(\ell_{12}=0, \ell_3=1)$ and $(\ell_{12}=1, \ell_3=0)$, where ℓ_3 denotes the angular momentum of the third quark relative to the centre of mass for the quarks q_1 and q_2. The Fermi hyperfine interaction for (q_1, q_2) is effective only in the term $(\ell_{12}=0, \ell_3=1)$. The $^4\underline{8}$ state has all quark spins parallel, so that the total hyperfine energy is simply $3\times\frac{1}{2}\times(1/4m^2) = +3/8m^2$. In the $^2\underline{10}$ state, the quark pair (q_1q_2) necessarily has I=1 (since $I=\frac{3}{2}$ holds for all three quarks), so that their spins are necessarily parallel in the configuration with $\ell_{12}=0$, by the Pauli principle (the "symmetric quark model" with singlet colour!); hence the total hyperfine energy is again $3\times\frac{1}{2}\times(+1/4m^2) = +3/8m^2$. The hyperfine interaction therefore does not separate the mean masses for the $^4\underline{8}$ and $^2\underline{10}$ submultiplets. This corresponds well to the physical situation, where the mean Δ^* mass is about 1660 MeV and the mean N^* mass for the three $^4\underline{8}$ multiplets is about 1680 MeV.

(iv) it is a general property of the $\underset{\sim}{70}$ representation that any SU(3)-invariant $\underline{S}_1\cdot\underline{S}_2$ will give equal spacing to the mass values for the $^2\underline{10}$, $^2\underline{8}$ and $^2\underline{1}$ submultiplets, in this order. This is qualitatively correct for the non-strange mean masses; with $^2\underline{10}$ at about 1660 MeV and $^2\underline{8}$ at about 1530 MeV, the mean ^{2}1 mass would then be expected to be at about (1400+150) $\approx$ 1550 MeV, allowing 150 MeV for the difference in the strangeness value, which is not far from spin-average mass for ΛD03(1520) and ΛS01(1405).

(v) The central potential V(r), given by (4.2) and sketched in Fig.6, has dV/dr>0. Hence the $\underline{\ell}_{12}\cdot\underline{S}^+_{12}$ term of (4.3) places the state of highest spin highest in mass. For the $(\underset{\sim}{70}, L=1^-)$ supermultiplet, this is in accord with observation for the $(3/2^-, 1/2^-)$ isosinglet pair (ΛD03(1520),ΛS01(1405)), which has by far the largest mass splitting.

However, in detail, the spin-orbit interaction in (4.3) does not lead to the mass splittings and mixing effects observed for the nine submultiplets of the $(\underset{\sim}{70}, 1^-)$. As found by Horgan[22] and later workers[24], a fit to these data appear to require that the $\underline{\ell}_{12}\cdot\underline{S}^{+}_{12}$ interactions should be dominated by large SU(3)-breaking terms.

[Subsequent to these Lectures, Isgur and Karl[25] have drawn attention to the role of the tensor coupling in the interaction. They lay great stress on the relationship of the hyperfine and tensor interactions, as being both consequences of the exchange of a vector gluon, and they showed that a good fit could be obtained to both the mass splitting and mixing effects deduced from the data for the N^* states assigned to the $(\underset{\sim}{70},1^-)$ supermultiplet, by retaining these two interactions as given in (4.3), while allowing the strength of the spin-orbit coupling to vary freely. In fact, they obtained a good fit for zero spin-orbit coupling. A possible rationale for this procedure arises if the confining potential V_{conf} in (4.2) has the form of a scalar coupling, rather than a vector coupling, as was assumed above. In this case, only the interaction $V_g \approx k\alpha_s/r$ would contribute appreciably to the spin-spin and tensor hyperfine couplings. while a scalar V_{conf} would contribute to the spin-orbit coupling as follows:

$$-\frac{1}{2m_1m_2}\cdot\frac{1}{r}\frac{dV_{conf.}}{dr}\left\{\left(1+\frac{(m_1-m_2)^2}{2m_1m_2}\right)\underline{\ell}_{12}\cdot\underline{S}^{+}_{12}+\frac{m_1^2-m_2^2}{2m_1m_2}\underline{\ell}_{12}\cdot\underline{S}^{-}_{12}\right\} \tag{4.6}$$

where we have retained some correction terms because they are relatively large. We note also that this spin-orbit coupling has the opposite sign to that given for a vector V_{conf} in expression (4.3), so that there is the possibility of cancellation between the spin-orbit couplings derived from V_g and V_{conf}, which might perhaps account for the above observation of Isgur and Karl. This possibility was put forth by Schnitzer[26] for the case of the $\bar{q}q$ system, as we mention below. It is being explored further for the full $(\underset{\sim}{70},1^-)$ supermultiplet by Isgur and Karl[27] and by Reinders[28] in different ways, with the present conclusion that a tolerable fit is obtained to both the mass spectrum and to the mixing matrices for states with the same J^P and isospin, where these are reliably known.]

Much progress has also been made in the interpretation of higher supermultiplets, corresponding to N=2 and 3 quanta of excitation, although not yet on the basis of QCD, only phenomenologically. For N=2, the supermultiplets $(\underset{\sim}{56},2^+)$ and $(\underset{\sim}{56},0^+)_2$ are already well established [24]. However, the interpretation of the mass spectrum does not appear to require any spin-orbit forces for $\ell_{12}=2$, a situation possibly connected with the above paragraph. There has been controversy concerning the existence of the supermultiplets $(\underset{\sim}{70},2^+)_2$ and $(\underset{\sim}{70},0^+)_2$. There is one 3-star state, NP1l(1780), whose mass suggests that it belongs to an N=2 excitation and which would then require assignment to the $(\underset{\sim}{70},0^+_2)$ supermultiplet, and there are four 2-star states, NF17(1990), ΛF05(2110) and ΣP11(2080) which fit naturally into the $(\underset{\sim}{70},2^+_2)$ supermultiplet, and ΛP01(1800) which fits naturally into the $(\underset{\sim}{70},0^+_2)$ supermultiplet. Cashmore et al.[29] have suggested that these states might all belong to the N=4 excitations $(\underset{\sim}{56},4^+_4)$, $(\underset{\sim}{56},2^+_4)$ and $(\underset{\sim}{56},0^+_4)$ and have shown that the data on the branching ratios and decay widths are not incompatible with such assignments and the data known for other baryonic resonances which certainly belong to N=4 excitations. They concluded that the data on baryonic resonances is consistent with the view that $\underset{\sim}{56}$ representations occur only for even N and even L, and that $\underset{\sim}{70}$ representations occur only with odd N and odd L, as also did Hey[30] more recently. However, this view takes no account of the systematics of the mass values observed for baryonic resonances and assigns baryonic states with widely differing mass values to the N=4 excitations; for example, the state NP11(1780) is then assigned to $(\underset{\sim}{56},0_4^+)$, whereas the state ΔH311(2420) belongs to the $(\underset{\sim}{56},4_4^+)$ supermultiplet. On the other hand, the existence of the N=2 supermultiplets $(\underset{\sim}{70},2_2^+)$ and $(\underset{\sim}{70},0_2^+)$ would imply the existence of 14 SU(3)-multiplets, whereas the states mentioned above exemplify at most four of them. To understand this situation, much further study will be necessary; on the experimental side, to locate and identify further resonances in this mass region, and on the theoretical side, to predict which of these resonance states should appear prominently and which may prove difficult to find experimentally because of small branching ratios for the entrance and exit channels most accessible for experiment.

A few N=3 baryonic states are known. The 4-star state ΛG07 (2100) and the 3-star state NG17(2190) are naturally assigned to $(\underset{\sim}{70},3_3^-)$, the Regge rotational recurrence of the $(\underset{\sim}{70},1_1^-)$ supermultiplet, although this has not been demonstrated; $\underset{\sim}{70}$ supermultiplets with L=1 (twice) and L=2 are also predicted. Two $\underset{\sim}{56}$ supermultiplets are also predicted, with L=1 and 3. It is of interest to note that the mass values can be predicted for some of the Δ^* states of the latter, from sum rules connecting them with mass values for the N=0 and N=2 states. The reason is that the even permutation symmetry required for the space wavefunctions of $\underset{\sim}{56}$ supermultiplets limits ℓ_{12} to even values, and hence, for N=3, to the values $\ell_{12}=0$ and 2. For other N=3 states, ℓ_{12} can have any even or any odd value $\ell_{12}\leq 3$. However, the qq interactions for the states $\ell_{12}=0$ and $\ell_{12}=2$ have already been determined phenomenologically from the well-known N=0 and N=2 $\underset{\sim}{56}$ supermultiplets. This leads to the prediction that the $(\underset{\sim}{56},1_3^-)$ supermultiplet should lie lowest of all the N=3 supermultiplets, and this appears to be borne out by the analysis of πN scattering data by Cutkosky et al.[31] establishing a ΔD35 resonance with mass 1925±20 MeV, which they have assigned to this supermultiplet. The sum rules mentioned above actually lead to the prediction[32] of mass 2088±25 MeV for this ΔD35 state, somewhat higher than the mass value reported. They also predict mass 2300 MeV for the state ΔD39, characteristic of the $(56,3_3^-)$ supermultiplet.

5. MESON SPECTROSCOPY WITH NON-CHARMED QUARKS

The mesonic states are regarded as $(\bar{q}q)$ systems, with internal orbital angular momentum L and internal quark spin S=0 or 1, their total angular momentum being given by the vector sum $\underline{J}=\underline{L}+\underline{S}$. The parity of such a state is $P=(-1)^{L+1}$ and it has generalized charge conjugation parity $\mathcal{C}=(-1)^{L+S}$.

With QCD, the $\bar{q}q$ potential has the same structure as that given by (4.2) and (4.3). The confining potential $\bar{V}_{conf}$ is not necessarily the same as V_{conf}, and the coefficient $\bar{k}$ now appropriate is $\bar{k}= -4/3$.

For L=0, there are 36 distinct meson states, comprising the three spin states for each charge state of the nonet ($\rho K^* \bar{K}^* \omega\phi$) of vector mesons, the charge states of the octet ($\pi K\bar{K}\eta$) of pseudoscalar mesons, and one further pseudoscalar state, the $\eta'(958)$ meson. In so far as SU(6) symmetry holds, the mesonic states contained in this vector nonet and pseudoscalar octet form the basis of an SU(6)x0(3) representation denoted by ($\underset{\sim}{35}$,$L=0^-$), while the 36th state constitutes the representation ($\underset{\sim}{1}$,$L=0^-$). For L=0, the only spin-dependent interaction is the Fermi hyperfine interaction and this has the correct sign to place the vector states at higher masses than their pseudoscalar counterparts.

For the L=1 configurations, the spin-orbit and tensor terms can come into play, whereas the hyperfine interaction plays little role since its leading term is proportional to $|\psi(r=0)|^2$, which is zero for $L\geqslant1$. The configurations with S=1 lead to nonets with $J^P=0^{++}$, 1^{++} and 2^{++}; those with S=0 lead to a $J^P=1^{+-}$ octet and a predicted 1^{+-} singlet state, not yet identified. With SU(6) symmetry, these three nonets and the octet constitute the basis states for the SU(6)x0(3) representation denoted by ($\underset{\sim}{35}$,$L=1^+_1$), the suffix 1 denoting that the state has one quantum of excitation energy, while the 1^{+-} singlet state would form the representation ($\underset{\sim}{1}$,$L=1^+_1$). We list the identified states in the following Table.

J^P	(I=1,s=0)	($I=\frac{1}{2},s=\frac{1}{2}$)	(I=0,s=0)		Comment	L·S	T
2^{++}	A2(1310)	$K^*(1421)$	f(1271)	f'(1516)	Ideal Nonet	+1	−1/5
1^{+-}	B(1229)	$Q_B(\sim1390)$	?	-	Octet?	0	0
1^{++}	A1(1100)	$Q_A(\sim1240)$	?	D(1286)	Nonet?	−1	+1
0^{++}	δ(976)	K(1250)	$S^*(993)$	ε(1200)	Nonet?	−2	−2
(1^{+-})	-	-	?		Singlet?	0	0

The Q_A and Q_B mass values are those given by Bowler[33]. The 2^{++} states form an almost ideal nonet. The f(1271) decay is dominantly to $\pi\pi$, although the mode $K\bar{K}$ is known with branching ratio 3%, while the upper state f'(1516) is known only in the mode $K\bar{K}$; also the f(1271) lies only 40 MeV from A2(1310). The other multiplets are incompletely known, but sufficient states are known to establish their existence, assuming the validity of SU(3) symmetry. The full set of expected I=1 states is known, corresponding to the predicted quantum numbers. At the far right of the Table are given the coefficients with which the spin-orbit and tensor interactions contribute to the mass values. We note:

(i) the mass ratio (A2-A1)/(A1-δ) = 1.7, close to the value 2 expected for LS coupling alone. This allows rather little role for a tensor interaction. However, the A1 mass is ill-defined and controversial[34,35].

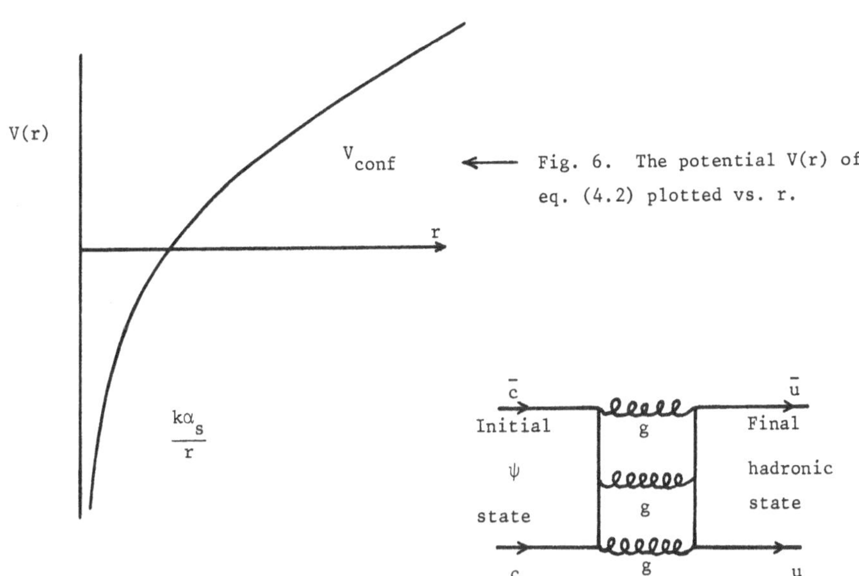

Fig. 6. The potential V(r) of eq. (4.2) plotted vs. r.

Fig. 8. Graph showing intermediate state ggg from $\psi = (\bar{c}c)$ to hadronic states.

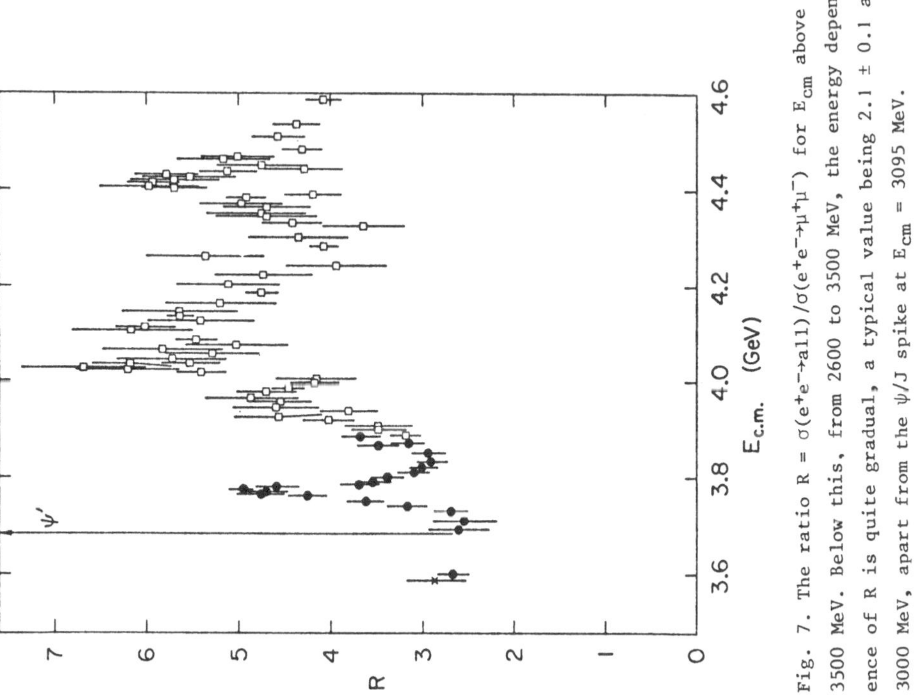

Fig. 7. The ratio $R = \sigma(e^+e^- \to all)/\sigma(e^+e^- \to \mu^+\mu^-)$ for E_{cm} above 3500 MeV. Below this, from 2600 to 3500 MeV, the energy dependence of R is quite gradual, a typical value being 2.1 ± 0.1 at 3000 MeV, apart from the ψ/J spike at E_{cm} = 3095 MeV.

(ii) the sign of the spin-orbit force from QCD would place the 2^{++} nonet at the highest mass, as is observed to be the case for the I=1 states and probably also for the s=±1 states.

Quite a number of mesonic states can be assigned to the configurations with N=2 quanta of excitation energy. These consist of the L=2 rotational excitation $(\underset{\sim}{35},2_2^-)$ and the first L=0 radial excitation $(\underset{\sim}{35},0_2^-)$. The L=2 and L=0 multiplets expected are now given, together with the established mesonic states having the appropriate quantum numbers.

L=2 Nonets	3^{--}	g(1680)	ω(1675)	L=0 Nonet	1^{--}	ρ'(1500)	ω'(1770)
	2^{--}	A3(1640)		Octet	0^{-+}	E(1416)?	
	1^{--}	ρ^*(1600)					
Octet	2^{-+}	?					

We note the following remarks:

(i) the two states assigned to L=0 are those established from their production in the electron-positron annihilation reaction,

$$e^+ + e^- \rightarrow "\gamma" \rightarrow \text{mesonic state } (J^P = 1^{--}, I=0 \text{ or } 1, s=0), \tag{5.1}$$

which proceeds through an intermediate one-photon state. Annihilation production of mesonic resonances through two-photon intermediate states is possible but will have a cross section at most $(\alpha/\pi)^2 \gtrsim 10^{-5}$ smaller, so that they can be neglected at present. The coupling $\gamma \rightarrow q\bar{q}$ is proportional to $\psi(r=0)$ for the state produced. Since $\psi(0)=0$ for L=2, it is natural to assign mesons produced through the reaction (5.1) to L=0 configurations. They necessarily have $J^P = 1^{--}$ and they are radial excitations of the ρ, ω and φ mesons. Above, we have labelled the observed states ρ' and ω' by one prime, since they are interpreted as the first radial excitations of the ρ and ω states. The states ρ^*(1600) and ρ'(1500) could well mix appreciably, since they are not far apart in mass, but this mixing can only be mediated by an interaction giving ΔL=2, i.e. a tensor interaction. There is no evidence for the occurrence of such mixing between these states.

(ii) the state E(1416) certainly exists, being known particularly for its decay modes $K\bar{K}\pi$ and $\eta\pi\pi$, the latter indicating $\mathcal{C}$ = + for its charge conjugation parity. Its spin-parity value is not yet determined, but the data are consistent with the values 0^-, 1^+, 2^-, etc. Its mass value is rather low (by more than 200 MeV) relative to the other 2^- mesonic states. It could perhaps be the missing 1^{++} meson in the $L=1^-$ table above, although its mass then appears rather high. Its assignment as 0^{-+} on the N=2, $L=0^-$ table above certainly appears the most plausible possibility, but there is no evidence to confirm this yet.

(iii) two SU(6)-singlet states are also expected for N=2, corresponding to the SU(6)xO(3) representations $(\underset{\sim}{1},2_2^-)$ and $(\underset{\sim}{1},0_2^-)$. The latter offers a further possibility for the assignment of E(1416), but no candidates are known for the state $(1_\omega, 2_2^-)$.

(iv) the I=1 mesonic states $\rho(773)$, A2(1310), g(1680) form a Regge trajectory with constant slope for $(\text{mass})^2$ vs. spin J. This is in fact the leading Regge trajectory for the I=1 mesons. It predicts mass 1980 MeV for the $J^{P\ell}=4^{++}$ I=1 meson. There have been reports[36] of an I=1 S(1930) meson, but this is not an established state. However, the I=0 $J^{P\ell}=4^{++}$ state h is firmly established, with mass 2040±20 MeV, which suggests the existence of an $L=4^{++}$ nonet in this mass region.

6. HADRONIC STATES WITH CHARMED QUARKS

The charmed quark, denoted by c, is assigned charge $Q_c = +2/3$ and has isospin I=0, being SU(3)-singlet; it is assigned the charm value C = +1. The charmed antiquark $\bar{c}$ has opposite quantum numbers Q = −2/3 and C = −1. The total charm of a system is then given by $C = N(c)-N(\bar{c})$. The attribute of charm was originally introduced by Bjorken and Glashow[37], as one possible means for the generalisation of SU(3) symmetry.

Its introduction today has two motivations:

(i) it allows the formation of 2 doublets of quarks,

$$\text{(a)}\quad (u, d\cos\theta_C + s\sin\theta_C) \qquad \text{(b)}\quad (c, -d\sin\theta_C + s\cos\theta_C) \qquad (6.1)$$

for the formation of weak interaction currents.[38] In these doublets (6.1), θ_C denotes the Cabibbo angle, known experimentally from the comparison of the strengths of the $\Delta s=\pm1$ leptonic decay modes (associated with the transitions $u \leftrightarrows s$ or $\bar{u} \leftrightarrows \bar{s}$) and the $\Delta s=0$ leptonic decay modes (associated with $u \leftrightarrows d$ or $\bar{u} \leftrightarrows \bar{d}$) with $\Delta Q = \mp 1$, for the semistable baryons and mesons. These doublets (6.1a,b) of the form (α,β) can each be used to construct a neutral current with the structure $(\bar{\alpha}J_\mu\alpha+\bar{\beta}J_\mu\beta)$. If the two neutral currents so formed from (6.1a) and (6.1b) have the same interaction strength, then their sum will have $\Delta s=0$ terms $u\to u$, $d\to d$, $s\to s$ and $c\to c$, but no $\Delta s=+1$ term $s\to d$. This cancellation is desirable because the experimental upper limits on the $\Delta s=\pm1$ neutral currents are extremely strong; for example, the $\Delta Q=0$, $\Delta s=\pm1$ decay $K_L^0\to\mu^+\mu^-$ has rate 0.2 sec^{-1}, whereas the $\Delta Q=-1$, $\Delta s=-1$ decay $K^+\to\mu^+\nu$ has rate 5×10^7 sec, more than 10^8 times faster than the former, whose measured rate can in fact be largely accounted for as a secondary effect involving the decay sequence $K_L^0\to\gamma\gamma\to\mu^+\mu^-$. The $\Delta Q=-1$ transitions which correspond to the doublet (6.1b) have the relative amplitudes

$$\text{(A)}\quad c\to d:\ -\sin\theta_C, \qquad \text{(B)}\quad c\to s:\ \cos\theta_C. \qquad (6.2)$$

The latter provides the dominant weak transition with $\Delta C=-1$, and involves the selection rule $\Delta s=-1$.

(ii) the ψ-family of mesons which have recently become established have a natural interpretation as $(\bar{c}c)$ bound states, known generally as the charmonium system. The mass observed for the lowest vector state, the ψ/J(3095) meson, suggests that the mass

m_c of the c quark should be greater than that for the (u,d) quarks by amount $(m(\psi/J)-m(\omega))/2$, leading to the estimate $m_c \approx 1.50$ MeV.

With the c quark and the ψ/J meson, the known nonet $(\rho\omega K^* \bar{K}^* \phi)$ of vector mesons must be extended by the following vector states, a D-doublet and an F-singlet with the structures

$$(D^+ = (\bar{d}c),\ D^0 = (\bar{u}c)),\quad F^+ = (\bar{s}c) \tag{6.3}$$

and their antiparticles $(\bar{D}^0, D^-)$ and F^-, in addition to ψ/J itself. This forms a total of 16 vector states with the structure $^3S_1(n=1)$. With QCD, we expect the $\bar{q}$-q interactions to be independent of quark type, except in so far as the non-relativistic reduction of these interactions introduces into them the mass values of the quarks involved. The latter vary so widely that there will be some gross violations of SU(4)-symmetry in the physical states. It is more useful to consider these systems as a set of 16 related states, rather than as belonging to the SU(4) representations $\underline{\underline{15}} + \underline{\underline{1}}$. Since the spin-dependence of the $\bar{q}q$ forces is of secondary importance, we may in fact consider the vector and pseudoscalar mesons as forming a set of 64 related states. With equal masses for the four quarks, these states would be based on the $(\underline{\underline{63}}+\underline{\underline{1}})$ representations of SU(8) symmetry, but the quark mass differences are so large that this SU(8) symmetry is badly broken, although the basic patterns of states still recognisably survive.

It would be desirable to have a unified treatment of the sixteen vector states, using the universal $\bar{q}q$ potential

$$V(q\bar{q}) = \bar{V}_{conf} + \bar{V}_g, \tag{6.4}$$

where $\bar{V}_g$ is given by the form (1.11). Since these $\bar{q}q$ systems are colour singlet, $\lambda^c(1)\cdot\lambda^c(2)$ takes the value $-4/3$, and we have again the form (4.2), but with V_{conf} replaced by $\bar{V}_{conf}$ and $k = -4/3$. However, a unified treatment would have to allow a relativistic treatment for one or both of the quarks, except for the $\bar{c}c$ system, in consequence of the low values for m and m_s. It is far from clear at present how the relativistic calculation should be carried out, for there are many ambiguities and difficulties, which we need not specify in detail here. In the meantime, non-relativistic methods can be used for the charmonium systems and that is where most calculational work has been centred, to date. Even the discussion of charmonium is not without its ambiguities, as we shall see.

For the $(\bar{c}c)$ system, we use the potential form obtained by making a non-relativistic reduction of the Dirac matrices $(\gamma_\mu(1)\gamma_\mu(2))$ with a radial potential $\bar{V}$ of the form

$$\bar{V} = (\bar{a}r+\bar{b})-(4/3)\alpha_s/r. \tag{6.5}$$

This leads to an expression $V_{q\bar{q}}$ of precisely the form (4.3), with V replaced by $\bar{V}$ and with $m_1=m_2=m_c$.

First, we consider the pattern of states in this potential. The ψ states are expected to be those which include a component of the state $n\,^3S_1$, since they are to be excited through the process (5.1). States dominantly of the structure $n\,^3D_1$ will generally have a $^3S_1$ component, since these are coupled by the tensor term in the interaction $V_{q\bar{q}}$ (cf. eq.(4.3)), and can therefore be directly excited. The lowest two states, $\psi/J(3095)$ and $\psi'(3685)$, are naturally identified with the n=1 and n=2 $^3S_1$-states and are therefore used as input data to determine the parameters of $\bar{V}$. As far as the rough location of the levels is concerned, the potential V_g is of secondary importance. A typical pattern of levels is that given by Kang and Schnitzer[39], as follows:

S	(3.095)		(3695)		4.20		4.58	4.95 GeV
P		3.456		3.964		4.40		
D			3.76		4.22		4.62	

Kang and Schnitzer used the parameters $\bar{a}$ = 0.30 GeV.f^{-1} and $\bar{b}$ = −1.72 GeV, with α_s=0; their quark mass was m_c=2.0 GeV. They also used the Klein-Gordon equation, but this is an unimportant refinement for the $c\bar{c}$ system. Eichten et al.[40] included V_g and took $\bar{a}$ = 0.2 GeV.f^{-1}, $\bar{b}$=0, with α_s = 0.2 and m_c = 1.6 GeV, fitting also the partial widths $\Gamma(\psi\to e^+e^-)$ observed for the ψ/J and ψ' states. Their calculation predicted 4.18 GeV for the n=3 S-state, 3.465 GeV for the n=1 P-state, and 3.765 GeV for the n=1 D-state, whence we see that these predictions do not depend critically on the parameters used.

The observed cross section ratio

$$R = \sigma(e^+e^-\to \text{all})/\sigma(e^+e^-\to\mu^+\mu^-) \qquad (6.6)$$

is plotted on Fig.7. The interpretation of the structure above $\psi'(3685)$ is not yet settled. It is natural to identify the sharp peak at 3.77 GeV with the n=1 D-state given by the above calculations. The broad hump from about 3.90 to 4.2 GeV appears to have appreciable structure, suggesting perhaps three separate resonance peaks within this mass range, and there is a well-defined peak at 4.41 GeV. Possibly the confining potential rises less rapidly than the linear form, in which case the higher states would have mass values lower than those predicted by Kang and Schnitzer.

The partial widths $\Gamma(\psi\to\ell^+\ell^-)$ give important evidence concerning the ψ states, since theyare related with the wavefunction $\psi(r)$ through the relation

$$\Gamma(\psi\to\ell^+\ell^-) = (16\pi\alpha^2/3)\ (Q_c/m_\psi)^2|\psi(0)|^2. \qquad (6.7)$$

The observed widths and the corresponding values deduced for $|\psi(0)|^2$ are as follows:

ψ mass (GeV)	3.095	3.685	3.77	4.0–4.2	4.41
$\Gamma(\psi\to\ell^+\ell^-)$ keV	4.8±0.6	2.1±0.3	0.37±0.09	∿1.8	0.44±0.14
$|\psi(0)|^2$ (GeV^3)x10^3	39±5	24±3	4.4±1.1	∿26	7.2±2

The low value for $\psi(3.77)$ is consistent with its interpretation as the n=1 $^3D_1$ state;

the value obtained requires about 20% admixture of the n=2 $^3S_1$ state. When α_s=0 in the potential $\bar{V}$, the use of a linear potential would predict the value $|\psi_n(0)|^2 = m_c\bar{a}/4\pi$, independent of n, as pointed out by Harrington et al.[41] and by Kang and Schnitzer[39]. The parameter values used by the latter lead to the value 48 for $|\psi(0)|^2$ in the units used above, which compares quite well with the value 39±5 observed for the ψ/J state.

It is of interest to compare these values for $|\psi(0)|^2$ with the values deduced for ρ(0.773), ω(0.781) and ϕ(1.020) from the empirical data for them[36]. In order to do this, we should first note that the coefficient Q_c^2=4/9 in eq.(6.7) should be replaced by 1/2, 1/18 and 1/9 for the cases of ρ, ω and ϕ. The values then obtained for $|\psi(0)|^2$, in the same unit as used for the ψ-mesons in the above table, are 2.9±0.3, 3.1±0.7 and 4.6±0.3, respectively. If a non-relativistic calculation were valid for the system ϕ=($\bar{s}$s), together with the same confining potential as used for ($\bar{c}$c), then the value $|\psi(0)|^2$ expected for ϕ(1.020) would be $(m_s/m_c) \approx 0.36$ times that for ψ(3.095), that is about 14, which is at least comparable with the empirical value of 4.6; the discrepancy remaining could well be due to the need for a relativistic treatment for the ($\bar{s}$s) states.

The hyperfine interaction in $V_{q\bar{q}}$ gives rise to a mass separation between the $^3S_1$ and $^1S_0$ state. The latter is usually identified with the X(2.830) state[42], observed in the transition sequence

$$\psi/J \rightarrow \gamma + X(2.830) \quad \llcorner\!\!\longrightarrow \gamma + \gamma \qquad (6.8)$$

This mass separation, ΔM = +265 MeV, appears unexpectedly large. The hyperfine term in V_g alone gives about +25 MeV. If V_{conf} is assumed to be of vector form, then it will also contribute to ΔM and the calculated value could become as large as +80 MeV[43]. We may note several other $^3S_1$-$^1S_0$ mass splittings, for comparison:

Mesonic pair	ρ-π	K^*-K	D^*-D (see below)
ΔM(MeV)	637	397	138
ΔM^2(GeV^2)	0.578	0.550	0.550

These data show an almost constant value for ΔM^2, on the other hand. If this value for ΔM^2 is used to estimate the mass for the $^1S_0$ state from the ψ/J mass, the result is 3.00 GeV, much closer to the ψ/J mass than that deduced from the γγ pairs in the decay mode (6.8).

With the Charmonium model, it is no surprise that γ-transitions should be observed following the excitation of the ψ'(3685) state, since the $\bar{c}$c calculations mentioned above predicted that its ^3P_J levels should lie between the n=1 and n=2 $^3S_1$ levels. The situation is summarized on Fig.10. The dominant ψ'→γχ, where the χ states are identified with the ^3P_J states of charmonium. The state χ(3450) shown on Fig.10 corresponds to very few observed events and has not yet been confirmed by later work.

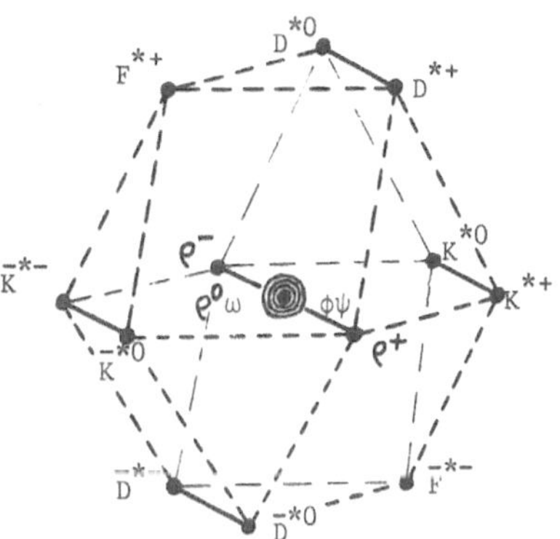

Fig. 9. The isospin multiplets for the 16 $^3S_1(\bar{q}_i q_j)$ vector states, where q_i= (u,d,s,c).

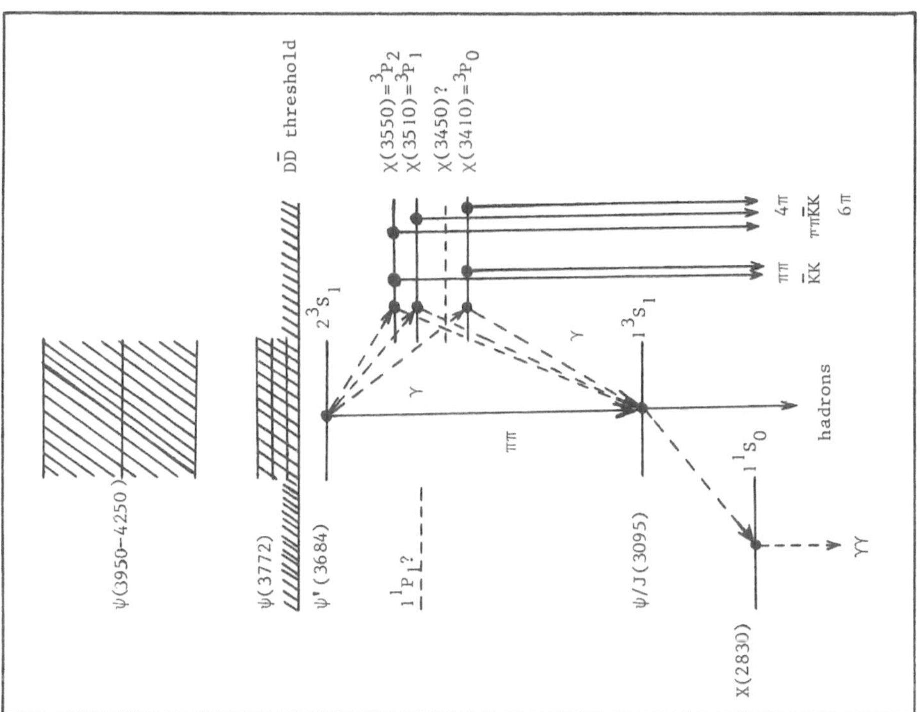

Fig. 10. The energy levels of Charmonium, showing the ψ and χ meson families and the known γ transitions between them. The other decay modes known for the χ states are indicated, together with the $(\bar{c}c)$ configurations with which they are now identified. The state X(2830) is also indicated, together with the γ transitions through which it has been observed; its identification is uncertain. We draw attention to the $D\bar{D}$ threshold, above which the ψ states have normal hadronic decay widths.

Hadronic decay modes have been observed from the other three χ-states, and the γ-transitions $\chi\to\gamma\psi/J$ have been observed from the upper two χ states. From the characteristics of these hadronic modes, and from the characteristics (angular correlations and relative rates) of the γ cascades $\psi\to\gamma\chi\to\gamma\gamma\psi/J$, it has been concluded that the χ states correspond to the configurations specified on Fig.9.

It is of interest to compare this P-state pattern with that known (cf. Sec.4 above) for the I=1 P states formed by the $(\bar{d}u)$ system, as follows:

$(\bar{d}u)$					$(\bar{c}c)$
	$A2(1310)^+$	$^3P_2$ — — —	$^3P_2$	$\chi(3554)$	
	$B(1229)^+$	$^1P_1$	$^3P_1$	$\chi(3510)$	
	$A1(1100)^+$	$^3P_1$ — — — ↗			
	$\delta(976)^+$	$^3P_0$ — — —	$^3P_0$	$\chi(3413)$	

The $^1P_1$ χ-state has not yet been observed. The γ-transition to it from the ψ' is forbidden by charge-conjugation parity and it is not yet clear how else it might be formed. For the $(\bar{d}u)$ systems, the spacing between the levels has been cited above (cf. Sec.4) as evidence that the $\bar{q}q$ non-central interaction is dominantly of spin-orbit character. For the $(\bar{c}c)$ system, this no longer appears to be the case. If there are both LS and tensor interactions then the spacing ratio is given by

$$R_1 = (^3P_2 - {}^3P_1)/(^3P_1 - {}^3P_0) = 2(a_{LS}-a_T)/(a_{LS}+5a_T), \qquad (6.4)$$

where a_{LS} and a_T are constants which characterize the strengths of these interactions.

However, Schnitzer[44] has pointed out that the LS and tensor interactions due to V_g cannot lead to the empirical value $R_1= +0.45$, for any choice of the parameters available, but he has shown[45] that it is possible to achieve this value for R_1 by adding an anomalous magnetic moment term to the quark-gluon interaction, as a purely ad hoc assumption. Schnitzer has also pointed out[45] that a scalar component in the confining potential would not add to the tensor interaction but would contribute to the spin-orbit interaction with the opposite sign; the net magnitude of a_{LS} would then be reduced and with it also the calculated value for the ratio (6.4). Other differences between the $(\bar{c}c)$ system and the $(\bar{d}u)$ system may then be attributed to the relativistic nature of the latter, as a question still to be settled.

It is of interest to note that the total spread δ of the fine structure for the $(\bar{c}c)$ system, $\delta=(E(^3P_2)-E(^3P_0))\approx 140$ MeV, is of the same order-of-magnitude as that (≈ 334 MeV) known for the $(\bar{d}u)$ system, in contrast with the situation noted above for the hyperfine structure. The observed total spread is readily obtained in calculations for the $(\bar{c}c)$ system; for example, according to Schnitzer[44], the potential of Eichten et al.[40] leads to δ=156 MeV, while that of Kang and Schnitzer[39] leads to δ=150 MeV.

We turn to discuss the partial widths for ψ/J and ψ' to decay to hadrons. These are remarkably small, being about 45 keV for ψ/J and about 100 keV for ψ'; in fact, these small widths were the most astonishing aspect of what emerged in the discovery and exploration of these new ψ phenomena. They are usually attributed now to the operation of the Zweig rule which allowed only final states whose mesons include the c and $\bar{c}$ quarks initially present in the ψ particle. However, the Zweig rule is required to be valid to rather a high degree of accuracy; to underline this, we recall that the decay process

$$\psi/J \to "\gamma" \to \text{hadrons}, \tag{6.5}$$

mediated by an intermediate photon, has a rate given by a partial width of about 12 keV, as can be deduced from the study of the ratio (hadrons)/$(\mu^+\mu^-)$ in e^+e^- interactions off the ψ/J resonance energy (in which case, both final states result only from an intermediate photon state) and the partial width $\Gamma(\psi/J\to\mu^+\mu^-)$, since this last process is also necessarily mediated by an intermediate photon. This width is only four times smaller than the partial width $\Gamma(\psi/J\to\text{hadrons})$ = 46 keV deduced for its direct decay to hadronic states, deduced from the total width $\Gamma(\psi/J\to\text{all})$ = 67±12 keV by subtracting the partial widths for the process (6.5) and for the leptonic decay modes $\psi/J\to\ell^+\ell^-$. It is of interest to note here that these direct hadronic decays are believed to obey isospin conservation, as well as SU(3)-symmetry to a lesser degree.

Since gluons couple strongly with quarks, and since gluons are neutral vector particles as are photons, why should the processes "$\psi/J\to$hadrons" not proceed much more rapidly than the process (6.5), through intermediate gluon states? The answer is believed to lie in the fact that gluons are the quanta of a non-Abelian colour gauge theory, and so have the property of Asymptotic Freedom, whereas photons are the quanta of an Abelian gauge theory, which does not have this property.

Let us consider this question for the ψ/J state. It cannot transform to a single intermediate gluon g, since this is colour octet, whereas ψ/J is colour singlet. It cannot transform to two gluons; when the state gg is colour singlet, it is even under $\mathcal{C}_g$="g-conjugation", whereas the colour singlet $^3S_1(\bar{c}c)$ state is odd under $\mathcal{C}_g$. Hence, the simplest intermediate gluon state allowed for ψ/J decay is ggg, as depicted on Fig.9. Since we are dealing with annihilation processes involving a large release of energy and momentum, the quark-gluon coupling constant α_s then appropriate is relatively small, so that it is reasonable to assume that the ggg intermediate state will also be the dominant contributor to the amplitude for $\psi/J\to$hadrons; the contributions from more complicated intermediate states will involve higher powers of α_s and will be correspondingly depressed. On this basis, we make a rough estimate for the ψ/J hadronic decay width as

$$\Gamma(\psi/J\to\text{hadrons}) \approx (\alpha_s^3)^2\ \Gamma_{\text{had.}} \approx 100(\alpha_s^3)^2 \text{ MeV}, \tag{6.6}$$

taking $\Gamma_{\text{had}}\sim$100 MeV as a typical decay width for normal allowed hadronic processes (such as $\rho\to\pi\pi$, $A2\to\rho\pi$, etc.). This estimate (6.6) can be in accord with the empirical value of about 45 keV, if we have $\alpha_s \approx 0.3$, a value which is still rather large relative to the electromagnetic coupling constant $\alpha \sim 1/137$. It is useful here to estimate roughly the decay width for the process (6.5) in the same spirit. The result obtained is

$$\Gamma(\psi/J \to "\gamma" \to \text{hadrons}) \approx \alpha^2 \Gamma_{had} \approx 5 \text{ keV}, \tag{6.7}$$

which is quite comparable with the empirical value of 12 keV, thus giving more credence to the line of argument followed here. Finally, we remark that these estimates imply that the Zweig rule concerning disconnected graphs must hold rather accurately since a $\psi-(\phi,\omega)$ mixing angle $\varepsilon(\psi)$ of as much as 1° could well contribute as much as 20 keV to the total width $\Gamma(\psi/J \to \text{hadrons})$. In the ψ state, the admixture of ϕ and ω components to the basic configuration $(\bar{c}c)$ can therefore have intensities of only less than about 0.05%.

Next, we consider the γ-transitions

$$\psi' \to \gamma + \chi(^3P_J) \to \gamma + (\gamma + \psi/J). \tag{6.8}$$

The rates observed for the first step in these cascades are completely normal (within a factor of 2) for "atomic E1 transitions" of a $(\bar{c}c)$ system, provided that the χ states have the spin values assigned to them on Fig.10. The observation of the decay of the $\chi(3.55)$ and $\chi(3.41)$ states to $\pi\pi$ and $\bar{K}K$ requires them to have even spin and parity, since they are isospin singlet states; the angular distribution observed for the γ-ray emitted in the transition $\psi' \to \gamma\chi(3.41)$, relative to the initial e^+e^- axis, corresponds to that $(1+\cos^2\theta)$ expected for J=0, while the isotropy observed for the γ-ray from $\psi' \to \gamma\chi(3.55)$ is not in disagreement with the expectation $(1+1/13\cos^2\theta)$ for J=2. The γ-decay widths are as follows:

$\psi' \to \gamma\chi$	$\chi(3.41)$	$\chi(3.45)$	$\chi(3.55)$
$\Gamma_{obs.}$ (keV)[46]	17±6	21±7	18±7
$\Gamma_{calc.}$ (keV)[47,48]	24(12)	35(14)	42(28)

where the $\Gamma_{calc.}$ given in brackets are the values calculated in a rather detailed treatment by Eichten et al.[48] which takes into account the coupling of these bound states with the open channels which exist not far above them.

For the second step in (6.8), the γ-decay is observed to occur in competition with hadronic modes of decay. These decay widths have also been calculated by Eichten et al.[48] in their detailed treatment, with the results

	$\chi(3.41)={}^3P_0$	$\chi(3.50)={}^3P_1$	$\chi(3.55)={}^3P_2$
$\Gamma_{calc.}(\chi \to \gamma\psi/J)$ keV	90	230	320
Exptl. branching fract$^{\underline{n}}$ for γ	3±3%	25±10%	12±7%

These results were surprising, since Γ_{had} is expected to be proportional to $|\psi(0)|^2$ and $\psi(0)=0$ for P-states. However, there will also be terms in Γ_{had} proportional to $|\underline{\nabla}\psi(0)|^2$ and we must conclude that the latter correspond to rather strong interactions (we may recall that for the $\pi^- P$ system, the strongest interaction processes are in the P-wave - for the S-wave $\pi^- P$ system, hadronic absorption (to $\pi^0 N$) and radiative transitions (to γN) are in competition (vide the Panofsky ratio 1.5)). If correct, the entries just above imply that the hadronic decay widths $\Gamma(\chi\to$had.) are 3±3 MeV, 0.7±0.3 MeV and 2.2±1.3 MeV for $\chi(3.41)$, $\chi(3.50)$ and $\chi(3.55)$, in turn. These P-state widths for the $(\bar{c}c)$ system are between one and two orders of magnitude larger than those observed for the S-state systems ψ/J and ψ'. This is partly comprehensible in terms of Asymptotic Freedom. The states $\chi(3.41)$ and $\chi(3.55)$ have $J^{P\mathcal{C}}=0^{++}$ and 2^{++}, respectively, and these allow the intermediate states gg. Corresponding to our rough estimate (6.6), we have here the estimate

$$\Gamma(\chi(^3P_0 \text{ or } ^3P_2)\to\text{hadrons}) \approx (\alpha_s^2)^2\Gamma_{had} \approx 100(\alpha_s^2)\ \text{MeV}, \tag{6.9}$$

neglecting any centrifugal barrier factor for J=2, which takes the value ≈1 MeV for the choice $\alpha_s \approx 0.3$ already indicated above. The $^3P_1$ state has the right g-conjugation parity $\mathcal{C}_g$=+ to allow a transition to the intermediate state gg. However, this transition is inhibited by an old selection rule first proved by Yang[49], forbidding a massive vector particle to transform into two neutral vector particles, each with transverse polarization. Since the possibility of longitudinal polarization depends on having $k(g)_\mu^2 \neq 0$ for at least one of the gluons, the transition amplitude $\bar{c}c(^3P_1)\to$ "gg"→hadrons depends considerably on the details of the final state reached and is therefore difficult to estimate. It is reasonable to expect some suppression in the hadronic rate for the $^3P_1$ χ-state relative to that for the $^3P_0$ and $^3P_2$ χ-states but it is difficult to say by how much. The value $\Gamma(\chi(3.50)\to$had.) deduced above does not appear unreasonable, in view of these remarks.

If the state X(2.83) is identified with the $^1S_0(\bar{c}c)$ configuration, usually denoted by η_c, the transition rate for $\psi/J\to\gamma X$ can be calculated directly as a $^3S_0\to{}^1S_0$ M1 spin-flip transition for the $(\bar{c}c)$ system. Assuming that the c quark has no anomalous magnetic moment, the calculated value is $\Gamma(\psi/J\to\gamma\eta_c)$=30 keV. In fact, an upper limit of 1.5 keV follows already from the MPSSSD experiment[50]. The DASP Group[51] have obtained the branching ratio

$$(X\to\gamma\gamma/X\to\text{all})\ \text{x}\ (\psi/J\to X\gamma)/(\psi/J\to\text{all}) = 1.4\pm0.4\text{x}10^{-4}, \tag{6.10}$$

although their observations have not yet been confirmed by other groups. Assuming X(2.83) is the η_c meson, a rough estimate of about 0.05 can be obtained for the first factor by estimating $\Gamma(\eta_c\to\gamma\gamma)$ by scaling up from the known width $\Gamma(\eta\to\gamma\gamma)$, and by estimating the rate $\Gamma(\eta_c\to$hadrons) from the expression (6.9) (since the transition $\eta_c\to$gg is allowed). This leads to the empirical estimate

$$\Gamma(\psi/J\to\eta_c\gamma) = 69\text{x}(2.8\pm0.8)\text{x}10^{-3}\ \text{keV} = (0.2\pm0.06)\ \text{keV}, \tag{6.11}$$

which is two orders of magnitude below the calculated rate. On the other hand, if the value $\Delta M \approx 25$ MeV calculated from V_g were correct, there would be no problem; the calculated branching ratio for $\psi/J\to\eta_c\gamma$ would then be only $3\text{x}10^{-4}$, and further, the detection of such a low energy γ-ray would be much more difficult. In this case, the state $\chi(2.83)$ would require some other identification. Confirmation of the result (6.10) is very desirable.

The discovery of the first D-meson was reported by Goldhaber et al.[52] in June, 1976. The present situation concerning the charmed mesons is as follows:

(i) the mass values given by Barbaro-Galtieri[53] for the D and D^* mesons, and by Yamada[51] for the F and F^* mesons, are given in MeV by the following table.

$J^P=0^-$	D^+ :	1868.3±0.9,	D^0 :	1863.3±0.9,	F^+ :	2030±60
1^-	D^{*+}:	2008.6±1.0,	D^{*0}:	2006.0±1.5,	F^{*+}:	2140±60.

We note that the Q-value for the strong decay $D^{*0}\to\pi^0D^0$ is only 7.7±1.7 MeV, while the decay $D^{*0}\to\pi^-D^+$ is probably energetically forbidden (Q=-1.9±1.7 MeV). The Q-values for the decay modes $D^{*+}\to\pi^+D^0$ and π^0D^+ are each about 5 MeV. The mass difference $(F^{*+}-F^+)$ is measured to be 120±40 MeV; the F^* meson is actually identified from the γ-ray emitted in the decay mode $F^{*+}\to\gamma F^+$.

(ii) the D-mesons are observed to decay to both $\bar{K}\pi$ and $\bar{K}\pi\pi$ final states, as well as to many others. Several individual branching ratios of interest are as follows[53]:

$D^0\to$	$K^-\pi^+$	2.2±0.6%	$D^+\to$	$\bar{K}^0\pi^+$	1.5±0.6% .
	$\bar{K}^0\pi^+\pi^-$	4.0±1.3%		$K^-\pi^+\pi^+$	3.9±1.0%
	$X^-_{\bar{K}}e^+\nu$	7.2±0.6%		$X^0_{\bar{K}}e^+\nu$	7.2±0.6%

The notation $X_{\bar{K}}$ means that all hadronic final states are included and that these are believed to include always a $\bar{K}$ meson; the final states $X_{\bar{K}}$ are dominantly $\bar{K}$ and $\bar{K}^*$. The final state $\bar{K}\pi$ necessarily has spin-parity limited to the series $0^+,1^-,2^+,..$ The phase space plot for the $K^0_s\pi^+\pi^-$ final state from D decay is rather uniform[54] and show none of the marked characteristics (in particular, zero intensity on the boundary of the plot) expected to hold if the $\bar{K}\pi\pi$ system had natural spin-parity 1^-, 2^+,...; for J=0, the $\bar{K}\pi\pi$ system necessarily has negative parity. The conclusion is that, whatever the spin value for D, the $\bar{K}\pi$ and $\bar{K}\pi\pi$ final states necessarily have opposite parity. Hence, parity is violated in the D-meson decay interactions and its decays therefore occur through weak interactions. The $\bar{K}\pi\pi$ distributions observed are quite consistent with J=0 and completely inconsistent with J=1.

(iii) the observation of the weak decay mode

$$D^+\to K^-\pi^+\pi^+ \tag{6.12}$$

is particularly important because it shows clearly that the selection rule $\Delta s=-1$ is associated with the weak transition $\Delta C=-1$, in accordance with the model of Glashow et al.[38] in which the dominant quark transition for this decay is $c\to s$ coupled with "Vac.$\to u\bar{d}$". There does exist the possibility of a quark transition $c\to d$ coupled with "Vac.$\to u\bar{s}$", which would lead to the decay mode $D^+\to K^+\pi^-\pi^+$. The model of Glashow et al. predicts that these latter transitions should have rate smaller by a factor $(\tan\theta_C)^4$

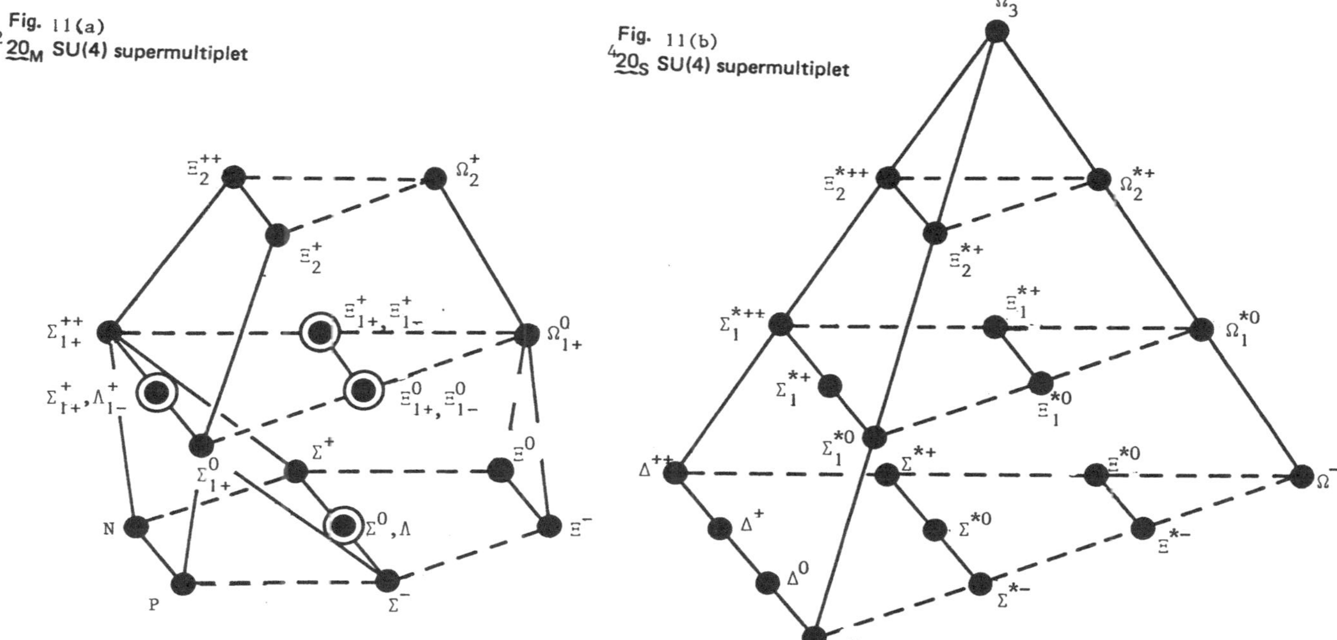

Fig. 11 displays all the charge states for the two SU(2) xSU(4) multiplets which comprise the lowest SU(6)x0(3) baryonic supermultiplet $(120,0_0^+)$, for (a) the $\underline{20}_M$ multiplet with spin-parity $(1/2^+)$ and (b) the $\underline{20}_S$ multiplet with spin-parity $(3/2)^+$. The four horizontal planes are for charm C = 0, 1, 2 and 3, and the system of naming the C > 1 states is discussed in the text. The axes in each of these planes denote isospin component I_3 and strangeness s. The bases of (a) and (b) reproduce the octet and decuplet patterns well known for the non-charmed baryons.

3×10^{-3} than the rate for $D^+\to K^-\pi^+\pi^+$, and indeed they have not yet been observed. There are also the possibilities of (a) $c\to s$ coupled with "Vac.$\to u\bar{s}$", and (b) $c\to d$ coupled with "Vac.$\to u\bar{d}$", predicted to have intermediate rates ($\sim\tan^2\theta_C$), but these modes do not have such distinctive signatures.

(iv) the total rate for the leptonic decay modes can be estimated simply, as being given by[55]

$$\Gamma(\text{charm}\to\ell\nu_\ell+\text{hadrons}) = f(z)G_F^2 m_c^5/192\pi^3, \qquad (6.13)$$

where the blocking factor $f(z=m_s/m_c) = 0.45$ for $m_c = 1.65$ GeV and $m_s = 0.55$ GeV. This estimate gives the rate $2\times10^{11}\text{sec}^{-1}$ for $D^+\to e^+\nu_e$ + hadrons; the measured branching ratio of 7.2% then leads to the estimate of 4×10^{-13}sec for the D-meson lifetime.

We turn now to consider briefly the charmed baryons. Again, only colour singlet configurations are considered as being relevant to the low mass region now available to experiment. The space wavefunctions for the three-quark systems are classified by their permutation symmetry S, M and A, as before, and these correspond to $\underset{\sim}{120}$, $\underset{\sim}{168}$ and $\underset{\sim}{56}$ representations for SU(8). The supermultiplets are now classified in terms of SU(8)xO(3). We shall confine our attention to the ground state supermultiplet $(\underset{\sim}{120},L=0_0^+)$, the extension of the SU(6)xO(3) supermultiplet $(\underset{\sim}{56},0_0^+)$. The $SU(2)_\sigma$xSU(4) content of this 120 supermultiplet is $({}^4\underline{\underline{20}}_S+{}^2\underline{\underline{20}}_M)$, where the suffix denotes the permutation symmetry of the SU(4) state; this suffix is redundant in the present case, because the S symmetry of the $\underset{\sim}{120}$ representation requires it to be the same as the permutation symmetry of the spin wavefunction (which is S for S=3/2 and M for S=1/2). All the substates of these two SU(4) representations are displayed on the three-dimensional Figures 11(a,b), which emphasize the high degree of symmetry which connects the properties of all these states in the limit where $m_c=m_s=m_u=m_d$.

Using the QCD potential V_{qq} with the parameters deduced from the discussion of the non-charmed baryonic states, and a charmed quark mass $m_c \approx 1.69$ GeV, De Rujula et al.[23] have calculated the masses of all the charmed members of the $(\underset{\sim}{120},0_0^+)$ supermultiplet. We now list all of the C≥1 states, naming them according to an extension of the notation of Lichtenberg[56].

s=0	$\Lambda_{1-}(2200)$	$\Sigma_{1+}(2360)$	$\Sigma_1^*(2420)$	$\Xi_2(3550)$	$\Xi_2^*(3610)$	$\Omega_3^*(4810)$
s=-1	$\Xi_{1-}(2420)$	$\Xi_{1+}(2510)$	$\Xi_1^*(2560)$	$\Omega_2(3730)$	$\Omega_2^*(3770)$	
s=-2	$\Omega_{1+}(2680)$	Ω_1^* (2720)				

In this notation, the number subscript is the charm C of the state. Its isospin I is specified by the Greek symbol used, being equal to the isospin of the non-charmed state specified by that symbol; for example, I=1 holds for the states Σ_{1+} and Σ_1^*, as for the non-charmed states Σ_0 and Σ_0^* already well known, and I=0 holds for the C=1 states Λ_{1-}, Ω_{1+} and Ω_1^*, the C=2 states Ω_2 and Ω_2^*, and the C=3 state Ω_3, as for the non-charmed states Λ_0 and Ω_0. The strangeness s for the state is then given by

$s=(s_0+C)$, where s_0 is the strangeness of the non-charmed state specified by the same Greek symbol; for example, the state Σ_{1+} has strangeness $s=(-1+1)=0$, as indicated on the display just above, while Ω_2 has strangeness $s=(-3+2)=-1$. The substates of each isospin multiplet then have the charge values

$$Q= I_3 + \frac{1}{2}(C+s+B), \qquad (6.13)$$

for $I_3=-I,-I+1,\ldots +I$. We have distinguished the S=3/2 charmed baryons by marking them with an asterisk; these states all belong to ${}^4\underline{\underline{20}}_S$ configuration of Fig.11(b). For the ${}^2\underline{\underline{20}}_M$ configuration depicted on Fig.11(a), there occurs two Ξ_1 multiplets. They belong to different SU(3) representations, so we distinguish them by adding a suffix ±, the + sign being for the state belonging to the $\underline{6}$ representation, the - sign being for the state belonging to the $\underline{\bar{3}}$ representation.

The unfortunate aspect of this notation is that it is violated by the notation we already use for the non-charmed baryons, specifically in that the $J^P=3/2^+$ state Ω(1672) is not denoted by the symbol Ω^* as this notation requires for spin-quartet states. We note that there is a $J^P=1/2^+$ state Ω_{1+} in the $\underline{6}$ representation within the ${}^2\underline{\underline{20}}_M$ multiplet, so that the distinction is logically necessary. Given this situation, it would be better to drop the asterisk notation completely, and to distinguish all states by using (2S+1) as an upper prefix, as we have distinguished the two $\underline{\underline{20}}$ baryonic representations which occur. This notation would involve some redundancy in a few cases, such as ${}^4\Delta_0$ or ${}^4\Omega_0$, but the alternative would be to rename the S=3/2 states using Greek letters different from those used for the S=1/2 states. We shall not go further into this matter here.

For the states of the $(\underline{120},0_0^+)$ supermultiplet, the mass values calculated by De Rujula et al.[23] range widely, from 939 MeV for the nucleons to 4810 MeV for the triply-charmed state Ω_3^{*++}. In order to emphasize the great lack of symmetry this means, all of the isospin multiplets of this supermultiplet have been plotted on Fig.12 according to their calculated mass values, together with all of the non-charmed baryonic states which have been established with mass values less than 2000 MeV. There will be charmed counterparts to all of the excited C=0 baryonic multiplets, so that there will be a rich spectroscopy to be found for each $C\geq 1$ a little above the masses calculated for the $(\underline{120},0_0^+)$ supermultiplet states and plotted on Fig.12.

Several baryonic states with charm C=1 have been reported from experiment already. The first information came from a neutrino-induced event observed by Cazzoli et al.[57] early in 1975 in a hydrogen bubble-chamber exposed to a neutrino beam with peak energy near 2 GeV from the AGS accelerator at Brookhaven National Laboratory. The event was fitted as

$$\nu_\mu + P \to \mu^- + \Lambda + \pi^+ + \pi^+ + \pi^+ + \pi^- \qquad (6.14)$$

with (Λππππ) mass 2426 MeV. Now that charm is established from the SPEAR experiments, there is little doubt that this represents the excitation of a charmed baryon through the weak interaction. The mass 2426 MeV lies close to 2420 MeV, the mass calculated

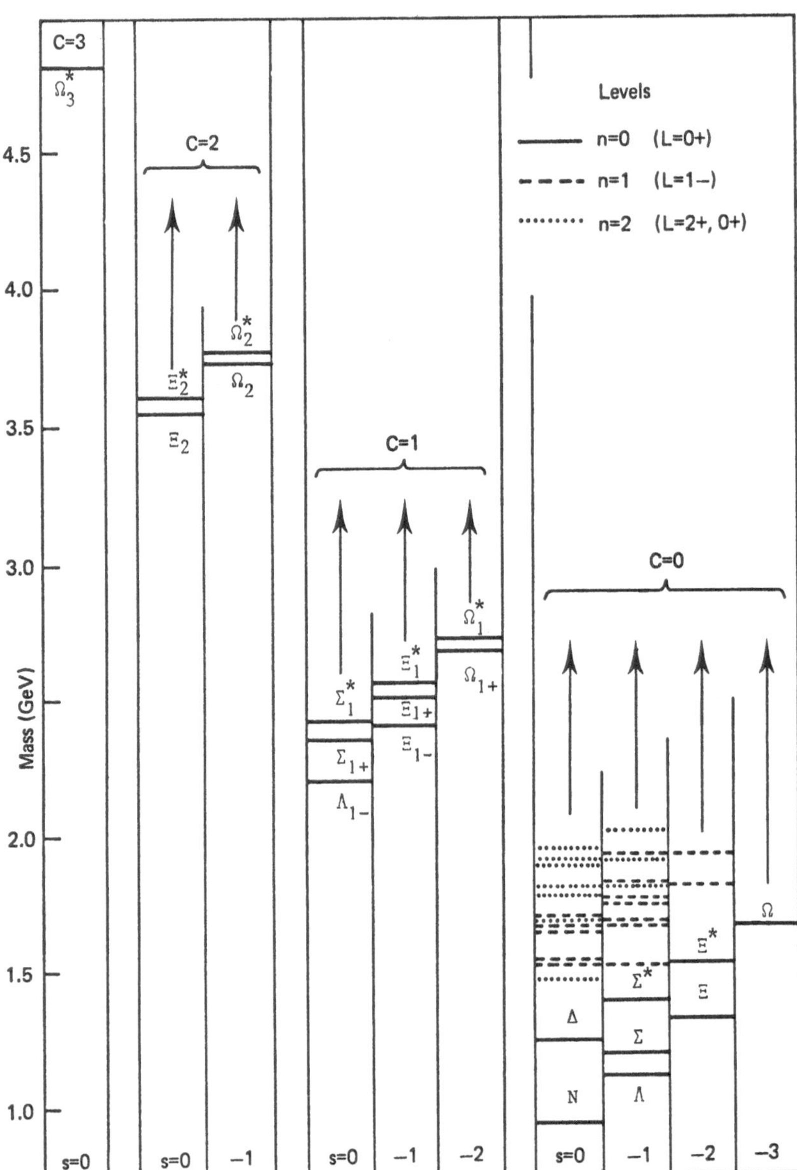

Fig. 12. The mass distribution of all the baryonic isospin multiplets is given as function of C and S. For C = 0, all the established states below 2.0 GeV are shown. For C ⩾ 1, the isospin multiplets shown are only those belonging to the $(\underset{\sim}{120}, 0_0^+)$ supermultiplet, the masses assigned to them being the values calculated by De Rujula et. al.[23] The notation used for the states is explained in the text.The states C ⩾ 1 marked with an asterisk have J = 3/2; the subscript + indicates a 6-representation of SU(3), the subscript - indicates a $\bar{3}$-representation.

for Σ_1^* by De Rujula et al. (see above Table). It is interesting to note that the event would then be interpreted as a transition from the P state, with spin-parity $1/2^+$ and in the $^2\underline{\underline{20}}_M$ multiplet, to the Σ_1^{*++} state, with spin-parity $3/2^+$ and in the $^4\underline{\underline{20}}_S$ multiplet,

$$\nu_\mu + P \rightarrow \mu^- + \Sigma_1^{*++}, \tag{6.15}$$

since this is a $\Delta C=+1$ transition of exactly the same class as is the $\Delta C=0$ reaction $\nu_\mu P \rightarrow \mu^- \Delta^{++}$ which occurs so strongly for $\nu_\mu P$ interactions at much lower energies. One of the three $(\Lambda\pi^+\pi^+\pi^-)$ masses for this event (6.14) has value 2260 MeV, which is not far from the mass of 2200 MeV calculated for Λ_{1-}^+, the lowest state among the C=1, s=0 baryons. If we accept this identification, then the first decay transition would be interpreted as

$$\Sigma_1^{*++} \rightarrow \pi^+ + \Lambda_{1-}^+, \tag{6.16}$$

a process directly analogous with the well-known strong decay transition $\Delta^{++} \rightarrow \pi^+ + P$. The state Λ_{1-}^+ can decay only through $\Delta C=-1$ weak interactions, among which the strongest transition is $\Delta s=-1$, as was anticipated in the weak interaction scheme of Glashow et al.[38] and as has been found to be the case for the decay of the D-mesons. The weak transition is apparently

$$\Lambda_{1-}^+ \overset{wk}{\rightarrow} \pi^+ + Y^{*0}, \tag{6.17}$$

where Y^{*0} denotes an excited C=0, s=-1 hyperon having the decay mode $Y^{*0} \rightarrow \Lambda\pi^+\pi^-$. Two identifications are possible for the pion emission sequence. The first gives 1927 MeV as the $(\Lambda\pi^-\pi_2^+)$ mass, and then 1380 MeV as the $(\Lambda\pi_2^+)$ mass. The second gives 1918 MeV as the $(\Lambda\pi^-\pi_1^+)$ mass, and then 1480 MeV as the $(\Lambda\pi_1^+)$ mass. We may note also that the $(\Lambda\pi^-)$ mass for this event is 1596 MeV. There are only two well-established s=-1 resonances in the mass region $\sim$ 1900 MeV, ΣF15(1915) and ΣD13(1940). Both of them are strongly inelastic resonances; for neither of them has the decay mode $\pi\Sigma^*(1385)$ been reported, although the mode $\bar{K}\Delta(1232)$ has been reported for the latter. Thus, the most plausible interpretation of the decay transitions following (6.17) is that Y^{*0} is the state ΣD13(1940), which decays in two steps

$$\Sigma D13(1940)^0 \rightarrow \pi^- + \Sigma^*(1385)^+ \rightarrow \pi^- + (\pi^+ + \Lambda). \tag{6.18}$$

This interpretation cannot be proven, on the basis of one event, of course, but it does illustrate nicely the kind of event which would result from the excitation of a C=+1 baryonic state, and the character of its sequence of strong, electromagnetic and weak transitions in the course of its decay. Neutrino interactions are rather favourable for the observation of charmed baryon excitation. The transition $\nu_\mu P \rightarrow \mu^- \Sigma_1^{*++}$ has a large matrix element, as neutrino transitions go; the fact that it is endothermic by about 1.5 GeV acts against it but this will be of lesser importance for much higher incident energies. Although the absolute cross section for charm excitation in neutrino-nucleon collisions is small, there will also be relatively little background. It seems reasonable to expect that rather clear data will be obtained in this way, in the course of time, which will lead us to some detailed knowledge of charmed baryon spectroscopy.

Further evidence for the state $\Sigma^{+}_{1-}(2260)$ has been presented by Knapp et al[58], from the study of multihadron final states due to an $\sim$ 50 GeV wide-band photon beam incident on a target at Fermilab. They found a clear peak in the $(\bar{\Lambda}\pi^{-}\pi^{-}\pi^{+})$ mass spectrum at 2260±10 MeV, with a decay width Γ=40±20 MeV consistent with their resolution. They also find evidence for a broad peak at about 2500 MeV for the $(\bar{\Lambda}\pi^{+}\pi^{+}\pi^{-}\pi^{-})$ mass, and they show that the $(\bar{\Lambda}\pi^{-}\pi^{-}\pi^{+})$ peak becomes very marked if the latter events are selected only from the events $(\bar{\Lambda}\pi^{+}\pi^{+}\pi^{-}\pi^{-})$ occurring within this 2500 MeV peak. In short, these states appear to have much the same properties and relationships as would the anti-particles of the two states $\Sigma^{*}_{1}(2460)$ and $\Lambda_{1-}(2260)$, in terms of which the event of Cazzoli et al. has been interpreted. It is not surprising that these states are seen only as antibaryons in these photoproduction experiments, for antibaryon events will have much less background; for final baryons, there will exist very many combinations $\Lambda(3\pi)^{+}$ in the many multihadron processes induced by photons which do not involve charmed baryons in any way and which are so much more readily excited as to swamp the charmed baryon signal.

No observations of charmed baryons have yet been reported from either hadronic interaction or electron-positron annihilation experiments. However, it is clear from the above remarks in this Section that we are now just at the beginning of an era of exploration of charmed hadron spectroscopy, analogous in principle (but different in detail) with the era of strange hadron spectroscopy which was in full swing by 1961 and which is still proceeding quite vigorously.

7. CONCLUSION

It will be useful to close with mention of some recent developments bearing on the parton model and QCD. These are on both the experimental and the theoretical sides, as follows:

(i) the observation that the empirical structure functions denoted previously by $F_i(x)$ are in fact dependent on the momentum transfer k_μ^2, so that they are better written as $F_i(x,k_\mu^2)$. This has become established for the deep-inelastic regime for eP and eN scattering[59], for μP and μN scattering[60] and for νP and νN inelastic scattering[61]. The trend of the scale invariance breaking is that $F_2(x,k_\mu^2)$ increases with increasing k_μ^2 at small x (x<0.2) and decreases at large x(x>0.2). This dependence is quite different in character from the effect of a form factor $f(k_\mu^2)$.

(ii) from the analysis of the perturbative series for the reaction amplitudes and the summation of the leading contributions in each order, and subsequently also by a construction based on the Renormalization Group Equations, Hinchliffe and

Llewellyn Smith[3,62] have shown that asymptotically free gauge theories leads to the following general form for the moments of the structure functions, as function of k_μ^2,

$$\int_0^1 x^n F(x,k_\mu^2)\, dx = M_n (\log(k_\mu^2/\Lambda^2))^{p_n}(1+0(\alpha_s)), \quad (7.1)$$

where p_n is given by

$$p_n = -\frac{4}{(33-3F)}\left\{1-\frac{2}{(n+1)(n+2)}+4\sum_{r=1}^{n}\frac{1}{(r+1)}\right\} \quad (7.2)$$

for QCD, where F denotes the number of flavours. Expression (7.2) gives $p_0=0$ and $p_n < 0$ for $n \geqslant 1$. The coupling constant α_s for QCD is given by

$$\alpha_s(k_\mu^2) = 12\pi/\{(33-2F)\ell n(k_\mu^2/\Lambda^2)\}, \quad (7.3$$

for $k_\mu^2 >> \Lambda^2$, where Λ is a parameter which has to be determined empirically; its value is believed to be approximately Λ=500 MeV. The coefficient M_n is calculable at present only for n=0. This form (7.1) is to be contrasted with the form

$$\int_0^1 x^n F(x,k_\mu^2)\,dx = M_n'(k_\mu^2/\Lambda^2)^{q_n}(1+0(\Lambda^2/k_\mu^2)), \quad (7.4)$$

which would hold for theories not asymptotically free.

We note that, for $n \neq 0$, eq.(7.1) predicts that all moments of the empirical structure functions approach zero as $k_\mu^2 \to \infty$. For n=0, eq.(7.1) states that the integral of the structure function approaches a calculable constant in the limit $k_\mu^2 \to \infty$. These predictions match the behaviour (i) noted empirically. With a positive definite structure function, the moments of the structure function can have this behaviour only if the structure function shrinks to the limiting form $M_n \delta(x+)$ as $k_\mu^2 \to \infty$, which is the trend shown by the data.

We note that the quark distrubition functions u(x),d(x),.. etc., are now to be regarded as functions of k_μ^2, with the notations $u(x,k_\mu^2)$, ... etc. The sum rules relating the integrals over structure functions with quark distribution functions hold as before, but they take the simple forms discussed in Sec.1 only in the limit $k_\mu^2 \to \infty$. The F_3 sum rule (1.5) is particularly insensitive to $\log(k_\mu^2/\Lambda^2)$ in this limit, because the gluon field contributes equally to $\bar{q}(x)$ and q(x), whereas F_3 gains its contributions from the combination $(q(x)-\bar{q}(x))$ in which these contributions precisely cancel.

As Llewellyn Smith[62] remarks, since the moments behave only as powers of log (k_μ^2/Λ^2), the structure functions $F_i(x,k_\mu^2)$ have their most rapid variation for the region of moderate k_μ^2, varying less and less rapidly as k_μ^2 increases. The density of charmed quarks,which is essentially negligible for the low k_μ^2 regime, increases rapidly at small x, with increasing k_μ^2, and could well contribute appreciably to the increase observed for F_2 at small x(cf. point (i) above). On the other hand, the prediction for the total neutrino cross sections is that σ^ν/E should decrease with increasing E, most rapidly in the region up to 50 GeV, and then more and more slowly as E increases beyond this.

These refinements will receive increasing attention in the next few years, as experimental information builds up in the high k_μ^2 region, with the use of the more intense and more energetic muon and neutrino beams which are now becoming available. We can see ahead many physical questions concerning the "elementary particles" whose understanding appears to depend on the colour gauge theory of hadronic interactions. Much has become understood already, in a qualitative way, using phenomenological approaches based on the general characteristics of gauge theories and on the particular properties of the gauge theory of colour and we may anticipate that much more will become understood through more detailed studies of the specific theoretical implications of particular gauge theories.

REFERENCES

1. V. Barger and R.J.N. Phillips, Nucl. Phys. B73 (1974) 269.
2. R. McElhaney and S.F. Tuan, Phys. Rev. D8 (1973) 2267.
3. I. Hinchcliffe and C.H. Llewellyn Smith, "Detailed treatment of scaling violations in asymptotically free gauge theories", submitted for publication (1977).
4. D.J. Gross and C.H. Llewellyn Smith, Nucl. Phys. B14 (1969) 337
5. D.H. Perkins, Contemp. Phys. 16 (1975) 173.
6. O.W. Greenberg, Phys. Rev. Letters 13 (1964) 598.
7. C.G. Callan and D.J. Gross, Phys Rev. Letters 22 (1969) 156.
8. See R.F. Streator and A.S. Wightman, "PCT, Spin and Statistics, and All That" (W.A. Benjamin Inc., New York, 1964).
9. M.Y. Han and Y. Nambu, Phys. Rev. 139B (1965) 1006.
10. R.H. Dalitz, in Hadron Interaction at Low Energies - Physics and Applications, Vol.1 (eds. D. Krupa and J. Pisut, VEDA, Slovak Acad. Sci., Bratislava, 1975), p. 145.
11. H. Lipkin, Phys. Letters 45B (1973) 267.
12. H.D. Politzer, Phys. Reports C14, No.3 (1974).
13. G.S. La Rue, W.M. Fairbank and A.F. Hebard, Phys. Rev. Letters 38 (1977) 1011.
14. G. Gallinaro, M. Marinelli and G. Morpurgo, Phys. Rev. Letters 38 (1977) 1255.
15. H. Fritzsch, M. Gell-Mann and H. Leutwyler, Phys. Letters 47B (1973) 365.
16. R. Blankenbeckler and R. Sugar, Phys. Rev. 142 (1966) 1051.
17. A.C. Butcher and J.M. McNamee, Proc. Phys. Soc. 74 (1959) 529;see also ref. 65.
18. F. Ajzenberg-Selove and T. Lauritsen, Nucl. Phys. A227 (1974) 1.
19. P. Swan, Proc. Roy. Soc. A. A228 (1955) 10.
20. J. Ribeiro, "Coloured Quarks and the Short-range Nucleon-Nucleon Interaction", D. Phil. dissertation, Oxford University (March, 1978).
21. J. A. Wheeler, Phys. Rev. 52 (1937) 1083 and 1107.
22. R. Horgan, Nucl. Phys. B71 (1974) 514.

23. A. De Rujula, H. Georgi and S. L. Glashow, Phys. Rev. D12 (1975) 147.
24. M. Jones, R. H. Dalitz and R. R. Horgan, Nucl. Phys. B129 (1977) 45.
25. N. Isgur and G. Karl, Phys. Letters 72B (1977) 109.
26. H. Schnitzer, Phys. Letters 65B (1976) 239.
27. N. Isgur and G. Karl, "P-wave Baryons in the Quark Model", submitted for publication (January, 1978).
28. J. L. Reinders,"Baryon Spectroscopy in the Non-relativistic Quark-Model with One-Gluon Exchange Potential",submitted for publication (1978).
29. R. J. Cashmore, A. J. G. Hey and P. J. Litchfield, Nucl. Phys. B98 (1975) 237.
30. A. J. G. Hey, Proc. Top. Conf. on Baryon Resonances at Oxford 5-9 July (1976), ed. R. T. Ross and D. H. Saxon (Rutherford Lab., Chilton, Didcot, 1976) p.463.
31. R. E. Cutkosky, R. E. Hendrick and R. L. Kelly, Phys. Rev. Letters 37 (1976)645.
32. R. H. Dalitz, R. R. Horgan and J. L. Reinders, J. Phys. G. 3 (1977) L195.
33. M. G. Bowler, J. Phys. G3 (1977) 775.
34. M. G. Bowler, M. A. V. Game, I. J. R. Aitchison and J. B. Dainton, Nucl. Phys. B97 (1975) 227.
35. J. L. Basdevant and E. L. Berger, Phys. Rev. D16 (1977) 657.
36. Particle Data Group, Rev. Modern Phys. 48 (1976) S1.
37. J. D. Bjorken and S. L. Glashow,Phys. Letters 11 (1964) 173.
38. S. L. Glashow, J. Iliopoulous and L. Maiani, Phys. Rev. D2 (1970) 1285.
39. J. S. Kang and H. J. Schnitzer Phys. Rev. D12 (1975) 841.
40. E. Eichten, K. Gottfried, T. Kinoshita, J. Kogut, K. D. Lane and T. M. Yan Phys. Rev. Letters 34 (1975) 369.
41. B. J. Harrington, S. Y. Park and A. Yildiz, Phys. Rev. Letters 34 (1975) 168.
42. W. Braunschweig et. al., Phys. Letters 67B (1977) 243.
43. H. J. Schnitzer, Phys. Rev. D13 (1975) 74.
44. H. J. Schnitzer, Phys. Rev. Letters 35 (1975) 1540.
45. H. J. Schnitzer, Phys. Letters 65B (1976) 239.
46. G. J. Feldman, in Proc. Summer Institute on Particle Physics (ed. M. C. Zipf, SLAC, Stanford, November 1976) p. 81.
47. R. H. Dalitz, Proc. Roy. Soc. A355 (1977) 601.
48. E. Eichten, K. Gottfried, T. Kinoshita, K. D. Lane and T. M. Yan, Phys. Rev. Letters 36 (1976) 500.
49. C. N. Yang, Phys. Rev. 77 (1950) 242.
50. H. F. W. Sadrozinski, Proc. 1977 Intl. Symposium on Lepton and Photon Interactions at High Energies (ed. G. Weber, DESY, Hamburg, December, 1977) p. 47.
51. S. Yamada, Proc. 1977 Intl. Symposium on Lepton and Photon Interactions at High Energies (ed. G. Weber, DESY, Hamburg, December, 1977) p. 69.
52. G. Goldhaber et. al. Phys. Rev. Letters 37 (1976) 255.
53. A. Barbaro-Galtieri, Proc. 1977 Intl. Symposium on Lepton and Photon Interactions at High Energies (ed. G. Weber, DESY, Hamburg, December, 1977) p. 21.

54. J. E. Wiss et. al., Phys. Rev. Letters 37 (1976) 1531.
55. M. K. Gaillard, B. W. Lee and J. L. Rosner, Revs. Modern Phys. 47 (1975) 277.
56. D. B. Lichtenberg, Lett. Nuovo Cimento 13 (1975) 346.
57. E. Cazzoli, A. Cnops, P. Conolly, R. Loutit, M. Murtagh, R. Palmer, N. Samios, T. Tso and H. Williams, Phys. Rev. Letters 34 (1976) 1125. Also private communication from Dr. N. Samios (1976).
58. B. Knapp et. al., Phys. Rev. Letters 37 (1976) 882.
59. R. E. Taylor, Proc. 1975 Intl. Symposium on Lepton and Photon Interactions at High Energies (ed. W. T. Kirk, SLAC, Stanford University, California, 1975), p. 679.
60. H. L. Anderson et. al., Phys. Rev. Letters 37 (1976) 4.
61. D. H. Perkins, P. Schreiner and W. G. Scott, Phys. Letters 67B (1977) 347.
62. C. H. Llewellyn Smith, "Deep Inelastic Phenomena", Lectures presented at the 1977 Cargese Summer Institute, to be published (1978).
63. G. R. Satchler, L. W. Owen, A. J. Elwyn, G. L. Morgan and R. L. Walker, Nucl. Phys. A112 (1968) 1.
64. R. A. Arndt and L. D. Roper, Nucl. Phys. A209 (1973) 447.
65. K. Wildermuth and W. McClure, "Cluster Representations of Nuclei", (Springer-Verlag, Berlin, 1966).

Selected Issues from

Lecture Notes in Mathematics

Vol. 431: Séminaire Bourbaki - vol. 1973/74. Exposés 436–452. IV, 347 pages. 1975.

Vol. 433: W. G. Faris, Self-Adjoint Operators. VII, 115 pages. 1975.

Vol. 434: P. Brenner, V. Thomée, and L. B. Wahlbin, Besov Spaces and Applications to Difference Methods for Initial Value Problems. II, 154 pages. 1975.

Vol. 440: R. K. Getoor, Markov Processes: Ray Processes and Right Processes. V, 118 pages. 1975.

Vol. 442: C. H. Wilcox, Scattering Theory for the d'Alembert Equation in Exterior Domains. III, 184 pages. 1975.

Vol. 446: Partial Differential Equations and Related Topics. Proceedings 1974. Edited by J. A. Goldstein. IV, 389 pages. 1975.

Vol. 448: Spectral Theory and Differential Equations. Proceedings 1974. Edited by W. N. Everitt. XII, 321 pages. 1975.

Vol. 449: Hyperfunctions and Theoretical Physics. Proceedings 1973. Edited by F. Pham. IV, 218 pages. 1975.

Vol. 458: P. Walters, Ergodic Theory - Introductory Lectures. VI, 198 pages. 1975.

Vol. 459: Fourier Integral Operators and Partial Differential Equations. Proceedings 1974. Edited by J. Chazarain. VI, 372 pages. 1975.

Vol. 461: Computational Mechanics. Proceedings 1974. Edited by J. T. Oden. VII, 328 pages. 1975.

Vol. 463: H.-H. Kuo, Gaussian Measures in Banach Spaces. VI, 224 pages. 1975.

Vol. 464: C. Rockland, Hypoellipticity and Eigenvalue Asymptotics. III, 171 pages. 1975.

Vol. 468: Dynamical Systems - Warwick 1974. Proceedings 1973/74. Edited by A. Manning. X, 405 pages. 1975.

Vol. 470: R. Bowen, Equilibrium States and the Ergodic Theory of Anosov Diffeomorphisms. III, 108 pages. 1975.

Vol. 474: Séminaire Pierre Lelong (Analyse) Année 1973/74. Edité par P. Lelong. VI, 182 pages. 1975.

Vol. 484: Differential Topology and Geometry. Proceedings 1974. Edited by G. P. Joubert, R. P. Moussu, and R. H. Roussarie. IX, 287 pages. 1975.

Vol. 487: H. M. Reimann und T. Rychener, Funktionen beschränkter mittlerer Oszillation. VI, 141 Seiten. 1975.

Vol. 489: J. Bair and R. Fourneau, Etude Géométrique des Espaces Vectoriels. Une Introduction. VII, 185 pages. 1975.

Vol. 490: The Geometry of Metric and Linear Spaces. Proceedings 1974. Edited by L. M. Kelly. X, 244 pages. 1975.

Vol. 503: Applications of Methods of Functional Analysis to Problems in Mechanics. Proceedings 1975. Edited by P. Germain and B. Nayroles. XIX, 531 pages. 1976.

Vol. 507: M. C. Reed, Abstract Non-Linear Wave Equations. VI, 128 pages. 1976.

Vol. 509: D. E. Blair, Contact Manifolds in Riemannian Geometry. VI, 146 pages. 1976.

Vol. 515: Bäcklund Transformations. Nashville, Tennessee 1974. Proceedings. Edited by R. M. Miura. VIII, 295 pages. 1976.

Vol. 516: M. L. Silverstein, Boundary Theory for Symmetric Markov Processes. XVI, 314 pages. 1976.

Vol. 518: Séminaire de Théorie du Potentiel, Proceedings Paris 1972–1974. Edité par F. Hirsch et G. Mokobodzki. VI, 275 pages. 1976.

Vol. 522: C. O. Bloom and N. D. Kazarinoff, Short Wave Radiation Problems in Inhomogeneous Media: Asymptotic Solutions. V. 104 pages. 1976.

Vol. 523: S. A. Albeverio and R. J. Høegh-Krohn, Mathematical Theory of Feynman Path Integrals. IV, 139 pages. 1976.

Vol. 524: Séminaire Pierre Lelong (Analyse) Année 1974/75. Edité par P. Lelong. V, 222 pages. 1976.

Vol. 525: Structural Stability, the Theory of Catastrophes, and Applications in the Sciences. Proceedings 1975. Edited by P. Hilton. VI, 408 pages. 1976.

Vol. 526: Probability in Banach Spaces. Proceedings 1975. Edited by A. Beck. VI, 290 pages. 1976.

Vol. 527: M. Denker, Ch. Grillenberger, and K. Sigmund, Ergodic Theory on Compact Spaces. IV, 360 pages. 1976.

Vol. 532: Théorie Ergodique. Proceedings 1973/1974. Edité par J.-P. Conze and M. S. Keane. VIII, 227 pages. 1976.

Vol. 538: G. Fischer, Complex Analytic Geometry. VII, 201 pages. 1976.

Vol. 543: Nonlinear Operators and the Calculus of Variations, Bruxelles 1975. Edited by J. P. Gossez, E. J. Lami Dozo, J. Mawhin, and L. Waelbroeck, VII, 237 pages. 1976.

Vol. 552: C. G. Gibson, K. Wirthmüller, A. A. du Plessis and E. J. N. Looijenga. Topological Stability of Smooth Mappings. V, 155 pages. 1976.

Vol. 556: Approximation Theory. Bonn 1976. Proceedings. Edited by R. Schaback and K. Scherer. VII, 466 pages. 1976.

Vol. 559: J.-P. Caubet, Le Mouvement Brownien Relativiste. IX, 212 pages. 1976.

Vol. 561: Function Theoretic Methods for Partial Differential Equations. Darmstadt 1976. Proceedings. Edited by V. E. Meister, N. Weck and W. L. Wendland. XVIII, 520 pages. 1976.

Vol. 564: Ordinary and Partial Differential Equations, Dundee 1976. Proceedings. Edited by W. N. Everitt and B. D. Sleeman. XVIII, 551 pages. 1976.

Vol. 565: Turbulence and Navier Stokes Equations. Proceedings 1975. Edited by R. Temam. IX, 194 pages. 1976.

Vol. 566: Empirical Distributions and Processes. Oberwolfach 1976. Proceedings. Edited by P. Gaenssler and P. Révész. VII, 146 pages. 1976.

Vol. 570: Differential Geometrical Methods in Mathematical Physics, Bonn 1975. Proceedings. Edited by K. Bleuler and A. Reetz. VIII, 576 pages. 1977.

Vol. 572: Sparse Matrix Techniques, Copenhagen 1976. Edited by V. A. Barker. V, 184 pages. 1977.

Communications in

Mathematical Physics

The journal is devoted to the following topics: General relativity, equilibrium and non-equilibrium statistical mechanics, foundations of quantum mechanics, classical and quantum mechanics of finitely many degrees of freedom, Lagrangian quantum field theory and constructive quantum field theory. Mathematical papers are accepted only if they are of direct relevance to physics.

Springer-Verlag
Berlin
Heidelberg
New York